# WHEWELL'S CRITICS

POZNAŃ STUDIES
IN THE PHILOSOPHY OF THE SCIENCES AND THE HUMANITIES

VOLUME 85



*Poznań Studies in the Philosophy of the Sciences and the Humanities*
is partly sponsored by SWPS and Adam Mickiewicz University

Address: dr Katarzyna Paprzycka · Instytut Filozofii · SWPS · ul. Chodakowska 19/31
03-815 Warszawa · Poland · fax: ++48 22 517-9625
E-mail: PoznanStudies@swps.edu.pl · Website: http://PoznanStudies.swps.edu.pl

## MONOGRAPHS-IN-DEBATE



**Katarzyna Paprzycka** (editor-in-chief)
Instytut Filozofii · SWPS
ul. Chodakowska 19/31 · 03-815 Warszawa · Poland
Katarzyna.Paprzycka@swps.edu.pl

*Monographs-in-Debate* is a new subseries of the *Poznań Studies in the Philosophy of the Sciences and the Humanities* book series. It publishes monographs that deal with the general area of interest of the *Poznań Studies* series. The special nature of the *Monographs-in-Debate* volumes arises from the recognition of the dialectical nature of philosophical work. Each volume contains a monograph, followed by peer commentaries and the author's replies.

POZNAŃ STUDIES IN THE PHILOSOPHY OF THE SCIENCES AND THE HUMANITIES, VOLUME 85
MONOGRAPHS-IN-DEBATE

John Wettersten

# Whewell's Critics

## *Have They Prevented Him from Doing Good?*

**Edited by**

**James A. Bell**

Amsterdam - New York, NY 2005

The paper on which this book is printed meets the requirements of "ISO 9706:1994, Information and documentation - Paper for documents - Requirements for permanence".

ISSN 0303-8157
ISBN: 90-420-1644-2

Printed in The Netherlands

## CONTENTS



**John Wettersten**
**WHEWELL'S CRITICS**
**HAVE THEY PREVENTED HIM FROM DOING GOOD?**



## COMMENTARIES

## REPLIES TO COMMENTARIES

# FOREWORD

A few biographical comments will tie Prof. John Wettersten's broader interests to this Whewell project. An orientation to the volume will follow.

The energy and interest that drove Prof. John Wettersten's analyses of reactions to Whewell's philosophy originated in a personal quest. A story will capture the gist.

John and I were graduate students and friends at Boston University. One animated disagreement stands out, this one over coffee in 1967. We were discussing intellectual intolerance in academia, especially in our chosen field of philosophy. I argued that the intolerance was inevitable but chaotically productive; he argued it was irrational and often dysfunctional.

At the heart of our discussion was the tendency of academics to cluster into schools of thought. With their thinking informed by a framework of ideas, they were often effectively blinded to the disadvantages of their own views as well as the advantages of competing schools. Thomas Kuhn's paradigmatic view was *à la* mode at the time, and it reinforced this outlook. I assumed that the dogmatism could only be countered effectively by a multiplicity of schools, with "purist" academics supporting the strengths of their school and emphasizing the weaknesses of competing schools. Despite the paradigmatic narrow-mindedness of many, I contended that progressive ideas – new insights and perspectives – would emerge from the debates, even if willy-nilly and unpredictably.

John demurred. He agreed that much of academia reflected Kuhnian ideas, but he did not believe it was inevitable. He recognized that traditions are deeply embedded in sociological contexts, but he did not believe they were totally a function of those contexts. His views fell within the "weak program" in the sociology of knowledge, even though that expression had not yet been coined. What is important here, though, is that John was in the early stages of his search for a better understanding of traditions, especially intellectual traditions. He argued that there should be guidelines to assess traditions critically, even from within, and to improve them in light of the criticism. Progress could thus be made rationally, or at least substantially so.

Prof. Wettersten's assessment of critics' reactions to Whewell's philosophy continues his search for those rational guidelines. His personal comments on the quest are found in a variety of places in the manuscript, more systematically at the beginning and end as well as in his replies to commentators. In an intellectual autobiography now being written, Prof. Wettersten goes into much more personal detail. The principal vehicle for his search here, however, is his penetrating analyses of the variety of responses to William Whewell. These responses were embedded in a number of different intellectual and cultural traditions, traditions which themselves were in flux to one extent or another. Prof. Wettersten filters his assessments of the critics' responses to Whewell through these factors. The result is a comprehensive discussion of the reasonableness – or unreasonableness – of those many responses. What makes the discussion even more fascinating, though, is a question that is never far from center stage: why were so many of the responses to Whewell's philosophy misconceived? A more contemporary correlate of that question: why have so many criticisms of Karl Popper's philosophy been misconceived?

Prof. Wettersten's manuscript is highly original. As a contribution to Whewellian literature, it explores the potential impact of Whewell's thinking and especially the resistance to his thinking. As a search for guidelines to assess traditions, it examines the potential and pitfalls of rational discussion when traditions are faced with bold and compelling challenges. As Guest Editor, I was privy to misreadings of this manuscript in one or both of these ways by a number of potential commentators. That should not be surprising, given its themes about misunderstanding, especially of original work.

The commentators' essays in this volume touch upon a variety of interpretative issues in Prof. Wettersten's manuscript. They enhance and extend a number of discussions in Whewellian philosophy, and they add perspectives that would otherwise remain tacit or unseen. Prof. Wettersten's replies carry this process further, and also give him opportunity to elaborate further on certain points in the manuscript. Far more than an add-on or denouement to the manuscript, then, the commentaries and replies are integral to the discussion. Do not miss them.

James A. Bell
*University of South Florida*
Box 3757
Carmel, California 93921
U.S.A.
E-mail: bell@luna.cas.usf.edu

John Wettersten

# WHEWELL'S CRITICS

## HAVE THEY PREVENTED HIM FROM DOING GOOD?

## PREFACE AND ACKNOWLEDGMENTS

The problem of how one can most productively study figures in the history of philosophy poses just as interesting problems as do the most interesting of these figures themselves. Interests may vary from mere curiosity about a past age, to historical views as live alternatives today, or to the way historical figures have shaped our heritage. For this reason the tasks of historians of philosophy are often separated, so that a study of a particular philosophy may not include a study of the influence of some thinker and a study of someone's influence may not include any attempt to interpret anew the content of this philosophy. There are, however, also attempts that seek to integrate these tasks such as E.A. Burtt's *The Metaphysical Foundations of Modern Science* (1954). In this book I have also tried to do that.

My interest in the history of the philosophy of science lies primarily in the study of the way philosophies of science have served to build our heritage: How have we come to think the way we do about science? My interest in this question is not merely historical, but also critical. I do not wish merely to know how we have come to think about science the way we do, but also whether this development has proceeded as one might wish, whether we can find times where other turns in the selection process of the various alternatives and the problems they have posed might have been better, and whether these developments need correction today.

I have worked under the presumption that, in order to think well about science today, it is not enough merely to know our philosophical heritage. We also need first to choose which traditions we wish to work in and then seek to critically appraise them. We need to do this because heritage sets the context in which today's problems our formulated. It serves as the background knowledge we unavoidably use in posing contemporary problems. The history of the philosophy of science cannot now be separated from the philosophy of science, as so many 20th century philosophers of science have vainly sought to do. In order to find and keep our bearings and pose our problems about science and society intelligently today, we need to critically examine the way the presumptions we make about science have been developed and defended. If we accept the background knowledge established today uncritically, we will be a slave to it. If we attempt to ignore it, we will do no better. Points of view which have been developed over centuries permeate our institutions and our thought. We can only control the influence of such ideas, if we begin to understand how they were developed and defended and how they came to shape the background knowledge we now use in thinking about science.

In order to preserve autonomy, then, we need to do more than ignore and/or merely accept our heritage as it has been taught to us. We need to study this heritage both respectfully and critically. We cannot live without the framework provided by heritage and we should honor it. But we should also seek to select it and to (re)shape it critically. A radical revision of our heritage lies beyond our capacity, but piecemeal changes and corrections are within our reach. This means that it is highly desirable to make regular attempts to re-think parts of it. In this way we may improve the intellectual background we unavoidably use when we set out to tackle the problems of the day.

It is with this in mind that I set out to study the philosophy of science of William Whewell. I wanted to defend his view with a new interpretation of his philosophy which shows it to be much more powerful, progressive and interesting than it has been taken to be. But I also wanted to show how this philosophy has been treated, how the discussion in the philosophy of science and some the important problems Whewell raised has proceeded, and why the time has come to critically reappraise not merely Whewell's philosophy but also this history and our standard interpretation of it. This aspect of our intellectual heritage lies, I think, heavily on current discussions in the philosophy of science. With a new perspective on it, we may, I hope, formulate our problems in the philosophy of science today somewhat better.

With these aims in mind I have tried to devise a method for the study of Whewell's philosophy which would integrate the study of it with the study of it's reception. In this way I hoped to throw some light on the tradition in the philosophy of science which rejected him as well as on his philosophy itself. My solution to the problem of how to proceed in my study of Whewell has been to study the discussion of Whewell's philosophy from the very first reviews of his two masterpieces until quite recent discussions of his philosophy. My aim has been to show not merely that Whewell has been misinterpreted, misread and rejected too quickly, but also how this process has led to the construction of an erroneous view of him and his competitors which permeates our discussion to this day. I wished to use the critical discussion of Whewell's critics to show how a poor interpretation of his philosophy and a far too negative attitude toward it has become a part of our heritage, how we need to change this heritage by changing our views both of Whewell and of those who rejected him, and how we may appreciate the positive contributions he has made to the history of the philosophy of science when we do so. The method of developing a new interpretation of a philosopher by a historical portrayal of the development of the discussion of him seemed especially appropriate for Whewell because there has been so little discussion of him that a virtually comprehensive portrayal of this development is possible.

Some readers of my manuscript have expressed dissatisfaction with this approach. They have sought here a new interpretation of his view presented in a standard way and nothing more. The knowledge of how this view was constructed has not seemed to them to be worth the effort it takes to understand it. The knowledge of the twists and turns the discussion of this view has taken seems to some of little importance. In response to this reaction I have presented my view of Whewell toward the end of this book, where I gather together the criticisms of him, sort out the good from the bad, and offer a new interpretation of his philosophy. Readers will, I presume, tend to look at this first, which I do not take ill, since I seldom read a book from cover to cover myself, preferring to skip here and there, to pose questions, and to find selections which I might enjoy. I hope, however, that the method I have employed has served the aim it was intended to, that is, to portray the development of standard interpretations of Whewell in order to show how and why we should change our views not merely of Whewell but also of the tradition which has treated him so badly.

Joseph Agassi first called my attention to Whewell's philosophy of science many years ago and we have discussed the material covered in this book on numerous occasions. I did not start any serious study of his philosophy, however, until it became clear to me, as I studied the roots of critical rationalism, that he had played an important role in the development all deductivist philosophies of science from the mid-nineteenth century to the present. I explained some of this development in my book, *The Roots of Critical Rationalism* (Wettersten 1992), which I was able to write due to a research grant from the Deutsche Forschungsgemeinschaft. I was given a chance to continue my studies of the relationship between methodology and psychology and of William Whewell in particular by the Twente Technical University. I am especially grateful for the support offered by Floris Cohen of that university. It was then that the idea of writing a critical, historical study of the reception of Whewell by the philosophical and scientific communities began to take shape. The main work on the present manuscript followed my stay at Twente and was considerably aided by continued collaboration with Agassi, though my view has, I think, come to differ somewhat from his, as I explain below. I am also grateful for helpful, critical comments on my manuscript from Maurice Finocchiaro, Michael Segre and Leszek Nowak.

John Wettersten
wettersten@t-online.de

# ANALYTICAL TABLE OF CONTENTS

# INTRODUCTION

## WHEWELL'S IMAGE AND IMPACT: TWO CONFLICTING TALES

Between the nineteen fifties and the nineteen seventies or eighties, problems in the philosophy of science took a drastic turn. Research moved from the nearly exclusive interest in the so-called logic of science or in the logical analysis of scientific theories to problems concerning the history of science. The possibility of reconciling descriptions of scientific practice with various theories of scientific research was explored. This change of direction seemed modern, new, challenging and refreshing. But it was not, as it may have seemed, the creation of new ways of doing the philosophy of science. It was the re-discovery of old ways with, of course, new innovations. Earlier thinkers have also sought to reconcile theories of scientific method with studies of the history of science. And these thinkers – William Whewell, Pierre Duhem, Emile Meyerson and others – received renewed interest.

This return to nineteenth century problems in the philosophy of science was not welcomed by all. Indeed, there is still a determined opposition which seeks to block its impact as much as it possibly can. And today, in the nineties, the opposition seems to have renewed success as the main themes in the philosophy of science return to analysis and those particular kinds of the philosophy of psychology which reinforce old positions. This attitude has a long tradition. In the past it has, indeed, been quite successful and may be so again. But now we see quite clearly that there are deep problems not merely about science, but also about how to study science. These problems are often not discussed, nor is their history part of our common heritage. We have consigned them to specialists. Indeed, the disputes in the past concerning how to study science have been declared settled. But they flare up again and again. Why?

There may be no especially interesting answer to this question. But there is a puzzle. How can an approach such as that of integrating the study of the history and the methodology of science, which is once declared completely superseded, arise again in a quite different context? Why has this occurred? Can we learn anything from the study of the history of the problem of how we should study science? There are, of course, a variety of approaches to the study of how we should study science and each is guided by a specific philosophy of science. We seem to be trying to catch our tails. But one approach which seems promising is to study the history of the problem of how to study science. How did it arise in something like its contemporary form? And, why was an approach, which seems so promising today, so roundly dismissed as it first emerged? How has this dismissal served to shape our contemporary problems? We may hope, perhaps, that a study of such questions as these may deepen our understanding of the problems we now face, whether or not we find thereby new solutions to them.

The *prima facie* answer to the question of where we should begin a study of approach to the philosophy of science in which the methodology of science and the history of science are integrated is not hard to find. It is with William Whewell. Although Whewell had been nearly forgotten in the ninety or so years between his death in 1866 and the rediscovery in the 1950s of the importance of the history of science, he is today common knowledge. And no wonder: he was the combination of historian and philosopher of science *par excellence*. But, if he was so modern, why was he almost forgotten? Why must we rediscover him? What was the fate of his views? How have they worked to help us formulate ours? Or, how have they failed to do so due to their rejection? We have many puzzles, it would seem, concerning this troublesome thinker. Let us look a little closer but still quite superficially at the views of Whewell's work.

Whewell established his reputation as a thinker of originality and power with his history of science (Whewell 1967a). In this history Whewell provided a systematic view of how science developed. He uncovered as well many new facts about the development of science which forced nearly everyone to revise their images of science. He has been more or less forgiven for the former which can, it has been thought, be ignored, because it was overambitious; his success in the later is admired and praised but, it has been claimed, Whewell contributed nothing new in principle – though much in scope. Moreover his history soon had to be revised and rendered up-to-date by more detailed, and thus piecemeal, studies. The systematic thinker Whewell did not have the patience or interest to do so. So specialists were needed who might take more care in reporting the facts. Historians, it was quickly established, should not suffer from the too-high ambition of bringing the whole of the history of science under one scheme – whatever merits Whewell's attempt to do so may have had.

The heart of Whewell's work, however, was his theory of knowledge. It was developed through studies of the history of science, which served as the touchstone of all of his philosophy – even for his views in ethics and theology. He deemed the central ideas developed in his philosophy to be of enormous importance and pleaded for indulgence, if he should keep repeating them in various fields and in various ways: they were, he thought, of the utmost importance for mankind. But no one shared his passion. His theory has been deemed eclectic and confused by virtually all commentators. Its publication evoked a lively and lengthy response from the leading thinkers in England, but it was quickly "established" that the theory suffered from such severe difficulties as to be hopeless. The established view of science as it was defended by John Herschel and John Stuart Mill was, it was thought, really much better. And since the end of this first debate – which began in the 1840s and lasted until Whewell's death in 1866 – Whewell's philosophy has, with a few minor exceptions, been simply ignored.

There has, however, been another response which is less well-known and even more important. This response is to seek to integrate those aspects of Whewell's theory which seem correct – above all his interesting observations concerning the history of science and the methods which have been actually used to guide scientific research – in different systems which are allegedly more coherent and/or more in accord with traditional views of science. The rationale for such attempts is the following. If one succeeds in taking particular aspects of Whewell's research, such as his view of the "colligation of facts" or of the "consilience of inductions" (I explain below), and builds these aspects into systems which are more coherent or more in line with established doctrine, these systems will be even stronger, because they will be more in accord with the facts. In order to undertake such an endeavor, no cognizance need be taken of Whewell's system or even of the way in which various parts fit together.

Although the thinkers who do this pay tribute to Whewell and invite further investigation of the author they implicitly accept as their teacher, there have, indeed, been very few thinkers who have taken Whewell's philosophical system seriously. But our current interest in him is incomprehensible, if we do not take note his system: we find him modern and a kindred spirit precisely because he sought (systematically) to unify the history and the philosophy of science. And today we seem still to deem systems suspect! Virtually no one has sought to describe those (new) problems which he uncovered and which led him to develop both his daring new theory of science and his path-breaking method for the study of science. Yet this latter approach is certainly crucial, if one wishes to appreciate the strengths of Whewell's view – in spite of its obvious defects. Whewell found no disciples or defenders. His closest friends, both intellectual and personal, very quickly adopted the view that his new proposals were no serious competitors to those hitherto under consideration. In England these were a Baconian view as it was developed by John Herschel and John Stuart Mill on the one hand and a Kantian view as developed by Mansel on the other hand.

The explanation for this phenomenon could be, of course, that Whewell's observations concerning science are instructive and insightful, even while the system in which he incorporated these observations is rather weak and confused. Perhaps, in seeking to revise and correct the philosophy of science of his day, Whewell really was forced into implausibilities and confusions. Perhaps, then, the desirable moves which Whewell made in his attempt to construct an alternative theory of science really should be incorporated in more standard, less adventurous views. It is, the argument continues, useless to follow Whewell in his abandonment of tried and true principles in order to construct a new systematic theory of science and/or a new methodology. The former leads to excesses and the latter is dangerous, since it is a threat to science.

This explanation has been the standard view from the publication of Herschel's influential review (Herschel 1841) – which quickly followed the publication of Whewell's two major works, his *History of the Inductive Sciences* (Whewell 1967a) and his *The Philosophy of the Inductive Sciences, Founded upon Their History* (Whewell 1967b) – to the present. Does our new appreciation for Whewell's program throw doubt on this view? Whewell did, of course, abandon tried and true standards for both science and the theory of science. But were the advantages of the changes he proposed and the disadvantages of alternatives, whether those of his contemporaries or those of later generations, given a fair hearing? Among Whewell's critics were some of the best, most courageous and most honest thinkers around. It would seem absurd to doubt that Whewell's theories were given not only a fair hearing, but an incisive and intelligent one as well. A review of the criticisms which were the foundation of the standard appraisal of his work would thus seem to be prima facie unnecessary, unless one had a special interest in the details of 19th century philosophy of science.

Nevertheless I would like to review the case to see what happened and why. Why did the best nineteenth century philosophers bring Whewell's theory of science into such miscredit? Why have philosophers of science ever since clung, with slight modifications, to the initial appraisals, even when we have today a new more sympathetic perspective on his work? There is no single overpowering reason for engaging in an exercise which throws into question the established judgments which have been crystallized out of intense discussion during a period of over one hundred years. Rather, there are puzzling discrepancies between the standard view according to which Whewell's philosophy is hopelessly confused and the impact it continues to have. These discrepancies render the standard view incomplete and confusing. I have already mentioned some of these quite briefly above. A review of such discrepancies here, then, may explain the interest of a study of how the standard appraisal of Whewell was constructed and maintained. Were mistakes in this appraisal made which limited Whewell's impact in unfortunate ways? Or are the discrepancies mere minor points which, when cleared up, will render the standard view of the matter complete? Can we learn from such a study how well or poorly our problems about how to do the philosophy of science have been formulated? Can such a study enable us to improve their formulation and/or our comprehension of them?

One discrepancy between the standard view, according to which Whewell's philosophy is hopelessly confused, and the impact of his philosophy is that the universal rejection of Whewell's philosophy did not – apparently – prevent this very philosophy from exerting influence in various fields and in various ways. Ideas which Whewell developed systematically for the first time turn up in other views. In the philosophy of science, for example, the importance of conjectures seems to become commonplace under Whewell's impact. In psychology the problem of perception was – apparently under the impact of Whewell's studies of observation – rendered quite intense: one could, it would seem, no longer claim without further ado that perception was built up out of individual sensations. The theory of inductive unconscious inference was introduced by Mill to explain away intransigent phenomena pointed out by Whewell and explained by his own theory of deductive unconscious inference. Mill's proposed alternative to Whewell was seen to be merely a name of a problem, that is, how can one show that the postulated reduction of perceptions to sensations really was possible? Claude Bernard's essay on methods in biology adopts virtually all of Whewell's methodological ideas. Through the work of Bernard, French philosophy of science was permanently changed – or so it would seem. Pierre Duhem followed Whewell and combined the study of the history of science with the philosophy of science and, like Whewell, developed a new comprehensive view of its history. He developed a philosophy of science which made that which Whewell deemed the third stage of the development of science – the extension and refinement of established ideas – the central task of all of science (Wettersten 1998). Whewell's influence can even be seen in philosophers such as Mach and the new-Kantians who see the need to account for the growth and discovery of scientific ideas, as well as the need to integrate such accounts into a theory of knowledge. Can such developments as these be properly understood without reference to the work of Whewell? And if we cannot adequately explain and understand these later developments without taking the impact of Whewell into account, can it really be the case that this source of so many modern ideas and problems is hopelessly confused? It may very well be the case; but no obvious means of reconciling his wide influence with his allegedly confused philosophy has been proposed by anyone.

A second discrepancy between the standard view of Whewell's philosophy as hopelessly confused and his impact – and thus a further reason for reviewing the criticism of Whewell and/or the process by which the standard view of his philosophy was constructed – is that a Whewell renaissance has recently begun. How can a hopelessly confused philosopher enjoy a renaissance one hundred and forty years after the publication of the first of his two major works? Indeed, he is still deemed confused, but this view merely hinders and does not block the interest in him. How, then, can we explain this renaissance? It will surely be useful to see the current discussion of Whewell in the light of a review of the tradition of Whewell criticism. Is this discussion of Whewell indebted to this tradition? If so, how can we reconcile that with the renewed interest in him? Or, are contemporary thinkers trapped by the possibly erroneous conceptions developed by this tradition? If so, we may still be puzzled by the new interest which the tradition alone can certainly not explain. Perhaps this recent discussion is genuinely new and represents an original contribution to our understanding of Whewell's philosophy, of Whewell's role in the history of philosophy, of his history of science and/or of the role his methodology has played in various sciences. But if so, why is this break with tradition not discussed? And how is this break with tradition to be justified? The discrepancy between his renaissance and the dim view of his philosophy raises many puzzles.

A third discrepancy between the standard view of Whewell's philosophy and its impact arises due to one reason for the Whewell renaissance. This renaissance is at least in part due to the success of Karl Popper's theory of science. Popper's theory is, like Whewell's, rejected more often than not, though unlike Whewell, Popper has found a large number of disciples, who are willing and able to defend the general direction of his views, even while changing and developing it. Whewell appears to be taken to offer an alternative to Popper. Now this fact – if it is a fact – is, of course, startling and a great complement for Whewell: to construct a view which is a live option after one hundred and thirty or forty years is an enormous achievement. But how can this development be possibly reconciled with the standard view according to which Whewell's view is hopelessly confused? Can it be the case that Whewell offers mere insightful observations that, even now after all this time, still need to be incorporated in modern views? This might be the case, but prima facie offers no satisfactory resolution of our puzzles, since it is Whewell's systematic theory which offers an alternative and which seems to need a new appraisal. Is it the case that Whewell is being introduced today as a new idea in order to block a still newer idea, a phenomena described by Gertrude Stein? (Wettersten and Agassi 1991; Stein 1946).

There is one further discrepancy between the standard view of Whewell's philosophy as confused and its impact. This last discrepancy is between the shallowness of the standard responses to his philosophy and the depth of the problems he put on the (hidden) agenda. Whewell's theory raised difficulties for its critics precisely because Whewell had the pretension to offer what he thought was a deep, powerful and unifying theory. But if it was deep, it was so precisely because it rejected standard traditional assumptions concerning science. His critics have said that one cannot do that and remain within the bounds of responsible, rational philosophy of science. But Whewell sought to be more careful and more responsible than any theoretician of science before him had been: he made a careful and systematic review of the facts of the history of science and demanded of himself that his theory remain in accord with them. Now, how do we judge such conflicting claims for responsible procedures? Whewell was even aware of this problem and offered his own answer. We test theories in science by their ability to explain facts in a unified way and in a way which makes it impossible that any alternative explanation of them should be possible. He repeatedly defended his positions by appealing to the facts, which his history of science had portrayed. His philosophy of sciences has the sub-title, *Founded upon Their History*. His own theory should obviously meet his own standards for science: he hoped to show that his theory and only his theory could account for this history. Now the tradition – implicitly – deems this answer inadequate. Although it is hardly discussed, the problems of rationality which it raises keep popping up. Can one respond to Whewell adequately without facing such problems of rationality? The discrepancy between the depth of the changes in problems introduced by Whewell and the shallowness of the considerations used to reject him – the failure to respect his own view of rationality – is puzzling.

My discussion thus far is quite superficial. There are, it would seem, some discrepancies between the way Whewell is traditionally presented and some aspects of the role which his theory has played in the history of philosophy and even in some of the sciences. And there are gaps here and there concerning, for example, the responsibility of Whewell's procedures which seem to have been rejected with no good reason. But this portrayal needs to be tested, and the discrepancies investigated, if we hope to clear away what may very well be minor points of difficulty or misunderstanding. The discrepancies may very well not be signs of deeper trouble. But even in this case it is certainly of some use to pursue them.

In order to examine these discrepancies, it would seem useful to first turn to the early criticisms of Whewell's theory. We may then see if they really do hit this theory in the way their authors claimed they do. This procedure may enable us to develop a picture of Whewell by starting with the views of his theory offered by his critics and, where necessary, correcting them or supplementing them. This procedure further recommends itself due to the fact that the appreciation of Whewell's system grew slowly and those systematic studies which exist built on the detailed criticism of his contemporaries: the detailed criticisms were integrated into systematic interpretations. How, then, was the traditional view of Whewell's philosophy developed? How did the discrepancies between Whewell's philosophy and its impact emerge? How can we explain them and/or explain them away? And how were the problems which Whewell proposed for consideration taken over, not taken over, reworked or suppressed by philosophers of science from 1840 until today?

# PART ONE

# THE BUILDING OF WHEWELL'S IMAGE

# CHAPTER 1

## IMMEDIATE REJECTION

The case against Whewell was quickly developed and well-defined. Whewell fought for his views until his death – after falling from a horse in 1866 at the age of seventy two – but already in his life-time, as he was Master at Cambridge, Mill's *Logic* had become the standard text in the philosophy of science. This already seems curious, since, if anybody knew Whewell's power as a thinker and the depth and breadth of his knowledge, it had to be his colleagues at Cambridge. Why did they, along with all the rest, turn their backs on his views so quickly? Whewell was, I should note, not the most popular person in Cambridge. He was reputed to have been a tactless man; it would seem likely that he offended more than a few of his colleagues – or that more than a few had taken offence. It is reported that the students went so far as to hoot and jeer from the balconies every time he walked into a college meeting – they relented and cheered him after the death of his wife and a long absence due to mourning. He wept. But he was also known and respected as the most powerful thinker around. Was he deemed lacking in tact merely because he was the dominating intellectual in Cambridge? I find the reports contradictory – which is not, of course, to say they are false, since people can be so. But we may still ask why his colleagues turned their back on him. Personal grudges or jealousy must have played a role. But here the same authors who describe his faults praise to the highest such qualities as his generosity. (He is reported to have given many loans without the least hope of repayment in his later days when, due to his marriages, he had become a wealthy man.) The personal picture becomes just as puzzling as the intellectual one. He was honored by all and the dominant intellectual at Cambridge who won every argument but he had no followers; he was tactless but he observed propriety carefully and he had deep friendships; he was extremely generous and not merely in what was for him a minor affair such as money and yet he was often even despised. Why then, was he rejected?

The response to Whewell's philosophy was quick and decisive and it came from all sides: Whewell's theory made innovations which could not, which should not be adopted. His views were not merely rejected but decisively so. What bothered Whewell's contemporaries so severely that they felt called upon to react so quickly and decisively? And did their reactions build a good foundation for a responsible appraisal of his work? Whewell had presented his central ideas in two major works, his *History of the Inductive Sciences* (1837; 1847; 1857), that is, natural science without mathematics, and his *The Philosophy of the Inductive Sciences* (1840; 1847 and a third edition in three volumes, *History of Scientific Ideas*, 1858; *Novum Organum Renovatum*, 1858; and *Philosophy of Discovery*, 1860). The immediate impact of his work was obviously powerful. One could see at a glance that he offered a clear, simple, systematic alternative to established views. He developed his views in well written treatises which showed enormous erudition. Such erudition alone had to be a great challenge. Indeed, science was, as everyone agreed, based on knowledge of facts and who knew more facts about the history of science than Whewell?

Now, by traditional standards, the conclusions Whewell reached were all wrong: induction, that is, the inference from facts to theories, played no role whatsoever in the growth of scientific knowledge. How could this be explained? How could one give credit to Whewell's knowledge and yet show that the rejection of the view which everyone knew to be true, i.e. that knowledge grew by inductive inference, simply went too far? Or, should one, indeed, accept such a deep change in the theory of science? At first blush Whewell's view could be just as little accepted as the power and erudition of its portrayal had to be honored. This is for us interesting and hopeful, since we find here that the very first commentators faced a problem which bothers us even today, that is, the discrepancy between his apparent power and his apparent inadequacy. Perhaps we may find here an explanation of how a confused view could have such a large impact. Perhaps, then, we will quickly find resolutions to some of the discrepancies in the pictures we find of Whewell.

In order resolve the discrepancy between Whewell's radical and allegedly confused departure from traditional philosophy of science, on the one hand, and the power and erudition of this work, on the other hand, a discrepancy which arose sharply and immediately, it must have seemed to some both desirable and necessary to show, not only that Whewell had rejected traditional views – that, after all, he did not need to be told – but also why this move was unacceptable. Critics needed to explain why superior theories of science could be obtained if one did not do that. Now it would seem that there are various possibilities for doing that, at which we will have to look. We find, indeed, at least five reactions to his work that sealed his fate and that, with a few exceptions, seem to have dominated the discussion to this day. Although most of these reactions object to his departure from standard views, they express different points of view and offer different criticisms. It would thus seem that we need to look at them separately first and then, perhaps, we may see how they contributed to the formation of the standard view of Whewell's work as hopelessly confused.

The first of these five reactions is that of John Herschel, who sought to show that Whewell's rejection of facts as the only source of knowledge led to excesses and brought him too close to Schelling and Hegel. The second reaction is that of John Stuart Mill, who sought to show that Whewell's methodology prevented him from accounting for proof in science and led to confusions concerning, for example, the difference between description and induction. A third reaction came from David Brewster, who claimed that his own biography of Newton had preceded Whewell in the fundamental point of view, but who also thought that Whewell had gone too far. Although Brewster perceptively saw the importance of Whewell's psychology, on which Whewell's methodology was based, and criticized it, the impact of his reaction was limited due to the overly partisan nature of his attack. Nevertheless it certainly contributed to the establishment of the point made by Herschel: they did not need to be told of the importance of hypotheses or conjectures or ideas for science. A fourth reaction was that of Henry Longueville Mansel. According to Mansel, Whewell's theory leads to excess and incompleteness, because it allows for the growth of categories beyond those laid down by Kant. (Even Kant is alleged to have gone too far.) A fifth reaction came from de Morgan who said two things: first, Whewell's theory of the intuitive recognition of necessary truth was inadequate, and second, Whewell did not, as his language indicated, offer a logic, nor was his subject induction. This has been often interpreted as the view that Whewell's theory was unacceptable because it conflicted with the principles of logic: scientific method required proof and Whewell had no theory of that. This was not, I think, the view of de Morgan who agreed with Whewell's appraisal of Bacon, who had great appreciation for Whewell's work, and who did not say that a theory of science ought to be a theory of proof. Exactly what he wished to say is not so clear.

These five reactions were not the only ones. One reaction, which is so far as I know anonymous to this day, was presented in two installments in the *Dublin Magazine* and among other things attacked Whewell's psychology (Todhunter suggests that the author was Archer Butler.) (Todhunter 1876, vol. I, p. 130). G. H. Lewes discussed Whewell's theory in his discussion of Kant (Lewes 1857) and Whewell responded in his *Philosophy of Discovery*. Spencer discussed his view and defended his use of the criterion of inconceivability (Spencer 1853, pp. 521-532); there is an anonymous review in the *Atheneaum* (Anonymous 1840) and his successor in the Chair of Moral Philosophy at Cambridge, Grote, published a discussion of Whewell in 1865 (Grote 1900, pp. 203-225). Todhunter has supplied an overview of the reactions to Whewell (Todhunter 1976, vol. I, pp. 129*ff*, 142*ff*). To this day there is no serious study of Whewell's character and life. Isaac Todhunter put together a two volume work on his life and work (Todhunter 1876) but it is rather bland and critical without taking Whewell's system seriously. We need to look at it as well. Since it contains letters and information concerning Whewell's intellectual development, it is extremely important as a source. But he apparently did the job, because no one else was willing to devote the time to a study of Whewell. There is a companion volume to Todhunter's work which concentrates on Whewell's private life by Mrs. Stair-Douglas, his niece, which was done at the behest of Whewell's family (Stair-Douglas 1881). It is also primarily a collection of correspondence, and thus also useful as a source and especially so if one wishes to get a flavor of Whewell's personality. But it is a mere compilation reflecting the insecurity of the editor in daring to offer an interpretation of the life and work of Whewell.

We must be surprised to note that there were no reviews favorable to Whewell's theory, unless we count de Morgan's as such as I will explain below. Whewell had to fight for his theory alone. It is incredible that a theory of science of such lasting interest could not find a single advocate even when its author enjoyed a high level of prestige and was known personally as a man of high integrity, deep understanding and wide knowledge by all the serious thinkers of his day. Were all of Whewell's critics too quick in their judgment? Did they take account of the strengths of Whewell's view? Did they fail to ask the question whether the change of principles proposed by Whewell would be desirable or not? Did they determine the views of Whewell's philosophy which followed? These are some of the questions we may hope to find answers to in a review of the early reactions to Whewell's philosophy.

Whewell completed his major works at a relatively young age – in his forties. He turned to other interests such as his concern with natural theology. But he worked steadily to revise, elaborate, defend and develop his views until his death. Although his enormous efforts did not lead to the acceptance of his views, I presume that he achieved a philosophical resignation in regard to the fate of his theory. His isolation from the younger generations – even at Cambridge, his own university which he loved – who followed above all John Stuart Mill must have been difficult. His later days apparently saw some happiness due to his two marriages, as well as deep unhappiness due to the death of each of his wives. But the reaction to his work must have laid on him heavily and made the loss of the consolation he had found in his marriages all the more bitter. From our perspective on the philosophy of science, in which the study of the history of science is exceedingly important, he must appear to be one of the most progressive and adventurous thinkers of his day; but he then had the image of a back slider. Issues with only a questionable bearing on his theory of science, such as his use of this theory to defend religion, no doubt played a role. A robust humanism of the sort offered by Mill could, perhaps, not then be countered with a conservative defense of religion. And a theory of science which leads back to the church would certainly seem out-of-date. But I am getting ahead of myself. Perhaps after a study of the reactions to Whewell's philosophy we may better understand his fate.

### 1.1. Sir John Herschel: The True Source of Knowledge Defended

One of the first reactions to Whewell's major works came quickly in a review published at first anonymously by Whewell's friend John Herschel, (Herschel 1841) to whom his *History* was dedicated. Certainly here we can expect to find a serious and sympathetic treatment of Whewell's work. Herschel was not merely a personal friend but a – if not the – leading scientist of the day and a serious student of scientific method. And indeed Herschel adopted a friendly and sympathetic tone which, while it does fit the style of the time in England quite well, also served him superbly in setting the task of his review. This task, he tells us, was no less than the appraisal of what a responsible thinker could accept from the (serious, challenging but sometimes questionable) work of Whewell. He begins,

> If the moral and intellectual relations of man have ever been justly regarded as transcending in importance all other subjects of human interest, the necessary dependence of his duties and responsibilities on his natural faculties must render it impossible to appreciate or define the one without entering into a close investigation and analysis of the other. And if, in the course of this inquiry, it appear, by reference to history and experience, that there exist in the intellectual constitution of our species springs of power and capacities of intelligence which have been but rarely drawn upon, and which have lain, as it were, torpid and dormant during long portions of history and among vast masses of population, it will become not less our interest than our duty to study with the most earnest solicitude the conditions under which the vigorous development and worthy employment of that power and those capacities can subsist. (Herschel 1841, pp. 142-143)

and, somewhat later,

> With the indulgence of such hopes and with the steadily increasing conviction of the possibility of their ultimate realization (of the extension of reason to new subjects of inquiry in the social and intellectual world), which every fresh advance in science affords, arises a necessity of occasionally, and indeed frequently, passing in review both the assemblage of the results obtained and the mode in which they have been obtained; with a view not only to the duly estimating the real value of our actual acquirements and the direction in which further progress appears most immediately practicable, but to the deducing from our experience of the progress already made, maxims and principles available in our future career. Science itself thus comes to be considered as an object not simply of philosophical interest but of inductive inquiry. (Herschel 1841, pp. 147-148)

and,

> Grave questions these, since, as we have already remarked, there are duties and responsibilities, individual and social, attached to their discussion. (Herschel 1841, p. 148)

Having thus set the stage for the development of his main theme Herschel set out to show that, although Whewell had, indeed, produced two impressive works, they had to be read carefully and with reserve and responsibility, for, they abandoned the secure foundation of science. He states,

> These (two views according to which we apply ideas formed antecedent to experience on the world on the one hand or simply distribute phenomena into groups according to analogy on the other hand) at first sight appear widely different, and indeed almost diametrically opposite views of the philosophy of knowledge; and we are thus, at the very outset of the subject, presented with two schools of such philosophy – that which refers all our knowledge to experience, reserving to the mind only a high degree of activity and excursiveness in collecting, grouping, and systematizing, its suggestions – and that, which assumes the presence of innate conceptions and truths antecedent to experience, intertwined and ingrained in the very staple and essence of our intellectual being, and commanding as with a divine voice, universal assent as soon as understood. The author of the very striking, profound, and in many important respects, original works of which we have undertaken to give some account, belongs to the latter of these schools; and, indeed, appears disposed to press its doctrines to a very far greater extent, and to place them in an infinitely bolder prominence than we have been at all aware of having been done before, except perhaps in the writings of some of the later German metaphysicians. (Herschel 1841, pp. 150-151)

In the beginning of the review, then, the specter of Hegel and Schelling, i.e. that of German metaphysicians, is raised. Herschel notes only much later in the essay that Whewell is also critical of German metaphysicians, i.e. Hegel and Schelling, even if he does tend to be too generous toward them (Herschel 1841, p. 230). Here, at the beginning of his essay, Herschel states his own preference as an alternative to the view of Whewell and German metaphysicians. He states,

> We confess in ourselves a leaning, though we trust not a bigoted one, to the other side. And this it is as well to notice at the outset, as it will occasionally tend to place us involuntarily in the apparent position of objectors to the form in which the matter of these works is propounded and treated; while yet we are impressed with a most hearty conviction of their substantive value and importance and a most genuine admiration of the extraordinary talent and boundless command of resources displayed in their conduct. (Herschel 1841, pp. 151-152)

The high praise which Herschel has for the works under review and the strict rejection of the point of view which is there developed creates a tension which needs reconciliation. And Herschel offers one immediately. He states,

> And after all, it seems far from certain that this opposition of views is anything more than apparent; for among the infinite analogies which may exist among natural things, it may very well be admitted that those only are designed, in the original constitution of our minds, to strike us with permanent force, to embody around them the greatest masses of thought and interest, to become elaborated into general propositions, and finally to work their way to universal reception, and attain to all the recognizable characters of truth, which are really dependent on the intimate nature of things as that nature is known to the Creator, and which have relation to their essential qualities and conditions as impressed on them by Him; so that the power bestowed on the mind of seizing on those primordial analogies, and its impulse to generalize the propositions which their consideration suggests, on the one view of the subject are equivalent to its endowment with a direct recognition of fundamental ideas and relations not derived from experience and the evolution from those ideas of necessary truths equally independent of experience in the other. And, perhaps, with this explanation both parties ought to rest content, satisfied that, on either view of the subject, the mind of man is represented as in harmony with universal nature; that, we are consequently capable of attaining to real knowledge; and that the design and intelligence which we trace throughout creation is no visionary conception, but a truth as certain as the existence of that creation itself. (Herschel 1841, pp. 152-153)

Herschel's apparently friendly offer of a compromise is, in fact, a statement of his own program. He offers here his own interpretation of the use of conjectures in science which he deems compatible with his own view, according to which all knowledge is derived from observation. This view is that the mind perceives the existence of analogies and does not impose ideas on the world. We have seen above that he deems himself forced to disagree with Whewell. He seeks to portray these disagreements as merely metaphysical and, therefore, as removable without touching anything of value in the empirical description of science. He states,

> As it is in the first of these books (Whewell's Philosophy) that Mr. Whewell develops and distinctly lays down those peculiar a priori views to which we have before alluded, and to which, as already said, we feel unprepared to yield entire assent, it will be necessary for us to examine rather in detail this part of his work, at the risk, it may be, of some degree of tedium to our non-metaphysical readers; though we shall endeavor, as far as possible, to divest our observations of technical metaphysical phraseology, which, sooth to say, we do not think that very obscure and imperfect science yet sufficiently advanced to indulge in otherwise than sparingly, and as it were emphasis gratis. (Herschel 1841, p. 180)

The first task which Herschel set himself in order to carry out his project of separating the false, metaphysical view from the valuable material found in the works under review was to review Whewell's History in order to appraise the mode of exposition on the one hand and the detailed results on the other hand. He presents this history as ambitious in its plan and yet sufficiently successful to deserve serious attention. His portrayals of it, however, follow the plan explained above: he explains how a sober reader acquainted with the facts would read it. And this is not how Whewell reads it. Whewell, for example, explains the failure of the Greeks to make progress as due to their study of language and their adherence to the concepts found in their own language. Since this language did not already contain the proper ideas for ordering the facts, they were limited by it and blocked from making progress. Herschel states that this is, to the reader familiar with what has been written on the subject, to be interpreted as the view that they had false classifications (Herschel 1841, p. 168).

A few pages later Herschel interprets a remark made by Whewell on an "inductive" act of ordering in terms of a year, a month, etc. as only expressing induction as he – and not as Whewell – conceives it (Herschel 1841, pp. 171-173). Whewell's passage on Kepler is mentioned without criticism, that is, the story of Kepler's use of conjectures and refutations is deemed no problem by Herschel for his own view of progress by inductive inference. But modern readers at least must find this puzzling. Joseph Agassi, in his article on Herschel's methodological treatise has shown what difficulties Herschel had on such points (Joseph Agassi 1971). And here Herschel gives no explanation whatsoever as to how they are to be reconciled or why conjectures are even needed. As to Whewell's treatment of science from Newton on, Herschel deems it a mere classification, which tolerably well suits the author's purpose (Herschel 1841, p. 182). But the enormous scope of Whewell's work cannot be brought together in one book and the "fact is that the eleven books, of which the remainder of Mr. Whewell's history consists, must rather be regarded as philosophical epitomes of their several subjects – outlines struck with a large and free hand, and destined to fix attention on leading features (though traced with perfect mastery and with consummate skill) – than as digested histories of the above enumerated branches" (Herschel 1841, pp. 183-184).

Whewell's history, then, offers a broad scheme and interesting discussions but no new or plausible view of the development of science – at least not when it is read "properly" and the metaphysical aspects or the peculiar views of induction are interpreted "correctly." Herschel thereby dismisses and explains away the core of Whewell's ideas. He does not note any serious problem that Whewell might even have attempted to solve. And problems that give his own view great difficulties are not mentioned. Herschel has thereby reduced Whewell's history to a failed attempt to provide an overview of the history of science which is misguided, but whose author has nevertheless managed to produce some good results.

Thus far we must be a little disappointed. The discrepancies between the standard interpretation of his work as confused, on the one hand, and its far-reaching impact, the renewed interest in his work and even the attempt to revive it, on the other hand, as well as the discrepancy between the shallowness of the responses to it and the apparent depth of the problems his view put on the (hidden) agenda are those discrepancies we would like to clear up. The basis of these discrepancies is the real or merely apparent power of Whewell's view. This real or apparent power is not really taken account of. It is neither explained nor (successfully) explained away. Herschel barely describes the plan of Whewell's *History* and it is certainly not explicitly evaluated. It is merely dismissed. But we do find an opposing point of view. Perhaps we will find this developed in the study of Whewell's *Philosophy of the Inductive Sciences*. And indeed, the most important critical treatment of Whewell's work comes in the first part of Herschel's treatment of Whewell's *Philosophy*. It is there that Herschel discusses the origins of ideas, that is, the central metaphysical point which separates his theory from Whewell's.

We may, unfortunately, be somewhat skeptical concerning Herschel's discussion right from the start. Herschel isolates the appraisal of Whewell's philosophy from that of the rest of his investigations, as if it had nothing to do with them, from, for example, the study of the history of science. Indeed, Whewell's descriptions of the process of induction, by which those ideas are found which unify facts, are accepted as quite precise and even "luminous" (Herschel 1841, p. 193). But this is clearly part of a strategy that is designed to separate the so-called metaphysical elements from Whewell's philosophy in order to save the proper and enlightening descriptions of science and its methods. These may then be treated as nothing particularly new. Rather, they are nice formulations and descriptions of what the educated reader might have known before. This strategy avoids coming to terms with the possible relevance of Whewell's historical considerations for his philosophical system.

Suppose, however, we try to take Herschel's approach on his own terms. How far can we go toward resolving our discrepancies with it? Whewell had four different kinds of arguments that were designed to show that ideas had to be imposed on the world. The first, somewhat more Kantian argument, was to argue that the necessity of knowledge required that ideas or knowledge not be derived from experience: no experience could justify, could allow us to deduce this necessity of our ideas, so, we presume a source of knowledge outside of experience to explain it. Whewell states,

> By the course of speculation contained in the last three chapters, we are again led to the conclusion which we have already stated, that our knowledge contains an ideal element, and that this element is not derived from experience. For we have seen that there are propositions which are known to be necessarily true; and that such knowledge is not, and cannot be, obtained by mere observation of actual facts. It has been shown, also, that these necessary truths are the results of certain fundamental ideas, such as those of space, number, and the like. Hence it follows inevitably that these ideas and others of the same kind are not derived from experience. For these ideas possess a power of infusing into their developments that very necessity which experience can in no way

> bestow. This power they do not borrow from the external world, but possess by their own nature. (Whewell 1967b, vol. I, p. 74)

Herschel seeks to meet this argument (He does not quote it, as I have done). by denying that necessity is at stake. Rather, a correct view of the matter would show that mere universality of scientific statements or the conviction of their universality is problematical. And this, he contends can be explained as a result of the

> nature of the inductive propensity – by the irresistible impulse of the mind to generalize ad infinitum, when nothing in the nature of limitation or opposition offers itself to the imagination – and by our involuntary application of the law of continuity to fill up, by the same ideal substance of truth, every interval which uncontradicted experience may have left blank in our inductive conclusions. What we contend for is, not that the propositions of geometry are other than necessary and universal, but that space being a reality (or a real condition), the mind, applying itself to that reality, discovers its properties by such application, which is experience, and in axiomatic propositions. For what, we may ask, can impress us with a sense of truth other than a clear perception of meaning? And what is a perception of meaning other than an intellectual experience of the real qualities and relations of the objects of our thoughts, as exemplified in special cases? (Herschel 1841, p. 188)

The second type of argument Whewell offered was that specific ideas, which we found in knowledge such as space, time and cause, could not be derived from experience but had to be imposed on it. Following Kant he deemed such ideas to be the conditions for the experience of objects, as we perceive them, and not a product of this experience (Whewell 1967b, vol. I, pp. 84*ff*, 128*ff*, 165*ff*).

Once again Herschel argues that these ideas are products of our experience. In a somewhat Kantian fashion he suggests, for example, that cause and time are products of internal observations and that space is merely a product of our observation of the nature of reality but, quite contrary to Kant, that ideas must be completely derived from observation.

The third type of argument which Whewell uses is more psychological, that is, he points out that we must learn how to perceive objects in space and we do this in an active, probing fashion and not by passively observing and making inferences. He can and does thereby make use of the then new and exciting physiological researches in vision conducted by Charles Bell concerning such phenomena as the probing nature of unconscious movements of the eyes (Whewell 1967b, vol. I, pp. 111*ff*).

This is ignored by Herschel. But it later became quite important in the dispute with Mill, who sought to develop Herschel's point of view, as well as in subsequent psychological research based on Mill's view.

The fourth type of argument offered by Whewell for the necessity of imposing ideas on the world is methodological. Whewell argued that before we have an idea we do not know which facts are relevant or even what the facts are. We cannot then merely abstract from the facts as they are given to us by simple observations. Rather, we need first to "colligate" them, that is, to bring some order to them. Only after this has been done can we see how they fit together (Whewell 1967b, vol. II, pp. 49*ff*; 1971, p. 263).

This procedure may appear to be circular, in that we presume a theory in order to prove it, but is not, because ideas that bring order to some facts may fail to do so with others. So, we may test them, refute them and move on to better ideas. Herschel ignores this aspect completely. We see already that we must go further to resolve at least some discrepancies. How does the tradition respond to this argument?

Herschel does not so much present new arguments concerning the issues at hand. Rather, he reviews the formulas of the standard inductivist, associationist view that had been developed by Bacon, Locke, Hume and others. These formulas were designed to show how any idea may be traced back to its source in experience alone. But they are not complete. It is asserted that every idea, say those found in geometry, can be reduced to experience, that is, their entire content consists of nothing outside of observations and directly observed relations between them. But the arguments are, of course, not complete: Herschel provides no method of carrying out such reductions. So, he has to fall back on transcendental arguments of the sort that science is not possible without such reductions.

Can Herschel make his point with transcendental arguments without such a method? We would expect him, given his point of view, to give us empirical arguments or inductive inferences to the truth of his position by showing us the facts. But he does not do this even in an essay in which it is maintained that this is the only legitimate form of argument. Whewell's alternative is not considered. We may, I think, be puzzled indeed concerning his own conception of what he was doing and the incompleteness of his discussion.

Perhaps we should not complain. This is, after all, a mere review and not intended as a substitute for a treatise explaining the origin of ideas such as those offered by Locke or Hume. Yet, it still appears somewhat superficial. For, it fails to note the strength of the Kantian arguments and ignores those concerning perception and, intimately related with these, those concerning the methods of science, that is, those referring to the probing activity of the mind in ordering experience with ideas – a theory which is found in Whewell's writings but not in Kant's. Herschel fails to note the strength of the Kantian arguments in that he presents mere sketches of how ideas might arise without filling in obvious gaps. The assertion, for example, that our idea of cause comes from the observation of acts of will (Herschel 1841, pp. 208-209) is obviously inadequate. Herschel thereby seeks to weaken the notion of cause and to render it specific to varying cases and thereby dependent on each case. But this will not, I think, save the day since it remains programmatic: a full demonstration of the derivation of all ideas from experience would require explicit derivation of these ideas from definite observations. But this would require that Herschel deal with precisely those arguments from psychology and physiology, which he has omitted from his discussion.

If Herschel wishes to satisfy himself with the mere expression of a point of view, this is certainly legitimate. And, indeed, this review is a fine statement of the sharp difference between Herschel and Whewell, which in the literature has been so often ignored – under the influence of Herschel, we might add. But he seems to deem it an argument. And this must leave us puzzled: if it is an argument, it is a modified transcendental argument of the sort proposed by Whewell and rejected by Herschel.

Herschel seeks to clarify the point of his discussion by saying,

> In making these remarks on Mr. Whewell's a priori doctrines, we are most anxious to be understood as limiting our disapproval strictly to the point of view from which he has contemplated his subject. In its handling there is everything to admire, nor are we aware that we have ever in the same compass encountered such a mine of recondite thought, searching inquiry, and pointed and brilliant illustration. But to these views he recurs again and again, and always with increasing decision, vives acquirit eundo, as if their force had grown upon him in their contemplation. (Herschel 1841, pp. 219-220)

And he continues in the same vein discussing both further aspects of Whewell's methodology such as colligation of facts and consilience of inductions and specific fundamental ideas. Having removed the erroneous metaphysics, however, he allows himself to be often appreciative, accepting most of what is offered with a few critical comments here and there, with a few reminders of the need for reduction of all ideas to experience.

The upshot, then, of Herschel's review was that, although most of Whewell's methodological and historical observations were to be accepted, these were deemed mere illustrations, sometimes detailed contributions to Herschel's own superior conception of science. Whewell's systematic or theoretical efforts were granted no significance whatsoever. But a Trojan horse had apparently been let in the door. The standard Baconian program defended in outline form by Herschel had to be developed to complete the responses Herschel proposed as proper. In order to do this one had to talk once more of the origin of ideas and of the systematic relation between this theory and the methodology of science. Herschel left these tasks for others to carry out. He was apparently satisfied with the response he gave. Perhaps he simply knew no better. We need to look further, however, to see if this point of view can be developed to resolve the discrepancies we find between the assertion of Whewell's theoretical insignificance, on the one hand, and his influence and interest, on the other hand. We do not need to look far. John Stuart

Mill found that he had to take up the challenge and sought to give more detailed responses.

## 1.2. John Stuart Mill: The Necessity of Strict Rules for Inductive Inference

In order to seek resolutions to our puzzles concerning the very bad opinion of Whewell's philosophy, which became established so quickly, on the one hand, and the lasting interest, influence and power of this view, on the other hand, we cannot avoid a close look at the work of John Stuart Mill. It was Mill's view, rather than Whewell's, which quickly became established; he was the new authority. Mill was also a man of unquestionable integrity, a firm believer in the power of reason, and a leader in battles for social justice and the respect for reason. If we may expect to find a fair and balanced treatment of Whewell anywhere, we may expect to find it in the work of Mill. Mill, we may presume, must have had powerful and good reasons for siding with and developing the view of Herschel rather than that of Whewell.

As Mill began his investigations into inductive inference, he had no intention of getting into the business of criticizing Whewell's philosophy. And, indeed, he wrote no essay on Whewell's work. Rather, he included his commentaries on Whewell's views in the various editions of his *Logic*, expanding here and deleting there, as the state of the discussion demanded. It might seem, then, that this dispute was not so important. Perhaps a mere review of more substantial treatments elsewhere. But, contrary to this impression, the battle with Mill was that discussion which settled the matter in favor of the traditional view. Surely, then, we may find here the best commentary on Whewell and the most powerful case for that interpretation of Whewell's work, which has come to be established. In order to see what this commentary is, we need to take a look at Mill's program, how it led him into the debate with Whewell, and what he had to say about Whewell's arguments.

### 1.2.1. *Mill's Program*

The program that Mill set himself in his Logic was to develop a complete and systematic view of the subject. Since, on his view, science proceeds by the application of those rules clearly and explicitly stated in this system for the first time, he purports to offer a complete and systematic view of scientific method as well. He states,

> This book makes no pretence of giving to the world a new theory of the intellectual operations. Its claim to attention, if it possesses any, is grounded on the fact that it is an

> attempt not to supersede, but to embody and systematize, the best ideas which have been either promulgated on its subject by speculative writers, or conformed to by accurate thinkers in their scientific inquires. (Mill 1973, vol.7, p. cxi)

and

> On the subject of induction, the task to be performed was that of generalizing the modes of investigating truth and estimating evidence, by which so many important and recondite laws of nature have, in the various sciences, been aggregated to the stock of human knowledge. (Mill 1973, vol. 7, p. cxii)

The view which Mill developed is the same as that which Herschel had earlier portrayed in his own treatise on methodology and in his review of Whewell's major works, that is, he develops the view that all inference is from particulars to particulars. Mill proposed to develop this view systematically so as to show its foundations. And this project should be carried through without any reference to philosophical or metaphysical disputes. He states,

> Logic is common ground on which the partisans of Hartley and Reid, of Locke and of Kant, may meet and join hands. Particular and detached opinions of all these thinkers will no doubt occasionally be controverted, since all of them were logicians as well as metaphysicians; but the field on which their principal battles have been fought, lies beyond the boundaries of our science. (Mill 1973, vol. 7, p. 14)

### 1.2.2. *The Conflict with Whewell*

Along the way and apparently quite without intent Mill came into conflict with Whewell. He had to, for, it was Whewell who had, more effectively than anyone else, pointed to the problems which confronted any attempt to reduce all knowledge to facts and inductive inferences from them. Already in the preface Mill notes the importance of Whewell for his own work and, indeed, in the two ways in which Herschel had noted his importance, that is, as a rich source and a hard competitor. He states,

> Whatever may be the value of what the author has succeeded in effecting on this branch of his subject, it is a duty to acknowledge that for much of it he has been indebted to several important treatises, partly historical and partly philosophical, on the generalities and processes of physical science, which have been published within the last few years. To these treatises, and to their authors, he has endeavored to do justice in the body of the work. But as with one of these writers, Dr. Whewell, he has occasion frequently to express differences of opinion, it is more particularly incumbent on him in this place to declare, that without the aid derived from the facts and ideas contained in that gentleman's *History of the Inductive Sciences*, the corresponding portion of this work would probably not have been written. (Mill 1973, vol. 7, p. cxiii)

Mill found himself quite quickly in conflict with Whewell for quite obvious reasons. Whewell denied the basic presupposition of Mill's *Logic*, that is, that all inference is from particulars to particulars. And Whewell had powerful arguments from metaphysics, from the analysis of scientific ideas, from psychology and from methodology for his own point of view. How, then, could Mill maintain that his own project was neutral to such fields, a study in logic alone and not a polemical treatise as well?

The conflict between Mill's plan to restrict himself to logic and to leave all metaphysics aside, on the one hand, and his polemics against Whewell, on the other, has been, of course, noticed. And we might expect that the conflict, and thus the importance of the discussion as a whole, is more apparent than real. But attempts to resolve it in favor of Mill's view by explaining it away seem terribly unconvincing. R.F. McRae, for example, in his introduction to Mill's *Logic* suggests that Mill distinguishes between propositions which are known directly and those which are known by inference and that his view rests on no assumptions concerning the former (Mill 1973, vol. 7, pp. xxiv-xxv). This is simply not true: Mill's logic rests on the view that we know facts directly by observations without use of theory. This is exactly what Whewell denies. Mill's *Logic* is, indeed, a systematic development of a particular view of the source of knowledge. It is an attempt to show that, and how, all knowledge may be obtained from this source. As such it is an attempt to deliver on the promises which Herschel had made, but which he did not or could not fulfill. Herschel had offered a program with suggestions as to how it should be, or must be, carried out. But he had not done so systematically. Mill's *Logic*, then, became the standard systematic development of this point of view and thereby the standard basis for the response to, and dismissal of, Whewell's philosophy of science. It would seem, then, that we need to look more closely at the conflict between Whewell and Mill and that we need to take it more seriously than the tradition to which McRae belongs has often done.

The most serious challenge to Mill's view came not merely from Whewell's philosophical system but more decisively from the historical studies which constituted the foundation of that system. Whewell maintained that the views defended by Herschel and Mill could not account for the facts of the history of science. Ideas could not be reduced to sensations and scientific discovery required the imposition of ideas, at first of guesses, on the facts. He made the first point in his response to Herschel (Whewell 1967b, vol. II, pp. 669-668) and the second in his response to Mill (Whewell 1971, pp. 238*ff*).

Good inductivists, such as Herschel and Mill were, could not ignore these claims about their position from such an authority as Whewell. In order to defend his position, then, Mill was led, apparently against his intentions, to give considerable attention to the alternative offered by Whewell and especially to an attempt to show that his view could account for the facts of scientific research. He saw the importance of Whewell's works as soon as they were published (1837 and 1840) and was thus led to add references and/or discussions of Whewell's work relatively late in the work on the *Logic* which had begun in 1830 (J. M. Robson, Textual Introduction in Mill 1973, vol. 7, p. lxiii and p. lxvi). It was in this later stage (1837 and later), after he had set the framework, that he dealt with the crucial questions of trains of reasoning, demonstrative science and induction, and then, it would seem, with Whewell, whose work on these points concerning the philosophy and history of science could no longer be ignored.

Mill's plan to defend the basic principles of his logic on the one hand and to systematically develop them on the other hand posed two sorts of problems. Commentaries on Mill's *Logic* have understandably tended to emphasize the second set of problems – those of developing his view – and to ignore the first set – those of defending his view – as much as possible. Whewell's systematic alternative quickly fell into such disrepute that no serious study of it was deemed desirable. If, however, we wish to see how this judgment was built and to solve our puzzles concerning it, we will have to reverse the procedure. Here, then, I will largely ignore such aspects of Mill's view as his four methods of induction or his inductive justification of induction and concentrate instead on his treatment of those problems which arose in his attempt to come to terms with Whewell's *History and Philosophy*.

### 1.2.3. *Mill's Response to Whewell*

Mill had a quite clear and simple view of all arguments. But it led to difficult problems. Mill defended the view that all deduction is merely shorthand for inductive inference, that is, a universal proposition in a proper argument is an induction from observed facts. All inductive inferences are arguments from facts to facts. A deduction, then, is an inference from some observed facts to some new one. Deduction is important in those sciences in which we have found principles and wish to apply them to new cases, since here deduction provides a useful shorthand, without thereby altering the fact that we are inducing to the new cases. It is clear that this view had to come into direct conflict with Whewell's view of mathematics and of science. Mill saw this and began the serious debate (Mill 1973, vol. 7, pp. 227-251).

Mill's responses to Whewell can be divided into his discussions of mathematics and proof, on the one hand, and of hypothesis and observation, on the other. In regard to mathematics and proof, Mill sought show that the study of mathematics was the study of facts alone by deeming definitions to be conventional and/or fictional and axioms to be based on induction. He deemed hypotheses to be regularly needed for discovery but always reducible to facts and thus provable by induction. Observation involved unconscious inferences and thereby went beyond what was observed, but this (unconscious) inference was always inductive and ultimately based on sensations.

Each of these responses failed for the same two reasons: facts and theories were not separated in the desired way and the needed theory of reduction was a mere program. Mill also had criticisms of Whewell's view and especially of his view of proof. Whewell's theory here faced severe difficulties, since it was based on a view of the intuition of truth, which did not seem at hand. He also defended the view that truth changed (Whewell 1971, pp. 250*ff*, 253, 492*ff* and p. 502).

Even here his view was not given its due, since the problems he raised were ignored. These were: What alternative to inductive inference can there be? And how can we account for the change of scientific theories found in the history of science?

### 1.2.3.1. *Mathematics and Proof*

The first aspect of the debate primarily concerns mathematics. There are three central issues: first, Is the subject matter of mathematics facts alone? second, How do we learn the axioms of mathematics? and, third, How do we know that they are necessary? or – as Herschel and Mill developed the debate – What does it mean to say that the axioms are necessary? At the end of this discussion Mill notes the identity of his views and those of Herschel. He states,

> The Quarterly Review [LXVIII,] for June 1841, contained an article of great ability on Dr. Whewell's two great works (since acknowledged and reprinted in Sir John Herschel's Essays [London: Longmans 1857] which maintains, on the subject of axioms, the doctrine advanced in the text, that they are generalizations from experience, and supports that opinion by a line of argument strikingly coinciding with mine. (Mill 1973, vol. 7, p. 248)

Mill begins the discussion of his central disagreements with Whewell by seeking to reconcile his own view that the source of all mathematical knowledge is facts with the apparent necessity to base geometry on definitions and axioms. On Whewell's view definitions and axioms are intertwined and together depend on fundamental ideas. But Mill can obviously not allow this: all mathematics must be based on induction and the result of an induction should be, say, an axiom. What role, then, do definitions play in mathematics? He suggests that they are merely hypothetical and, therefore, that they may be allowed into mathematics without a foundation in induction. This is not to say that they are arbitrary. On Mill's view they would seem to be determinations of the objects which mathematicians study and as such should not contain falsehoods (Mill 1973, vol. 7, p. 228). But operations with fictitious objects are not forbidden (Mill 1973, vol. 7, p. 229). The need for definitions, then, do not show that some type of entity other than facts is the subject matter of mathematics.

This argument is only part of the story and preparatory for his discussion of other, more important disagreements. Mill deems axioms to be needed for the attainment of mathematical knowledge. Are they based in induction? And is induction sufficient to prove such (necessary) propositions? In regard to the first question Mill adopts Herschel's thesis: the axioms of geometry are learnt first by experience, that is, by induction. The extension of this knowledge beyond experience is due to imagination. Whewell's objection, that we cannot directly compare ideas to reality, since we use the ideas to perceive reality, and hence must know their truth in a different way, is dismissed in the same way as Herschel had before, merely on the basis of common sense: we see the world, they say, the way it is. They offer, in effect, a transcendental argument for their view.

Mill sets the terms of the debate for the third question – How do we know that axioms are necessary? – by arguing as follows:

> Although Dr. Whewell has naturally and properly employed a variety of phrases to bring his meaning more forcibly home, he would, I presume, allow that they are all equivalent; and that what he means by a necessary truth, would be sufficiently defined, a proposition the negation of which is not only false but inconceivable. I am unable to find in any of his expressions, turn them what way you will, a meaning beyond this, and I do not believe he would contend that they mean anything more. (Mill 1973, vol. 7, p. 238)

By setting the terms of the debate concerning our knowledge the necessity of axioms in this way Mill is enabled to restrict the debate to the question of how the inconceivability of the rejection of scientific ideas is to be explained. He has here two strategies. One is, of course, to explain inconceivability on his own view. He states,

> The account which I should give of this progress of the scientific mind is somewhat different. After a general law of nature has been ascertained men's minds do not at first acquire a complete facility of familiarly representing to themselves the phenomena of nature in the character which that law assigns to them. The habit which constitutes the scientific cast of mind, that of conceiving facts of all descriptions conformably to the laws which regulate them – phenomena of all descriptions according to the relations which have been ascertained really to exist between them; this habit, in the case of newly discovered relations, comes only by degrees. So long as it is not thoroughly formed, no necessary character is ascribed to the new truth. But in time, the philosopher attains a state of mind in which his mental picture of nature spontaneously represents to him all the phenomena with which the new theory is concerned, in the exact light in which the theory regards them: all images or conceptions derived from any other theory, or from the confused view of the facts which is anterior to any theory, having entirely disappeared from his mind. The mode of representing facts which results from the theory, has now become, to his faculties, the only natural mode of conceiving them. It is a known truth, that a prolonged habit of arranging phenomena in certain groups, and explaining them by means of certain principles, makes any other arrangement or

explanation of these facts be felt as unnatural: and it may at last become as difficult to him to represent the facts to himself in any other mode, as it often was, originally, to represent them in that mode. (Mill 1973, vol. 7, pp. 245-246)

The second strategy is to show some problems which result from the acceptance of Whewell's view that inconceivability should serve as a criterion of truth. He doubts that inconceivability should be so important since this is an affair of the mind (Mill 1973, vol. 7, p. 238).

He supports his view by pointing to changes in the history of science. This should be Whewell's strength, and in most respects it is, but here Mill can nevertheless use it with some effect against him. On the basis of his historical investigations Whewell finds that he must assume that necessary truth changes and grows. He bites the bullet and maintains that a theory can be true and later not true, can be at one time inconceivable and later (or earlier) conceivable. He defends this view by explaining that two views which seem irreconcilable may be so developed so as to remove the conflict (Whewell 1971, p. 251) and/or one may explain another at a deeper level (Whewell 1971, pp. 502-503).

This position is hard to defend and Mill has here an opportunity. But he still makes it too easy for himself. He ignores the problems which Whewell seeks to solve and thereby the strengths of Whewell's solutions. There is a gap: How can Mill explain changes in the history of science? Are false theories not part of science? In this case not only Descartes but also Kepler or Galileo must go. If they are part of science, can scientific theories be false? What then of induction which is the basis of all science because it offers proof? These questions were simply suppressed and as a consequence the relative ability of Whewell's theory *vis à vis* Mill's theory to solve particular problems was never properly discussed. Perhaps here we find the beginning of our answer to the puzzles concerning Whewell's poor reputation and lasting influence: Did he seek to solve problems – even poorly – which others failed, or refused, to even take notice of?

### 1.2.3.2. *Scientific Method: Hypotheses and Observation*

Although Mill deems the methods of mathematics to be identical with those of science, he handles the theory of the methods of empirical sciences separately. The superiority of his view as a theory of physical science needs to be demonstrated by showing that it can give a better, more complete account of the facts of scientific research. In addition to dealing with those problems which arise within Mill's own system such as the explication of specific forms of induction, Mill had not only to show that his view could account for those factors used by Whewell to support his own theory but also to show that his account led to a better view of science. In order to show the superiority of his own view, Mill raised one fundamental objection to Whewell's theory: this theory cannot account for proof in science. He states,

> Dr. Whewell's theory of the logic of science would be very perfect if it did not pass over altogether the question of Proof. But in my apprehension there is such a thing as proof, and inductions differ altogether from descriptions in their relation to that element. Induction is proof; it is inferring something unobserved from something observed: it requires, therefore, an appropriate test of proof; and to provide that test, is the special purpose of inductive logic. When, on the contrary, we merely collate known observations, and, in Dr. Whewell's phraseology, connect them by means of a new conception; if the conception does serve to connect the observations, we have all we want. As the proposition in which it is embodied pretends to no other truth than what it may share with many other modes of representing the same fact, to be consistent with the facts is all it requires: it neither needs nor admits of proof; though it may serve to prove other things, inasmuch as, by placing the facts in mental connexion with other facts, not previously seen to resemble them, it assimilates the case to another class of phenomena, concerning which real Inductions have already been made. Thus Kepler's so-called law brought the orbit of Mars into the class ellipse, and by doing so proved all the properties of an ellipse to be true of an orbit: but in this proof Kepler's law supplied the minor premise and not (as in the case of real inductions) the major.
>
> Dr. Whewell calls nothing Induction where there is not a new mental conception introduced, and everything induction where there is. But this is to confound two very different things, Invention and Proof. (Mill 1973, vol.7, pp. 304-305)

Mill has here a program for showing the superiority of his account of the facts of scientific research. On this program the task of developing a true theory of science consists in showing how scientific knowledge is generated and proved from observations alone. In order to accomplish this task Mill needs to show, contrary to Whewell, that there is an independent basis of facts from which one may make inductions, that inductions provide proofs and that Whewell's view does not. I will not here discuss the second task, since it constitutes the positive aspect of his program and since, from the publication of Mill's work to the present, philosophers have continually tried to improve Mill's work to fulfill this claim. This exercise has been justified on the grounds that there is no alternative – Whewell cannot offer a theory of proof – and that there must be a basis in fact for scientific inference. It would be equally foolish to defend Whewell's theory of proof by intuition as to criticize Mill's theory of inductive inference.

I may note here that Whewell seems to offer two criteria or signs of truth of scientific theories. One is that the theory is confirmed by, to use modern terminology, independent tests (Whewell 1967b, vol. II, pp. 65*ff*). Mill correctly rejects this as proof. But he fails both to see the methodological importance of such tests and to appreciate Whewell's argument (Mill 1973, vol. 7, p. 500). According to this argument no example in the whole history of science can be found, in which a theory has made a correct prediction about a new class of events, and then has been subsequently shown to be false (Whewell 1967b, vol. II, p. 67).

The apparent first criterion is, however, no independent criterion, but merely a special case of the second, which is his theory of proof. The second criterion of truth or method of proof is that a theory is proven to be true when it can explain all the facts in a specific field combined with the realization that no other theory can. Mill rejects this view on the ground that Whewell himself has shown, contrary to his own view, that even exploded scientific hypotheses may always or nearly always be modified to account for the facts. He here refers to Whewell's discussion of the modification of Descartes' theory of vortices to reconcile it with Newton's (Mill 1973, vol. 7, pp. 502-503). He also objects once more to Whewell's appeal to inconceivability even in those cases where no alternative seems available (Mill 1973, vol. 7, p. 503).

These are no doubt serious objections. Yet Whewell's theory was stronger than Mill conceded. Both parties to the dispute held that science in fact attained the truth by proving its theories true. Whewell argued that his hypothesis could explain the facts, facts which were then agreed upon, even though we are no longer agreed about them today, and no other theory could. If one correctly saw the facts of the history of science, one would see the necessity of his explanation. In response to Mill he repeatedly pointed out the historical adequacy of his view and the inadequacy of Mill's. The central example was that of Kepler, but he returned to the theme with other examples as well (Whewell 1971, pp. 251, 253, 254, 255, 264 and 273).

Let us turn, then, to the discussion of one of those facts Whewell found it necessary to account for, that is, to the question of the nature of the observational basis of science. Is it dependent on ideas or independent of them? Whewell thought that his historical studies as well as other arguments demonstrated such dependence. His most forceful statements, though by no means his only ones, of this dependence are, perhaps, his discussions of the antithesis of philosophy.

There are two central aspects to Mill's discussion of the observational basis of science. On the one hand, Mill sought to show that observations used in scientific inductions could be gathered, at least in principal, independently of the use of the theories. He conceded that, as a matter of fact, hypotheses regularly play crucial roles in gathering facts. He sought, however, to explain this phenomena away as a mere technical matter of efficiency. He further sought to show that even though normal observation was not observation merely but (unconscious) inference as well, the foundation of his view and that of scientific theories in direct observation was firm.

The two central tasks which Mill sought to carry out, that is, to explain away the role of hypotheses in fact gathering and the role of ideas in perception, presume that Whewell's views on these matters are true. Mill states in regard to the first: "It is allowable, useful and often even necessary, to begin by asking ourselves what cause may have produced the effect, in order that we may know in what direction to look out for evidence to determine whether it actually did" (Mill 1973, vol. 7, pp. 498-499). But he demands that in the end we reduce the law to particulars or that we induce it from the facts which we may find by deduction or ratiocination. The specific methods of induction combined with the independence of the knowledge of the facts should secure the truth of the propositions so induced.

Since the role of hypotheses even in the gathering of facts had been granted by Mill, he had to show that this method of procedure or of discovery does not violate or undermine his own demand for inductive proof. In order to accomplish this task he maintained that it was possible to describe facts in various ways. But such alternative descriptions were not yet theories subject to induction. They could all be true – as Whewell had erroneously maintained for theories – but could be happy or not in that they would serve as good or poor bases for inductions.

This issue was crucial in the debate between Mill and Whewell and it is here that we may hope to find a foundation for the very poor view of Whewell's work. For, Whewell's whole system would, indeed, collapse, if Mill could carry out the task which he set himself. Could, then, Mill succeed in demarcating descriptions from inductions in the way he intended to? Could he show that various (true) descriptions of the same phenomena were possible and also different from any induction? Or was Whewell correct in maintaining that this distinction could never be properly drawn, since descriptions had the character of theories?

Mill chose for his central example the discovery by Kepler that the motion of the planets was elliptical. This, he said, was a description and to be distinguished from induction. He sought on this basis to construct a consistent interpretation of Kepler's research as the following quotations make clear. He states,

> The descriptive operation which enables a number of details to be summed up in a single proposition, Dr. Whewell, by an aptly chosen expression, has termed the Colligation of Facts. In most of his observations concerning that mental process I fully agree, and would gladly transfer all that portion of his book into my own pages. I only think him mistaken in setting up this kind of operation, which according to the old and received meaning of the term, is not induction at all, as the type of induction generally; and laying down, throughout his work, as principles of induction, the principles of mere colligation. (Mill 1973, vol. 7, p. 294)

This distinction should apply to Kepler, that is, Kepler's description of the path of the planets was "another and better mode of representing the same observations (as the theory that they were carried round by wheels)" (Mill 1973, vol. 7, p. 295).

The case, Mill contends, is directly analogous to that in which a navigator sails around an island and thereby discovers that it is such and so describes it (Mill 1973, vol. 7, pp. 295-296). He continues,

> Subject to the indispensable condition which has just been stated, I do not conceive that the part which conceptions have in the operation of studying facts, has ever been overlooked or undervalued. No one ever disputed that when we include a multitude of things under a general expression, there is implied in the expression a conception of something common to those things. But it by no means follows that the conception is necessarily pre-existent, or constructed out of the mind by its own materials. If the facts are rightly classed under the conception, it is because there is in the facts themselves something of which the conception is itself a copy; and which if we cannot directly perceive, it is because of the limited power of our organs and not because the thing itself is not there. (Mill 1973, vol. 7, p. 296)

and,

> Yet is a fact surely, that the planet does describe an ellipse; and a fact which we could see, if we had adequate visual organs and a suitable position. (Mill 1973, vol. 7, p. 297)

Now it might appear that the difference between descriptions and inductions is clear and that we have found a firm basis for the negative appraisal of Whewell's work. But then we discover that descriptions may also be hypotheses and, indeed, in need of verification. He states,

> And lastly, we must add to these all hypothetical modes of merely representing, or describing, phenomena; such as the hypothesis of the ancient astronomers that the heavenly bodies moved in circles; the various hypotheses of excentrics, deferents and epicycles, which were added to that original hypothesis; the nineteen false hypotheses which Kepler made and abandoned respecting the form of the planetary orbits; and even the doctrine in which he finally rested, that those orbits are ellipses which was but an hypothesis like the rest until verified by facts. (Mill 1973, vol. 7, pp. 494-495)

Mill has now asserted that descriptions are unique and may all be true, that they are known by observations and that they at times need to be verified. The confusion becomes apparent as Mill later tries to resolve the conflict. He states,

> We observed, in the second chapter, that the proposition "the earth moves in an ellipse", so far as it only serves for the colligation or connecting together of actual observations, (that is, as it only affirms that the observed positions of the earth may be correctly represented by as many points in the circumference of an imaginary ellipse, is not an induction, but a description: it is an induction, only when it affirms that the intermediate positions, of which there has been no direct observation, would be found to correspond to the remaining points of the same elliptic circumference. (Mill 1973, vol. 8, p. 647)

We find, then, that Kepler's theory is both a description and an induction. The theory as description should, he continues, be the basis for an induction. But his distinction has collapsed just where he needed it most, that is, in those cases where research begins with hypotheses. This is, as Whewell's historical research had shown, the normal and most important type of scientific research. It appears, then, that our search for a resolution to our puzzles cannot be found in this aspect of Mill's work. Indeed, it seems to sharpen it, since Whewell's most illustrious opponent, a man of the highest integrity, has not adequately answered Whewell on this fundamental point of disagreement. Indeed, this kind of attempt to save inductivism was largely dropped. Since there has, however, been a recent attempt by Gerd Buchdahl to save Mill's view to some degree, as I explain below, we may, perhaps, reserve our judgment. But for now we have no answer.

The second central aspect of Mill's discussion of the observational base of science, that of the role of ideas in even simple observation or perception, may arouse our curiosity even more. This issue is just as crucial for the two points of view, since Whewell's view was based in his psychology as well as his views of scientific method. Here as well he had pointed to facts which seemed to render his view necessary. If the view of Whewell's work which came to be established is to be deemed well founded, his critics must certainly have good alternatives to Whewell's position. In order to pursue the inquiry we may turn to Book IV, Of Operations Subsidiary to Induction, where Mill continues his discussion of induction and even after the development of his view of this subject had been completed. Did he see the difficulties he was in? Mill discusses the conditions for induction, or, "those operations of the mind, which are either necessarily presupposed in all induction, or are instrumental to the more difficult and complicated inductive processes" (Mill 1973, vol. 8, p. 641).

He is, it is easy to see, above all concerned to meet Whewell's challenge to his central doctrines: it was Whewell who argued most powerfully that Mill's methods could not be those used in science, since they presumed a false psychology and/or theory of observation. Mill's theory of four methods of proof presumed just that which was most difficult, that is, the identification of the relevant facts and their relations (Whewell 1971, pp. 238*ff*). In seeking to answer this criticism Mill not only grants Whewell the basic tenets of his methodology. He grants him basic portions of his psychology of perception as well. He begins by noting that "What we are said to observe is usually a compound result, of which one-tenth may be observation, and the remaining nine-tenths inference" (Mill 1973, vol. 8, p. 642).

The concession that observations is nine-tenths inference may seem advantageous to Mill, since he needs to explain both the secure foundation of science in observation as well as errors of observation. If observation is nine-tenths inference, errors may then be explained as errors of inference (Mill 1973. vol. 8, p. 642). And observation, strictly speaking, may contain no error and be that basis for science whose existence is denied by Whewell.

The inference which we purportedly find in observation must be, of course, subject to the rules of induction in order to serve their proper role in Mill's system (Mill 1973, vol. 8, p. 644). But when we look for an argument or evidence that this is so, we find nothing but a transcendental argument. Mill must postulate that there is in normal observations something more than mere (veridical) observation in order to account at all for the facts of perception emphasized by Whewell. His system allows for only one answer: humans make inductions on observations. This is the psychology of learning which he has uncritically taken over from the Baconian tradition. Now he cannot show that this is the case. He has no empirical psychological study which would support it. So, he must argue that it must be the case, that is, that without it we cannot explain science. And this begins to appear suspicious, since it just his theory of science which was at issue. Is Mill arguing here in a circle from his own methodological theory to a psychological analysis which, given this theory, must be true without an independent appraisal of the facts?

It is hard to see that Mill has here an answer which is convincing to anyone not already convinced that all psychological processes must be somehow reduced to observations and inductions from them. We once again fail to find that explanation of the weakness of Whewell's system which was so convincing to them and which to our eyes might seem powerful – even it should happen to be false. And, indeed, we find Mill somewhat confused himself after this long effort to explain away the troublesome facts brought forward by Whewell. He goes so far to say,

> We thus see that it is impossible to express in words any result of observation, without performing an act possessing what Dr. Whewell considers to be characteristic of Induction. There is always something introduced which was not included in the observation itself; some conceptions common to the phenomenon with other phenomena to which it is compared. An observation cannot be spoken of in language at all without declaring more than that one observation; without assimilating it to other phenomena already observed and classified. (Mill 1973, vol. 8, p. 645)

But this is "an operation which precedes all induction, and supplies it with its materials. It is a perception of resemblances, obtained by comparison" (Mill 1973, vol. 8, p. 645).

We here find Mill confused as to whether perception or observation is an inference, and thus induction, or mere perception of resemblance and no inference. He gives us two answers to the same problem, but not does tell us why he needs two nor under what conditions one or the other explanation is to be deemed correct. It would seem that he needs both. Observation must be inference in order to explain away alleged content not obtained by induction. But unconscious inference which cannot be made explicit seems a poor foundation for science: error in observations has been explained as due to erroneous (unconscious) inference. Such error must be somehow excluded from science.

The problem which Mill here faces and the confusion into which he falls are identical to those discussed above in regard to descriptions and inductions. He needs both descriptions and pure observations to establish the basis of science his theory requires. But he can not hold the two apart, so he ends up with two inconsistent explanations for the same phenomena. We have not yet found the answer to our puzzles concerning the power of Whewell's view on the one hand and its decisive rejection on the other hand. We have what amounts to a mere assertion that Whewell's view must be false. We do not find convincing answers to the problems Whewell's view raised.

The Baconian tradition, we might note, chose to concede that perception involved unconscious inference and left the direct perception of resemblance as the alternative aside. This was done quietly, but we might conjecture that the reason for it was that the view of direct perception of resemblances was not compatible with the simultaneous assertion of unconscious inference. Helmholtz is often given credit for the theory of unconscious inference, even though both Whewell and Mill to say nothing of Schopenhauer had proposed such theories before. Perhaps the reason for this is that he offered both an inductivist view in contrast to Schopenhauer and Whewell and one which did not incorporate the alternative of direct perception of resemblance: his view seems identical to Mill's in regard to unconscious inference, but simpler and more coherent because he leaves aside this second view.

We might conjecture that Helmholtz offered a much better alternative and that we might find there the answer to the question of how and why the opinion of Whewell's view was formed in the way it was. This cannot, of course, be strictly true since, although Helmholtz's work overlaps that of Whewell and of Mill, it is somewhat later and played no role that anyone has mentioned in this discussion. And this hypothesis is puzzling for other reasons as well. Helmholtz offered an alternative answer to some problems raised by Whewell. But these answers are not part of a system. They are separate answers to specific problems in perception and perhaps in the necessary principles for science – of causality, for example. They touch the central issues in the debate between Whewell and Mill quite directly but they offer no systematic alternative. They might be deemed refinements to Mill's system.

Now if we deemed Helmholtz's view a refinement of Mill's and thus a systematic alternative to Whewell the logic of the argument would be somewhat different. Although Helmholtz uses the same transcendental argument as that used by Mill to argue for his inductive theory of unconscious inference, he does not at the same time argue for direct observation but rather for the incredibly high reliability of unconscious inference. It may thus, we may presume, serve as a foundation of science. Helmholtz himself does not spell these things out and, I think, for good reason: he had no desire to open up this debate in Germany. But even if we deem Helmholtz as a competitor, we do not come farther with our problem, since the very fact that we would have to reconstruct such a debate to see how it may have come about shows the power of Whewell's view and could not possibly show why it was then so quickly rejected. And even the details which Helmholtz discussed were not so clear: Hering had offered steady competition in the theory of perception and later views such as those of the Würzburg school and the Gestalt school indicate the difficulties a Helmholtzian view faced. This is all very interesting and I have described it elsewhere (Wettersten 1992) but it does not seem to help us here.

I come, then, to the end of my discussion of the first reaction to Whewell. Herschel and Mill uncovered problems in Whewell's view of necessity and of justification. They apparently tried to take seriously his defense that his view could explain the growth of science better than others by explaining away his alleged advantages. They sought also to meet him on the field of psychology in the same way. But in each case they offered merely a program. When they tried to develop this program they lapsed into inconsistencies, they found gaps they could not fill and, as a result, used transcendental arguments of a sort described by Whewell and rejected by them. They could apparently hold to their program only by ignoring the even deeper problems which their own view faced. They also did not argue that the problems faced by their own view were more interesting or challenging or offered more hope than Whewell's. The agenda seems to have been set not by the logic of the criticism but by the desire to adhere to established views even in the face of difficulties. All this would seem to indicate that our puzzles concerning Whewell are more serious rather than less due to our examination of the responses of Herschel and of Mill. The problems which Whewell raised and the gaps in the answers which were given to them seem to show how serious his opposition was and not, as we sought, the reasonable grounds for his rejection as a confused thinker. There were further criticisms of Whewell's philosophy, however, which also quickly appeared and we need to turn to them as well to understand the rejection of Whewell's philosophy.

### 1.3. Augustus de Morgan: Playing on the Sidelines

One of the most likely places to find serious answers to our questions is, perhaps, in the comments of Augustus de Morgan. He was not such a direct party to the dispute as were Herschel, Mill, Mansel and Brewster. Yet he was a very sophisticated reader who had thought about the issues at hand and one who, we would suspect, would not be frightened by the kind of innovations which Whewell hoped to introduce. And, indeed, we do find at the beginning of the first of de Morgan's two essays a broad basis of sympathy in the recognition of the inadequacy of Bacon's theory and, we may infer, the need for some new alternative which overcomes the deficiencies found there. He states, "We cannot find our casus belli in Dr. Whewell's estimate of Bacon's method: for the main difference between him and us on this point . . . lies in the amounts of development" (de Morgan 1859, p. 43).

De Morgan does not find this alternative in Whewell's work but he is highly appreciative of the description of scientific discovery which Whewell has offered. He states, "It is a most excellent thing that there exists a work so well calculated to show the student how human progress has been made, as the one now before us" (de Morgan 1859, p. 44).

He also deems him the leader of the day in the field of the psychology of research. He states,

> . . . Whewell is, in a very special manner, the psycho-physical, or physico-psychological writer or our day in England. We do not know of anyone else who combines the same amount of . . . thought and reading in mathematics, physics, and metaphysics. Still less do we know of any one who has laboured in the same manner to connect physics and psychology. (de Morgan 1859, p. 43)

Whewell apparently hoped for an encouraging critique from de Morgan. The first of his two reviews was apparently written at the request of Whewell (Todhunter 1876, vol. 2, p. 416) and concerns Whewell's *Novum Organum Renovatum*. There is one point of dissent from Whewell's essay which is dealt with. It concerns the question of the proper provinces of logic, of induction and of the theory of discovery. De Morgan presents his problem as a terminological one when he states,

> There is a desire to have the word logic applied to a science which takes in much which the word excludes and excludes much which the word takes in. Dr. Whewell, who makes all physical discovery go under the logic of induction, and Mr. John Mill, who puts all discovery of truth under general logic, are to us equally employed in straining the term beyond its proper meaning. (de Morgan 1859, p. 43)

This terminological problem is alleged to be important since confusion in the use of terms give rise to confusion in the content. And this applies to Whewell's work as well. He states, "But when Dr. Whewell, or any one else, persists in confounding induction with the determination of the general notion under which the induced particulars are to be brought, he perpetuates a confusion which has already done a great deal of mischief" (de Morgan 1859, p. 44).

De Morgan's contribution, then, is to distinguish logic and induction from discovery or other means of finding the truth. We have met a similar distinction earlier in Mill between invention and proof and we will meet something similar below in more recent work where the context of discovery is distinguished from the context of justification. (Joseph Agassi has informed me that it was La Place who introduced this famous distinction in his *System de Monde*. It became, however, of importance in the 19th century as a means of establishing the view that Whewell's theory was incomplete, that it was a theory of discovery merely and no theory of proof.) It was said against Whewell that he has a theory of scientific discovery, but not one of proof or of justification. For justification, many have claimed, we need induction in the old sense, that is, inference from particular statements to a general one. And thus it might seem that de Morgan's view is designed to reinforce this judgment. It has been so read. But is this de Morgan's opinion? He seems to leave the door open for Whewell, since one could deny that induction as it has been conceived is actually used in science (de Morgan 1859, p. 44).

In order to find a perspective on de Morgan's attitude here it may be useful to digress for a moment to note his view of Herschel who wished to base all science in observation via induction (de Morgan 1815, vol. I, pp. 80*ff*). In a discussion of Bacon de Morgan himself digresses to discuss Herschel's *Discourse on the Study of Natural Philosophy*. He observes that Herschel places at the head of his book "the well-known first aphorism of the *Novum Organum*: 'Man being the servant and interpreter of nature, can do and understand so much, and so much only, as he has observed in fact or in thought of the course of nature: beyond that he neither knows anything nor can do anything' " (de Morgan 1915, p. 80).

But this book, de Morgan says, contains "notions of discovery far beyond any of which Bacon ever dreamed; and this is because it was written after discovery, instead of before." Sir John Herschel, in his version, has avoided the translation of *re vel mente observaverit*, and gives only "by his observation of the order of nature." In making this the opening of an excellent sermon, he has imitated the theologians, who often employ the whole time of the discourse in stuffing matter into the text, instead of drawing matter out of it. By observation he (Herschel) means "the whole course of discovery, observation, hypothesis, deduction, comparison, etc." (de Morgan 1915, pp. 80-81)

He continues,

> In another work, his treatise on Astronomy, Sir John Herschel, after noting that a popular account can only place the reader on the threshold, proceeds to speak as follows of all the higher departments of science. "Admission to its sanctuary, and to the privileges and feelings of a votary, is only to be gained by one means – sound and sufficient knowledge of mathematics, the great instrument of all exact inquiry, without which no man can ever make such advances in this or any other of the higher departments of science as can entitle him to form an independent opinion on any subject of discussion within their range." How is this? Man can know no more than he gets from observation, and yet mathematics is the great instrument of all exact inquiry. Are the results of mathematical deduction results of observation? We think it likely that Sir John Herschel would reply that Bacon in coupling together *observare re* and *observare mente*, has done what some wags said Newton afterwards did in his study door – cut a large hole of exit for the large cat, and a little hole for the little cat. But Bacon did no such thing: he never included any deduction under observation. (de Morgan 1915, pp. 81-82)

De Morgan thought that science had become mathematical and that any serious person had to take that adequately into account (de Morgan 1915, p. 2). And neither Bacon nor Herschel, when he appeals to Bacon, had done that. De Morgan rejects the program of Herschel and Mill of reducing all knowledge to observations and inferences from them. He suggests, indeed, the same view of Herschel as that found in Agassi's path-breaking article of over a century later (Agassi 1971b). Herschel had included hypotheses in science and this is progress, but clung to the old Baconian views with which this innovation could not be reconciled. De Morgan, then, does not call for induction or a theory of induction along the lines of Herschel and Mill. Just what he hoped for is not so clear. He sets desiderata which a theory must meet, above all it must account for mathematics. (The empiricist theory of mathematics of Herschel and Mill is apparently not even taken seriously.) But how that is to be done is by no means clear. We are still left with puzzles about what to do with Whewell.

I end my digression, then, on de Morgan's view of Herschel and induction. It helps see, perhaps, better what de Morgan's view of induction was not, than what it was. De Morgan presents his view of Whewell's attempt at a theory of induction as follows.

> Accordingly, when Bacon, rejecting and contemning the old logic, proposed a substitute, he did not declare against the name of logic but rather intimated that what he had to propose might be called by the same name. It is not he but his followers who, refusing to part with the venerable word, have explicitly propounded the logic of induction as a science contrasted with the old logic. Dr. Whewell very decidedly takes this position: and though we do not quarrel with any of his conclusions, we are entirely opposed to the use which he makes of the words logic and induction. (de Morgan 1859, p. 43)

Now de Morgan does assert that "Kepler and his contemporaries discovered that Mars moves in an ellipse, that the earth moves in an ellipse, &c.: and thence inferred, by what is most technically called induction that the planets moved in ellipses" (de Morgan 1859, p. 44).

De Morgan thus held that three aspects of the study of knowledge, those of discovery, of logic and of induction should be separate. Concerning logic, he says, there is growing consensus "that logic means the study of the necessary forms of thought, the analysis of the modes under which all men must think" (de Morgan 1859, p. 43).

Whewell, he suggests, has not described this but rather discovery. Induction also occurs in science on de Morgan's view but he does not appraise its significance. Perhaps it occurs only in specific frameworks which would make little difficulty for Whewell's view since it would then be a mere means of refinement of fundamental ideas. De Morgan says after all that he does not here quarrel with any of Whewell's conclusions. Logic must be separated from both discovery and induction, since it concerns necessary truths. These are not to be obtained by Whewell's methods of discovery nor, I presume, by induction.

De Morgan's effort here seems merely aimed at proper demarcation and Whewell treated it as such in his response when he said, "But I do not wonder at your denying these devices (methods by which scientific discoveries are made) a place in Logic; and you will think me heretical and profane, if I say, so much the worse for logic" (Todhunter 1876, vol. II, p. 417).

Nevertheless Whewell came into conflict with de Morgan on one important point. He maintained that his method could enable one to discover necessary truths. On de Morgan's suggestion, then, it might be a part of logic. This issue was the single point of concern of de Morgan's second review, his review of *On the Philosophy of Discovery*. (The title of this book was apparently due to de Morgan's criticism of *Novum Organum Renovatum*. *On the Philosophy of Discovery* appeared in 1860, de Morgan's review of *Novum Organum Renovatum* in Jan. 1859. In the preface to his book which is I presume – following Blanché – falsely dated Feb. 8, 1856, Whewell states that he did not seek a philosophy of induction but a philosophy of discovery and for this reason changed the title; Whewell 1871, p. v).

In his second review, de Morgan (1860) emphasizes his agreement with Whewell's theory that necessary truths grow. He agrees with him that the necessary truths of mathematics may not appear so to the uninitiated but will to those properly educated and "We agree also with Dr. Whewell, and so far differ with most of his opponents, that knowledge of necessary truth may be progressive" (de Morgan 1860, p. 502).

De Morgan's example of the possibility of drawing any map with four colors – a proposition known to map makers without knowledge of its necessity – comes from mathematics rather than science however. He comments on Whewell's claim to have proved the necessity of the proposition that all matter is heavy so:

> But it is quite possible to conceive a ball of something hard enough to knock our heads off, taking its way along the earth's surface, performing that feat upon some human beings – say Dr. Whewell and ourselves, to treat both sides of the argument impartially – and then, though palpably of the consistency of iron, stopping quietly in mid-air, poised like the earth in space, without falling to the ground. This, we say is a possible conception: all we can say against it is, that there is no such kind of matter within the history of experience. Dr. Whewell conceives himself to have demonstrated that all matter is heavy: to have demonstrated that no such ball exists. (de Morgan 1860, p. 501)

and,

> The matter stands thus: Dr. Whewell has, in his own mind, arrived at a distinctness – come to a clearness, as the Puritans used to say – respecting the idea of matter, which makes him feel that there is no difference in quality between the propositions that all matter is heavy and that two straight lines cannot enclose a space. He tells us that we shall arrive at the same, when our idea of matter is as clear as his. We look round for a second to his motion, and we do not find one. (de Morgan 1860, p. 501)

De Morgan concedes that there is a sense in which, given some propositions, say some of the laws of motion, others would of necessity be forced on one (de Morgan 1860, p. 502). He there refers favorably to Whewell's distinction between the necessary and the empirical parts of a proposition. But the new aspect pointed to has, he says, no name, although it should. This new conception indicates that the laws of thought have been put in happy correspondence with the laws of nature.

De Morgan's comments are highly appreciative and more genuinely so, I think, than Whewell's other critics. His view has apparently been read however as reinforcing the criticisms of Herschel, Mill and Mansel, even though he disagrees with these other thinkers on important points. It may seem to do this by denying that the idea of necessary truth can be extended to natural science, as Whewell wished to do, and by, therefore, demanding a new theory of the proof of scientific theories which, if Whewell's philosophy is interpreted as a mere theory of discovery and nothing more, is lacking. There thus appears to be some consensus here as to where Whewell's confusions lie and that these must be grounds for the rejection of central portions of his system.

But even this consensus does not seem to answer our puzzle about the judgments of Whewell's philosophy. De Morgan's criticisms were by no means a call to simply set Whewell's philosophical views aside, but may just as well, even better, be seen as a call to improve them. If they were properly demarcated and expressed in a more appropriate language they would obviously have, on de Morgan's view, much to teach us. Such new views might, of course, also be problematical. De Morgan is not convinced by Whewell. But he sees nothing better, or mentions nothing better, among those alternatives then available. And Bacon and Mill, he notes, also suffer difficulties. It is ironic that Whewell's response, so much the worse for logic, has been interpreted so negatively instead of the basis of agreement between de Morgan and Whewell to demarcate the fields of logic and scientific method in a more happy way.

De Morgan's reviews left open, perhaps, a choice of how one should interpret the problems of Whewell's system. These problems could be interpreted as mere grounds for rejection or they could be taken up as problems which needed to be solved on the way to a still better theory of science. There was little discussion of this point between them though they were friends. De Morgan's remarks were appreciative, intriguing and amusing, even when they did not lead to clear cut answers to some questions we might wish to put to him. He discusses Whewell elsewhere as follows.

> The Master of Trinity was conspicuous as a rough customer, an intellectual bully, an overbearing disputant: the character was as well established as that of Sam Johnson. But there was a marked difference. It was said of Johnson that if his pistol missed fire, he would knock you down with the butt end of it: but Whewell, in like case, always acknowledged the miss, and loaded again or not as the case might be. He reminded me of Dennis Brulgruddery, who says to Dan, Pacify me with a good reason, and you'll find me a dutiful master. I knew him from the time when he was my teacher at Cambridge, more than forty years. As a teacher, he was anything but dictatorial, and he was perfectly accessible to proposal of objections. He came in contact with me in his slashing way twice in our joint lives, and on both occasions he acknowledged himself overcome, by that change of manner, and apologetic mode of countenance, which I had seen him employ towards others under like circumstances. I had expressed my wish to have a thermometer of probability, with impossibility at one end, as 2 and 2 make 5, and necessity at the other, as 2 and 2 make 4, and a graduated rise of examples between them. Down came a blow: "What! put necessary and contingent propositions together! It's absurd!" I pointed out that the two kinds of necessity are but such extremes of probability as 0 and infinity are of number, and illustrated by an urn with 1 white and n black balls, n increasing without limit. It was frankly seen, and the point yielded; a large company was present . . . I have said that Whewell was gentle with his pupils: it was the same with all who wanted teaching: it was only on an armed enemy that he drew his weapon. (De Morgan 1915, vol. II, pp. 246-247)

De Morgan leaves all options open with pregnant remarks. We seem to be close to finding an answer to our puzzle concerning the poor opinion of Whewell, since it was the decision to take the option of rejecting Whewell's system and the problems which it raised which was crucial in sealing the fate of Whewell's philosophy for the next one hundred years. But just as we come close to an answer it slips away from us. We find no discussion whatsoever of the alternatives of tackling the problems

Whewell's system raises. De Morgan hints with a friendly wink at what might be done. But then he trots away just as we wish to ask for more. We find a policy of accepting the problems growing out Mill's system as the proper agenda and rejecting those which grew out of Whewell's, even though a thinker of such importance as de Morgan found such interesting things to say in his comments on Whewell. I will leave this topic of the reasons for the choice of an agenda for now but I will need to return to it below.

### 1.4. David Brewster: Getting Stuck in the Middle

One of the leaders of the prompt and negative reaction to Whewell's work was David Brewster. He remained Whewell's somewhat embittered opponent until Whewell's death and the sharpness of his tone combined with his attribution to Whewell of petty motives may make us skeptical that we may here find those reasons which led to such a sharp rejection of Whewell's philosophy. But it would no doubt be sensible to look both at the causes of Brewster's antagonistic attitude as well at his substantial critiques to improve our understanding of the reaction of Whewell's contemporaries to his work. The study of Brewster's reaction is also interesting because he was a competitor to Whewell. Brewster had already published his *The Life of Sir Isaac Newton* in 1831. Brewster there sought to learn about scientific method by studying how Newton had proceeded. It is only by the study of the biographies of eminent scientists, he said, that one may learn how research is conducted successfully (Brewster 1831, p. 336). And this process is, he says, by no means that described by Bacon. Rather,

> The impatience of genius spurns the restraints of mechanical rules, and never will submit to the plodding drudgery of inductive discipline. The discovery of a new fact unfits even a patient mind for deliberate inquiry. Conscious of having added to science what had escaped the sagacity of former ages, the ambitious spirit invests its new acquisition with an importance which does not belong to it. He imagines a thousand consequences to flow from his discovery: He forms innumerable theories to explain it, and he exhausts his fancy in trying all its possible relations to recognized difficulties and unexplained facts. The reins, however, thus freely given to his imaginations, are speedily drawn up. His wildest conceptions are all subjected to the rigid test of experiment, and he has thus been hurried by the excursions of his own fancy into new and fertile paths, far removed from ordinary observations. Here the peculiar character of his own genius displays itself by the invention of methods of trying his own speculations, and he is thus often led to new discoveries far more important and general than that by which he began his inquiries. For a confirmation of these views, we may refer to the History of Kepler's Discoveries; and if we do not recognize them to the same extent in the labours of Newton, it is because he kept back his discoveries till they

> were nearly perfected, and therefore withheld the successive steps of his inquiries. (Brewster 1831, pp. 336-337)

We see already that Brewster had adopted a point of view which in some basic elements was the same as that later expounded by Whewell in his History of 1837. Brewster, by the way, generously thanks Whewell for providing him with information used in his biography (Brewster 1831, p. vi).

Brewster did not wait long to respond to Whewell's History. Already in 1838 he published a long review of it (Brewster 1838). And his judgment is severe. He states,

> In concluding our notice of Mr. Whewell's volumes, we must regret our inability to speak more favourably of a work composed with so much haste, and breathing so sectarian a spirit. To the character of an independent history of the inductive sciences, or of a "comprehensive survey of the existing state of human knowledge," it has no claim: nor, looking to its ulterior aim, can we permit ourselves to believe that on such a pedestal the stately column of a higher philosophy can be reared. To us, indeed, such an attempt appears utterly hopeless; and we cannot conceive how a Newton or a Laplace could give laws to inventive and original minds, still less can we imagine that heaven-born genius can be either checked or directed in its flight by those who have never imbibed its spirit or wielded its power. (Brewster 1838, p. 150)

This judgment of Whewell's sectarian spirit had at least two reasons – or motivations. The first was Whewell's total neglect of Brewster's own work which Brewster points out, giving a long quote to prove his case. (The quote from Brewster's Life, given above, is a part of this reference; Brewster 1831, pp. 125-126). The second is Brewster's jealous if not chauvinistic defense of the rights of Scotsmen. He states,

> These and other extraordinary omissions and oversights which we have already pointed out, are doubtless owing to our author's rapid march, as well as to his impatience of enquiry. But in taking a general glance at the materials of his work, we think we can perceive system even in his errors; and we do not hesitate to say, that the generalization which he has most successfully pursued is that of grossly neglecting the claims of the philosophers and authors of Scotland. We shall not enquire how far the animadversions of Playfair, of Leslie, of Brougham, and of other eminent Scotsmen, may have excited unfriendly feelings in our sister universities; and still less shall we enquire to what degree personal and even political feelings may have mingled their poison in this injustice towards our intellectual home. (Brewster 1838, p. 147)

The obvious self-interest of Brewster's point of view leads us to suspect him of exaggeration and partisanship of his own. And Brewster devotes the major portion of his review to the discussion of Whewell's omissions, his mistakes and false judgments of the merit or priority of various scientists. He aims to show that Whewell's work was sloppy and done in haste; it was uncritical in its use of secondary sources; it ignored excellent histories which had preceded it; it went beyond the capacity of the author to judge the worth of specific pieces of research; and it was, of course, biased against the Scotch.

Brewster's criticisms and judgments themselves are sometimes rather exaggerated. He complains, for example, that Mrs. Somerville, who was known for a popular scientific text, had been poorly treated by omission. But in other cases he points to errors in judgment, such as Whewell's giving equal credit with Charles Bell to a man named Mayo. Whewell later made a separate study and modified his view; Todhunter says that in the end Whewell had published three different appraisals of this event (Todhunter 1876, vol. I, pp. 116-117). Brewster also complains that in some cases Whewell's *History* is superficial as in the case of Pythagoras. He has an interesting discussion of the case of Galileo, in which he defends the view that Galileo had a friendly reception from the Vatican but forced the case needlessly with his use of Simplicio as a character representing the Pope. He carries the discussion farther. He also points to a variety of errors which he believes himself to have found in Whewell's work.

It would obviously take me too far afield to go into the details of these various controversies. In a letter to the editor of *Edinburgh Review* dated Oct. 28, 1837 Whewell rejected all the criticisms with the exception of two: he agreed to re-examine the question of the contributions of Mayo and of Bell and he conceded that he had not given adequate recognition of the debt of Albazon to Ptolemy. He also responded in kind to some derogatory remarks about Cambridge:

> This wretched rant is the echo of a slavish tradition, handed down from the brighter and prouder days of the Edinburgh Review. That work, at all periods, has spoken of the English Universities with equal bigotry and ignorance; and the shallow and vulgar conceit thus generated, which tarnished some of the fairest pages of the better scientific critics of its former times, is well fitted to make it contemptible, now that it has no longer any such to boast of. (Todhunter 1876, vol. I, p. 116)

Early in his intellectual career Whewell had the reputation of one who worked in excessive haste. I am not sure just when it first appeared and perhaps the *History* was the prime cause. Whewell's alleged reluctance to rework the text to remove errors such as there were in succeeding editions (there were three editions) and to thank his predecessors for instruction and his critics for corrections did not help. It is hard to believe however, that errors or weak discussions alone, which could hardly be avoided in such a large undertaking and which were obviously exaggerated by some of his critics, were alone the cause of Whewell's difficulties in finding appreciative responses to his point of view. There were passages which everybody, including Brewster, highly praised and it was the most comprehensive text available. It was immediately translated into German where it was apparently well known. Brewster gave it all he had but made too many mistakes himself and exaggerated too often to damage the history so severely that it could not remain a serious challenge to any historian or philosopher of science. If we want to understand the rejection of Whewell it might be useful, then, to turn to the discussion of his philosophy as well which, with few exceptions such as Todhunter, was held in even less esteem.

Brewster wrote a second long review, this time of the *Philosophy*, shortly after it appeared (Brewster 1842). He had already begun the discussion of Whewell's philosophy in his review of his *History*. He there noted that Whewell's purpose was not to collect facts but to develop a point of view concerning them (Brewster 1838, p. 116). Brewster hoped to show, right at the beginning of his review, that Whewell, in his philosophy of science, has nothing new to say. To maintain, for example, that Aristotle did not have the appropriate idea to explain why the light which passed through a figure of any shape whatsoever, if it traveled far enough, would produce a circle, is only to say that he made a false conjecture and did not have sufficient knowledge of scientific method to subject it to tests. We all know, that is, Brewster knew, as he points out, that conjectures were crucial in scientific research. This discussion of an example of marginal importance at best led to a long and bitter fight as I explain below. Brewster here quickly dismisses Whewell's philosophy as trivial and that was, perhaps one cause for the following antagonism. He does not mention Whewell's theory of the stages of the development of science nor does he evaluate the broader frame which Whewell offers other than to offer some detailed complaints. Let us turn then to his review of Whewell's Philosophy.

Brewster quickly picks up in his review of Whewell's *Philosophy* where he had left off in his review of the *History*. He states at the beginning,

> It is with considerable reluctance, and not without some pain, that we sit down to record our opinion of the work before us. But if the task (of appreciating and relating the labours of scientists and the reformers of science) is performed by an individual whose speed neither time nor tide can stay – whose movements are neither equable nor straight-forward, who comes into perpetual collision with those who are struggling in the same cause, and thus creates a general mutiny in the "tented field" – it becomes the duty of the critic to interpose his authority, however feeble, in attempting to give peace to a community thus divided against itself. (Brewster 1842, pp. 265-266)

and,

> Our disappointment . . . has been great: and so great has been the disappointment of others, that competent judges regard the present work as the Philosophy of scholastic Metaphysics rather than of physical Science; and question the ability of any natural philosopher to analyse its materials, and pronounce upon its merits. That this opinion is a correct one, is proved by two of the ablest and most partial notices of it; in which it is substantially treated as a metaphysical work, where all the thorny questions of speculative philosophy are discussed and decided. Distinctions which language can with difficulty mark, speculations which strain the human mind in its attempt to grasp them, and positions on which no two writers are agreed, are thus made the foundation of

> Inductive Science, and the noble pillar of physical truth is thus erected on piles driven into the quicksands of mysticism and sophistry. (Brewster 1842, p. 266)

We find two central complaints which we have already seen in Herschel's review, that is, that Whewell threatens the foundation of science and that he overreaches in his attempt to describe and to put in one schema the whole of science. We may hope, then, to find further support of these charges which obviously played such a large role in the rejection of his view and especially by the scientists. We also note that Whewell was taken as a revolutionary and as such was not treated well. Stiff opposition from those such as Brewster who thought he failed to participate properly in the cooperative endeavor of furthering science was the norm. But how is this to be judged? And is it mere personal or theoretical as well? Brewster first said Whewell's view was nothing new and now it is a serious threat. (Herschel did the same in a very different tone.) Let us look for some resolutions of these puzzles in his arguments. Brewster opens the attack with an attempt to show Whewell's lack of competence. His tactic is to show the inappropriateness of one of Whewell's numerous suggestions for terminological innovations ('dipolarization' instead of 'depolarization' in certain cases). Though Whewell often answers Brewster's charges in the second edition of his *Philosophy*, he never discusses this point and leaves his proposal as it stands. Though Whewell had very important things to say about language – he defended the use of theory impregnated terms as a necessary means of scientific progress even in those cases where the theory is false (with some truth content) (Whewell 1858, pp. 294*ff*, esp. p. 299) – this suggestion concerning 'depolarization' is merely given in a paragraph and as far as I can tell the proposal was defensible whether or not it was the best. Surely this kind of discussion is no foundation for an appraisal of Whewell's philosophy. Let us look further.

The next section of Brewster's review (Brewster 1842, pp. 268-283) is, indeed, by far more interesting and goes to the heart of the matter. Brewster here discusses the psychological-cum-physiological foundation of Whewell's theory. This important foundation of Whewell's theory was sometimes treated as mere metaphysics – by Herschel, Mill and Brewster – and sometimes as psychology – by Herschel, Mill and Brewster. The problem lay in Whewell's assertion of the necessary role of the activity of the mind in perception: a theory of the activity of the mind was metaphysics and unacceptable; a theory of perception was psychology and a foundation of science. And here Brewster sides with Herschel and Mill: the observation of facts does not require a special activity of the mind. In order to show this and to establish the foundation of science in the observation of facts – even when observation requires genius to be of use in science – Brewster attacks Whewell's psychology and the physiology –

above all that of Charles Bell – which Whewell had used to support his own view of the matter (Whewell 1967b, vol. I, pp. 119*ff*).

Whewell's philosophy of science rested on the assumption that the mind was active and contributed to the content of thought. In his theory of perception he sought to show that perception could not be explained without – in modern terminology – a feedback mechanism. We have impressions to be sure, but we continually test and improve these impressions. In this way we learn to understand and perceive – the two processes are not to be sharply distinguished – the world (Whewell 1967b, vol. I, pp. 111*ff*). Now, how does this process take place? Whewell found here his most important source of support from the physiological researches of Charles Bell. And the most important of these was that the eyes were in constant motion and that this motion was crucial for the perception of objects in space. It was a mechanism by which the eyes could continually test and improve perceptions. But this process would not work, Bell conjectured, if there were no impressions of the movement of the eyes. The movement of the eyes had to be coordinated with the impressions which the eyes received if this movement should be useful in perception. So, Bell postulated, we are in some sense conscious of this movement and he called this consciousness a sixth sense. Whewell does not limit the role of such a sixth sense in perception to sight, however. He maintains as well that we do not perceive space by touch alone but here as well rely on a muscular sense. Whewell here appeals to the theory of Brown (Whewell 1967b, vol. I, pp. 118*ff*).

Now Brewster will have none of that and proceeds to criticize all three thinkers. He first discusses Brown's physiological theory of the sixth or "muscular" sense and complains that it is too vague – according to Brown, Brewster states, the whole muscular frame is an organ of sense – while casting aspersions on his mastery of "physical truth." He fails, however, to even mention the physiological theory of Bell according to which there were nerves to carry the sensations of muscle movements to the brain. And Whewell does not commit himself to such a theory but, following Bell, asserts that perception cannot be explained without including in this explanation the factor of the awareness of muscle movements. Perhaps in his discussion of this psychological hypothesis we will find something more interesting.

Brewster has one argument. He postulates individuals with no muscle movement and maintains that they would have the same perceptions as those who do. In the case of touch the movements are obtained by some mechanical or electrical devices and in the case of sight only one case is considered: a person with no sense other than sight is set on a stool and revolved: what does he perceive? (The case of a man with only the one sense of sight was introduced by Reid and used and criticized by Whewell as Reid had maintained that such persons would see the world as curved and thus have, for example, a different geometry in which, say, lines which intersected once must intersect again.) In the second edition Whewell responds (Whewell 1967b, p. 121). He argues that a person with no sense, but sight who was rotated on a stool, would simply see a series of pictures. He would have no conception of being back at his point of origin when he was. And only by action in space could he form a conception of such pictures covering a solid body. I do not know how to judge this thought experiment. Whewell's conjecture as to what would be seen seems to me quite plausible and the argument as a refutation quite weak. Let us proceed but we will come back to the problem of the perception of objects in space below. Brewster turns to another view which Whewell took over from Bell according to which there is a spot on the retina which is particularly sensible. The searching motion of the eye is then explained as an attempt to receive the image of an object on this sensible spot. Brewster maintains that there is no sensible spot, that the spot referred to is less sensible than all others and that the sensibility increases as the line of vision moves away from this spot. And in the second edition Whewell notes this last point and implicitly concedes that it may be correct (Whewell 1967b, Part I, p. 123) though no mention of Brewster is made either here or in other corrections which are made due to his criticisms or in other responses to them. Whewell here maintains that "the facts are very curious, however they may be explained, but they do not disturb the doctrine delivered in the text." It is hard to follow Whewell here since just which "doctrine delivered in the text" is intended is hard to say. If the doctrine of the existence of a sensible spot is intended, then the objections are certainly relevant and require response. If the doctrine of a specific movement of the eye toward a specific spot is intended, then this doctrine is effected, if not overthrown. Whewell was anything but generous in his response and this weakens his case. But his theory does not depend on this physiological hypothesis and both thinkers seem to have known that. Brewster continues,

> But the grand error committed by Mr. Whewell is his implicit adoption of the doctrine, "that the idea of position or relation in space is not give by the retina, but by the muscular sense". If he had been aware of the objections against this doctrine already published, and never answered, he ought to have brought them under the notice of his readers, even though they should not have influenced his own decision. (Brewster 1842, p. 277)

Brewster maintains that the view that the motion of the eyes is needed to perceive up and down is absurd and empirically refuted. Whewell, however, does not hold this view. Rather, he holds that we cannot even sensibly pose the question as to why we see things as up or down: it is a given of our means of perception that objects are so perceived and by means of the retina. Those who pose this question confusedly try to compare the image on the retina with what we see. Brewster argues that we need to consider the angle of the rays entering the eye to explain the matter (Brewster 1842, p. 281). And Whewell agrees with Brewster's observations but finds them unnecessary to handle the question of upright or inverted images (Whewell 1967b, Part I, p. 294).

But Brewster now believes to have found an even better criticism of Whewell's theory. He maintains that, if vision alone can give us no perception of position and relations in space, we can also not perceive form with sight alone. Whewell limits the perception of form or of space which is acquired with the so-called sixth sense to large objects or extensive spaces. And, Brewster continues, "Now, we ask, and we ask it with a feeling of some triumph, what is the size of the object, or the extent of the space, where the power of sight abjures the assistance of the senses, and exercises its own right of judgment?" (Brewster 1842, p. 278).

In the next edition Whewell answered so, "The expression of the first edition was 'large objects and extensive spaces.' In the text as now given, I state a definite size and extent, within which the sight by itself can judge of position and figure" (Whewell 1967b, Part I, p. 125).

He here parts company with Bell who made no such limitation. The limitation within which the eye can judge position and figure is to objects whose whole can be seen with one position of the eyes. Brewster effectively criticizes Whewell's theory of how we see with two eyes or how we unite two images into one. Whewell here appeals to the view that there are corresponding points on the two retinas which Brewster merely observes is outdated. In the following edition Whewell says that

> perhaps it is better to avoid the assertion that the law of corresponding points is "a distinct and original principle of our constitutions," as I said in the first edition. The simpler mode of stating the law of our constitution appears to be to say, that each eye determines similarly the position of objects; and that when the positions of an object, as seen by the two eyes, coincide (or nearly coincide) the object is seen as single. (Whewell 1967b, Part I, p. 295)

Through his discussion of Whewell's theory of perception Brewster forced Whewell to make corrections or merely to withdraw to a vaguer theory. One can hardly say that Whewell's physiology of perception which, following above all Bell, should exhibit the physiological mechanisms which aided the activity of the mind in perception, was adequate. The path breaking research of Wheatstone (1838) was not integrated with that of Bell; some points were simply in error and some incomplete. And Whewell's responses to this critique were, as always in the discussion with Brewster, given in a begrudging spirit which effected their content: mere modifications were given of texts with no serious discussions of problems.

The weaknesses of Whewell's physiology of perception should not be overestimated, however. It took account of some of the most important research in the physiology of perception, that of Charles Bell, and integrated it in a broader theory of knowledge. His own view of the need for the activity of the observer in order to observe was by no means refuted. He gave, indeed, powerful arguments for the role of the observer and pointed to examples such as the false perception of the direction of a turning windmill which had to be taken into account. And Brewster made no such attempt. Some objections made by Brewster were serious and interesting, but not grounds for simple dismissal. They were the basis for a discussion, but we do not find them sufficiently developed to question the theory that perception is active, which is the view Whewell needs for his philosophy.

Brewster has, of course, entirely other reasons for rejecting Whewell's philosophy. He once more goes to work on the accuracy of Whewell's text. This time the object of his complaints is Whewell's discussion of scales of color. Whewell discusses the problem of whether the scale of colors produced by a prism may be used as a measure. It would seem not, since the colors cannot be numerically and thus exactly identified. But, says Whewell, there are possibilities due to the existence of definite lines discovered by Fraunhofer and that thin plates of specific thickness have definite colors, say a plate of air at a thickness of twelve millionths of an inch is orange. Brewster complains that the attribution is wrong since Wollaston is neglected and the thickness is wrong – it should be eight millionths of an inch – light cannot be used since the atmosphere creates variations. And Whewell gives no explanation of what a chromatometer is good for.

Whewell silently changes his text in the second edition. He adds a reference to Wollaston, he changes the numbers, he explains that white light is meant and that he did not intend to give a complete account, and adds that he had had no chance to study those books which Brewster points out he has ignored (Whewell 1967b, Part I, pp. 328*ff*). Brewster has one internal criticism which is that Whewell maintains that the Fraunhofer lines make a chromatometer for specific purposes possible, that they do not allow for numerical measures and that numerical measures are necessary for all purposes. Whewell implicitly concedes that Brewster is right on all counts. But they are all minor: Whewell can and does change his text and his theory is not touched by it. Perhaps his credibility was. It is hard to judge. Whewell was, of course, petty in his response in that he failed to acknowledge the need for corrections given Brewster's criticisms.

We look further for serious difficulties and we come once again to the discussion of Aristotle which we met for the first time in Brewster's review of Whewell's *History*. The issue here discussed is crucial even if the example is not. And Brewster says that he did not want to debate the issue with this example of Aristotle but, as I will explain shortly, with one from the researches of Hooke, Newton and Young. The problem, just as in Brewster's review of the *History*, is whether Whewell's theory of the success and failure of scientific research according to which success is achieved when appropriate ideas are found and not achieved when they are not found is adequate. Whewell had tried to show with many examples that, only when an appropriate idea is found, can phenomena be successfully explained and Brewster complained that this was trivial. It is not quite as trivial as it now appears however. Bacon, the tradition, and leading theorists such as – a bit later – John Stuart Mill had maintained that success or failure depended on observation and the correct use of inductive methods. Whewell said that did not matter, and was indeed of no use if one did not have the appropriate idea and superfluous at best if one did (Whewell 1971, pp. 263-264). Where do such thinkers as Herschel and Brewster stand on this issue? Whewell quotes Herschel in his answer to Mill to support his own case. Yet Herschel does not side with Whewell as he makes amply clear in his review. But neither he nor Brewster side with Bacon either, since they acknowledge the importance of hypotheses. Nevertheless, they stay as close as they can to Bacon in that they maintain that observations are independent of ideas and that factual discoveries are of great and independent importance. And Brewster will show that by criticizing Whewell's examples and introducing some of his own.

What Brewster wants most to do is to show that the observation of key facts is a crucial step in scientific discovery. This was the view he proposed in his *Life of Sir Isaac Newton*. Such factual discovery may, contrary to Bacon, require genius to make but it is not dependent on prior ideas and in this sense Brewster remains Baconian. In order to support his view he has already attempted to criticize the psychology according to which ideas are needed for perception. Here he wishes to argue from the history of science. He proposes to show that the attainment of an appropriate idea is neither sufficient nor necessarily prior to the perceptive observation of some key fact in scientific discovery. We may have an appropriate idea and pass over it if we have not made the right observations. And we may not find the appropriate idea until we have made the proper observations.

In order to prove the first point Brewster points to the case of Hooke's conjecture that thin plates are colored due to interfering pulses reflected from the surfaces of the plate. Newton, he says, rejected this idea, applied another and misled philosophers for a century and the idea of Hooke was nearly forgotten. But, after Young had demonstrated the interference of pulses in the phenomenon of inflexion or of diffracted pencils – rays of light converging to or proceeding from a point – one could see that Hooke's idea was correct. In his review of the *History* Brewster summarizes his point of view so,

> The discovery of a new phenomena will necessarily give rise to various conjectures as to its cause; and the number and nature of these conjectures will depend on the fancy, the knowledge or the ingenuity of the individual. But the true philosopher will submit each conjecture, as it is formed, to the scrutiny of observations and experiments; and even if he should find them all delusive, he will probably discover, in the researches to which they have given rise, some new fact or principle which may finally lead him to the solution of his problem. (Brewster 1838, p. 119)

In his second review he states,

> Dr. Hooke did apply the distinct and appropriate idea of interfering pulses. Newton rejected it as not appropriate, and substituted another much less appropriate. Dr. Young subsequently discovered the fact of the interference of diffracted pencils, and this fact furnished the appropriate idea which was desired. This is an exact illustration of our position, which, we should hope, will be regarded as unquestionable. (Brewster 1842, p. 288)

The point of difference between the two thinkers concerns the question: are factual discoveries made prior to or independent of ideas. They both allow for conjectures but Brewster contends that conjectures are designed to explain the facts we have discovered – they do lead also to new factual discoveries – whereas Whewell maintains that we need ideas of some sort before we can make the factual discoveries. What does Whewell say in response to Brewster's challenging example? Nothing. Brewster complains about Whewell's failure to take up his challenge. But Whewell did respond to Brewster's criticism of his analysis of Aristotle's explanation of the phenomena of light passing through openings, of whatever shape whatsoever, forming circular images if the light is allowed to proceed far enough. Aristotle had maintained, so Whewell, that this phenomena is due to the circular nature of light. The correct explanation, he says, is that light travels in rays and the sun is circular. Because Aristotle did not have the correct conception of rays he did not form a correct idea. In his first review Brewster explains Aristotle's failure as due to his failure to test his imperfect ideas. This failure was due to his lack of knowledge of scientific method and due his lack of knowledge of the fact that light travels in rays. He accepts, in short, Whewell's analysis but translates it into his own language to make the discovery of facts prior.

In the first edition of his *Philosophy* Whewell responded and quite sharply indeed. Whewell maintained that, if one has the appropriate idea, one needs no experiments: it can be obtained with geometrical reasoning alone. And he adds in a pique, "Is the Reviewer's mind in that stage of speculation in which the truths of geometry are considered as best proved experimentally?" (quoted in Brewster 1842, p. 288).

Brewster was obviously angry and went to work. He sought out the passage in Aristotle for the first time and could not find it. He did find a discussion of the problem. But he found there that Aristotle used a quadrangular rather than a triangular aperture, as Whewell had maintained, and that Aristotle had given no explanation such as that attributed to him by Whewell. Still worse, Aristotle gave a new explanation which Brewster himself now endorses: the image produced is always in conformity with the aperture but the rays of light falling on the edge of the figure are feeble and are not perceived. He asks,

> Do not these observations demonstrate our views, and overturn those of Mr. Whewell? For the soundness of them, we appeal to our fellow-labourers in science. We appeal even to Mr. Whewell the philosopher against Mr. Whewell the legislator of science. We appeal to the undergraduates against the Master of Trinity. (Brewster 1842, p. 290)

How did Whewell respond to all this? He gave the candidates for the Smith Prize at Cambridge in 1842 the task of refuting Brewster's view that the rays of light produced the image defined by the aperture at all distances. He corrected the passage in the second edition of his *History* to give a correct account of Aristotle's view and he deleted the reference to this example in his *Philosophy* (Todhunter 1876, vol. I, pp. 116-117). A Dr. Anderson defended Whewell's version of the true explanation and rejected that of Aristotle and Brewster. Brewster then denied that the opinion to be refuted in the Smith's Prize examination was his. And, Todhunter tells us, there is a draft of a reply by Whewell in his papers.

If Brewster had been a moderate critic he may very well have been a very good one. But he often damaged himself with his excessive judgments of Whewell's work. In the case of Aristotle, he more or less uses Whewell's theory to refute Whewell, that is, he says that Aristotle and not Whewell had the appropriate idea. And this would seem to show that one needs, as Whewell said, the appropriate idea. Issues were lost in the desire to catch the opponent making mistakes – on both sides. Brewster's criticisms were sometimes on the mark and his examples, especially that of Hooke, Newton and Young quite nice. They sometimes did force modifications in Whewell's presentation and should have forced some in his theory. How, for example, can he explain that true or appropriate ideas are sometimes rejected? He could have done so easily but did not. He could have said, namely, that the idea as it was first proposed had not been sufficiently refined by empirical research such that its truth could be seen. And the alternative was not refuted, as it should have been, because it was, as he did in fact hold, protected by ad hoc measures. But the discussions degenerated into pettiness and backbiting comments from both sides.

We have analyzed a good portion of Brewster's attack on, and alternative to, Whewell's theory. The issues discussed were often important and Brewster's criticisms were both detailed and at times guided by leading ideas. And still we must be dissatisfied. We find mistakes in Whewell's text to be sure, which Brewster tries hard to make the most of. But these are often harmless against Whewell's theory and even against his *History*. There is apparently one serious and central issue which concerns the nature and role of factual discoveries in science. It is raised once more in a discussion of accidental discoveries in science: for Whewell there are none, since all factual discoveries depend on ideas; for Brewster they are crucial, since an important part of science is the independent discovery of facts. I will not analyze this in detail. It does not, I think, lead to more satisfactory results as those already discussed: Whewell points out the role of ideas and Brewster must concede that one must be informed. And, he says, once more with perception, that one person may make an observation and another may see its true significance. Whewell does not take an interest in or respond to such criticism. Toward the end of his essay Brewster quotes Whewell to the effect that rules for research would not help such researchers as Kepler, Fresnel or Brewster and declares that Whewell has given way here and is in agreement with Brewster. But Whewell also never sought to reduce the process of discovery to the following of some rules.

Do we find here, then, the reasons for the rejection of Whewell? We find that Whewell did, indeed, treat some thinkers such as Brewster somewhat poorly in not acknowledging the progress that they had made in methodology by, say, in the rejection of Bacon and the substitution of a method of hypothesis. We find further that Whewell did make some detailed mistakes in his historical studies. Still further we find that he did ignore some good criticism such as the objection that he idealized science or the objection that the history of science exhibited complexities, which his view did not take account of. Brewster pointed out such difficulties in his example of Hooke, Newton and Young.

But we still do not find that crucial or decisive objection. We do not find it because Brewster wishes to maintain both that he is in basic agreement with Whewell on methodology and at the same time criticize him sharply. On the one hand differences are merely verbal; on the other hand Brewster is clearly right and Whewell wrong. If this is merely a matter of style, he could say so but does not. If it is merely a matter of accuracy, he could say that, but who would then compete with Whewell in scope? And if it is a matter of theory and not just of detail, then these theoretical differences would have to be spelled out. But Brewster did not do that; perhaps he simply wished to avoid "metaphysics." The crucial points of the role of ideas in perception and in scientific research are dealt with, but Brewster offers no clear alternative: he asserts that facts are discovered independently but calls Whewell's conceptions facts or his facts appropriate ideas and admits background knowledge even in "independent" discovery of facts. The discussion of psychology and physiology of perception shows difficulties in parts of Whewell's view but no convincing alternative to explain away the facts which led Bell and Whewell to the view that the mind must be active in perception. And there were many problems for the denial of this view: perception was, as all admitted, very complex. And Bell's theory of eye-movements at least deserved an answer from the traditional theorists. Brewster came quite close to raising the level of discussion. He attempted serious criticism based on both the facts of perception and the history of science. But his apparent hostility and jealousy led him to continually overshoot; his contemptuous remarks about Whewell made a reasonable discussion difficult; and he made mistakes himself. The discussion which might have taken place in which each thinker might have substantially improved the views of the other hardly took place. Brewster had severe difficulties of his own such as that caused by his desire to move away from Bacon without moving too far so as to endanger science: he sought to save in his theory the role of the independent discovery of facts, but just like Herschel could not integrate this view with his more liberal attitude toward the role of hypotheses. The appeal to the conservative view, however, was crucial and, it would seem, effective. I find it hard to judge the short term effect of such a hard fought and often petty battle as that which took place between Whewell and Brewster. But there was quite clearly an Edinburgh-Cambridge fight in which, I presume, Brewster could expect support in Edinburgh. The long term impact is also hard to judge. The interesting points Brewster had to make were apparently lost, though Brewster may have reinforced the standard view with some effect.

### 1.5. Henry Longueville Mansel: Kant by English Rules

The rejection of Whewell's theory by the inductivists Herschel and Mill on the grounds that it (1) ignored the true views of the origins of ideas and of scientific method and (2) could not account for proof was complemented with an attack from the deductivist side developed by Henry Longueville Mansel. Mansel was then engaged in an attempt to make Kant acceptable in England. Since Mansel's analysis comes from a Kantian and thus systematic point of view, we may hope here to find some insight into the reasons for the rejection of Whewell's philosophy. Although Whewell's philosophy was rejected by the inductivists Herschel and Mill on the grounds that it was confused, they did not adequately take Whewell's system or its problems into account. They quarreled instead over his departure from their first principles. We may hope to find in

Mansel a critic more sympathetic to Whewell's Kantian and systematic aims and thus a better analysis of the strengths and weaknesses of Whewell's attempt. And here we must surely find the criticisms of his system which became so influential.

Mansel has further appeal – especially to his English audience – because he sought to state Kant's views in a clear and simple language and to correct what he deemed to be errors in them. We may expect to find in his analysis not merely sympathy but clear analyses as well. He hoped to render a Kantian point of view defensible by analyzing the psychological foundations of logic and science. Mansel thought that there was a need for a similar analysis of Whewell's theory: Whewell's use of Kant had to be, so Mansel, corrected in order to make it immune to the attack from Mill. He states,

> It is much to be regretted that Dr. Whewell, who has made good use of Kantian principles in many parts of his "Philosophy of the Inductive Sciences," has not more accurately observed Kant's distinction between the necessary laws under which all men think, and the contingent laws under which certain men think of certain things. His neglect of this distinction has given a seeming advantage to the empirical arguments of his antagonist, Mr. Mill, who is thus enabled apparently to decide the question at issue by what is in reality no more than argument ad hominum. (Mansel 1851, p. 309)

The advantage which Whewell thereby renders his empiricist opponents is, according to Mansel, due to his theory of the necessity of laws which should be deemed contingent and his resulting view of the proof of this necessity by use of the criterion of the inconceivability of their denial. Mansel endorses Mill's criticism of this aspect of Whewell's theory in spite of his Kantian position (Mansel 1851, p. 309).

Mansel believed that a more careful and correct demarcation between those principles, which are a priori laws of the mind and therefore necessarily imposed on all experience, on the one hand, and those principles, which are true but contingently so, that is, the world just happens to be that way, on the other hand, will enable one to show that there are, indeed, a priori principles without any need to appeal to the inconceivability of their denial in order to prove their necessity. He suggests,

> But on psychological grounds it would seem as if the point of separation between a priori principles and empirical generalizations ought not to be very difficult of determination. The difference is not one of degree but of kind; and the separation between the two classes of truths is such that no conceivable progress of science can ever convert the one into the other. That which is inconceivable, not accidentally from the peculiar circumstances of certain men, but universally to all, must be so in consequence of an original law of the human mind: that which is universally true within the field of experience indicates an original law of the material world. No transformation of one into the other is possible, unless the progress of science can change mind to matter or matter to mind. It is therefore incumbent on the philosopher

who would extend mathematical certainty to the domain of physical science, to confirm in every instance his theory by a psychological deduction of his principles, as Kant has done in the instances of Space and time. (Mansel 1851, p. 310)

Mansel sought to apply to Whewell's theory the same program he used to analyze Kant's system. He sought to determine by a psychological analysis whether the principles which Whewell deemed necessary and a priori were really universal laws of the operations of all minds or only a generalization – perhaps a true one. This was needed, he thought, because only then, only if one limited the a priori, necessary categories, could one avoid Mill's criticism. Now Whewell himself, of course, never thought such a tactic to be necessary. He had other responses to Mill which he deemed adequate. He sought primarily to show that his view of inconceivability was not merely psychological; the inconceivability of which he spoke was dependent on the nature of ideas. And it could account for differences in appraisals of what is inconceivable and what is not: only a properly educated mind could see the properties of the ideas which rendered them necessary (Whewell 1852). So, Mansel's remedy was, from Whewell's point of view, not needed. Whewell explained.

Mr. Mansel does not agree either with Mr. Mill or Dr. Whewell; he adheres to the original Kantian thesis, that Space and Time are sources of necessary truths, but denies the office to the other Fundamental Ideas of Dr. Whewell. In reading what has been said by Mr. Mill, Mr. Mansel, and other critics, on the subject of what I have called Fundamental Ideas, I am led to perceive that I have expressed myself incautiously, with regard the identity of character between the first two of these Fundamental Ideas, namely, Space and Time, and the others, as Force, Composition, and the like . . .

The special and characteristic property of all the Fundamental Ideas is what I have already mentioned, that they are the mental sources of necessary and universal scientific truths. I call them Ideas, as being something not derived from sensation, but governing sensation, and consequently, giving form to our experience; – Fundamental, as being the foundation of knowledge, or at least of Science. And the way in which those Ideas become the foundations of Science is, that when they are clearly and distinctly entertained in the mind, they give rise to inevitable convictions or intuitions, which may be expressed as Axioms; and these Axioms are the foundations of Sciences respective of each idea. The Idea of Space, when clearly possessed, gives rise to geometrical Axioms, and is thus the foundation of the Science of Geometry. The Idea of Mechanical Force, (a modification of the Idea of Cause,) when clearly developed in the mind, gives birth to Axioms which are the foundation of the Science of Mechanics. The Idea of Substance gives rise to the Axiom which is universally accepted, – that we cannot, by any process, (for instance, by chemical processes) create or destroy matter, but can only combine and separate elements; – and thus gives rise to the Science of Chemistry.

Now it may be observed, that in giving this account of the foundation of Science, I lay stress on the condition that Ideas must be clearly and distinctly possessed. The Idea of Space must be quite clear in the mind, or else the Axioms of Geometry will not be seen to be true: there will be no intuition of their truth; and for a mind in such a state, there can be no Science of Geometry. A man may have a confused and perplexed or a vacant and inert state of mind, in which it is not clearly apparent to him, that two straight lines cannot enclose a space. But this is not a frequent case. The Idea of Space is much more commonly clear in the minds of men than the other Ideas on which science depends, as Force or Substance. It is much more common to find minds in which latter Ideas are not so clear and distinct as to make the Axioms of Mechanics or of Chemistry self-evident. Indeed the examples of a state of mind in which the Ideas of Force or of Substance are so clear as to be made the basis of science, are comparatively few. They are the examples of minds scientifically cultivated, at least to some extent. Hence, though the Axioms of Mechanics or of Chemistry may be, in their own nature, as evident as those of Geometry, they are not evident to so many persons, nor at so early a period of intellectual or scientific culture. And this being the case, it is not surprising that some persons should doubt whether these Axioms are evident at all; – should think that it is an error to assert that there exist, in such sciences as Mechanics or Chemistry, Fundamental Ideas, fit to be classed with Space, as being, like it, the origin of Axioms. In speaking of all the Fundamental Ideas as being alike the source of Axioms when clearly possessed, without dwelling sufficiently upon the amount of mental discipline which is requisite to give the mind this clear possession of most of them; and in not keeping before the reader the different degrees of evidence which, in most minds, the Axioms of different sciences naturally have, I have, as I have said, given occasion to my readers to misunderstand me. I will point out one or two passages which show that this misunderstanding has occurred, and will try to remove it.

The character of axiomatic truths seen by intuition is, that they are not only seen to be true, but to be necessary; – that the contrary of them is not only false, but inconceivable. But this inconceivableness depends entirely upon the clearness of the Ideas which the axioms involve. So long as those Ideas are vague and indistinct, the contrary of an Axiom may be assented to, not because it is possible, but because we do not see clearly what is possible. To a person who is only beginning to think geometrically, there may appear nothing absurd in the assertion, that two straight lines may enclose a space. And in the same manner, to a person who is only beginning to think of mechanical truths, it may not appear to be absurd, that in mechanical processes, Reaction should be greater or less than Action; and so, again, to a person who has not thought steadily about Substance, it may not appear inconceivable, that by chemical operations, we should generate new matter, or destroy matter which already exists. (Whewell 1852, pp. 8-10)

and he explains further,

Though Mr. Mansel does not acknowledge any subjective conditions of intuition besides Space and Time, he does recognize other kinds of necessity, which, I should equally refer to Fundamental Ideas: because they are, no less than Space and Time, the foundations of universal and necessary truths in Science . . . To these Principles he ascribes a "metaphysical necessity," the nature and grounds of which he analyses with great acuteness. But what I have to observe is, that whatever differences may be pointed out between the grounds of the necessity, in this case of metaphysical necessity, and in that which Mr. M. calls mathematical necessity which belongs to the Conditions or Ideas of Space and of Time; still, it is not the less true that the Ideas of Substance and of Cause, do afford a foundation for necessary truths, and that on these truths are built Sciences. (Whewell 1852, p. 14)

Whewell makes amply clear that for him fundamental ideas are necessary, that they have differing grounds, but are united by the fact that they are seen by the sufficiently educated mind and only by the sufficiently educated mind to be necessary, and that they build the foundations for various sciences. For Whewell, then, there is no need to distinguish between the necessity of space and time and the necessity of other ideas. There is also no need to distinguish between a Kantian and a Platonic view: in all cases of fundamental ideas, the ideas grow, are learned by those who study an area and are seen by those to be true and necessary, to see that only these ideas can describe the world as it is. No idea is a mere form which organizes phenomena; all fundamental ideas describe the world truly.

Further, the restrictive alternative offered by Mansel involved sacrifice. Whewell sought to explain the growth of scientific knowledge and deemed inductivism or empiricism inadequate to account for that. Mansel was here no better off: he had to rely on induction and empiricism to account for the growth of our knowledge of "empirical generalizations" just as much as Herschel or Mill. And they could not account for growth, for necessity or for proof such as it in fact occurred in science, as we have seen above in the discussion of their attempts to account for those facts which Whewell pointed to in support of his own theory.

If we are primarily interested, however, in the discovery of the basis for the negative judgment of Whewell's philosophy, it is certainly desirable to overlook these anticipated difficulties in Mansel's own view of science and see if he has offered good reasons for a more limited view of a priori ideas or principles. It is here where we may expect to find Whewell's confusion, that is, in the use of Kantian ideas in contexts where they cannot be consistently made to apply. Mansel seeks to carefully avoid such confusions by demarcating with his new method the proper province of logic (Mansel 1851, pp. 283-284). His aim is clearly quite broad since the proper demarcation of logic is the basis for a proper estimation of the limits of our knowledge in science, metaphysics or ethics. (Mansel's approach is perhaps quite close to that of Jacob Friedrich Fries who also sought a new more psychologically oriented method of identifying a priori categories; see Wettersten 1995, pp. 137-138; 1992, pp. 153-159. Mansel refers to Fries occasionally, but I am unsure as to what debt Mansel has to Fries. Fries' disciple Apelt commented on Whewell as well and I will discuss his view below.)

Mansel is just as opposed to Schelling and Hegel as any other proper Englishman. But, like Whewell, he believes that the strict empiricism of Herschel and of Mill goes too far: some a priori principles exist. His problem, then, is to identify and to demarcate (some of) them, so as to provide the necessary correction to Herschel and Mill, but not to go too far in the direction of Schelling and Hegel. I suppose, but can find no direct quotations to support my supposition, that Mansel believes that

Whewell has quite unwittingly already done this. This problem of demarcation was common to Whewell and to Mansel so one may compare their approaches to it. Mansel sought, as did Herschel, Mill and Whewell, to analyze the content of thought. This analysis of the source of ideas serves then as the foundation of the theory of the limits of knowledge. Mansel contends, just as Whewell does, that there are two sources of knowledge. On Mansel's view these two sources are, on the one hand, the formal conditions of experience – and therefore of knowledge – and the content of knowledge. Now, since all content of knowledge is given by presentations, the (psychological) problem of the foundation of logic is to identify and demarcate that which we obtain by mere presentation and that which we use to give form to presentations in order to achieve representations (of objects) (Mansel 1851, pp. 7*ff*). He states, "The possibility, therefore, of any branch of scientific inquiry depends upon the psychological question, how many presentative faculties has man? Every such faculty may furnish distinct materials for thought" (Mansel 1851, pp. 38-39).

Now Mansel is not sure of the extent of that which is presented to us. It is a task for psychology to determine it, however, and he proposes to make a beginning. He argues that sensations alone are by no means sufficient to supply the full content which we in fact find in thought (Mansel 1851, pp. 42-43). Mansel has, therefore, a program which is identical to that of Herschel and Mill in that he sets himself the task of identifying the source of the content of all ideas in that which is given to man from without. But his program allows for modes of "presentation" other than sensation and is thus much broader in this respect as well as in his supposition of the existence of a priori forms. His results are that space and time are "presented" to man and the mathematical principles which give them form are a priori and synthetic. In regard to mathematics, then, the difference between Whewell and Mansel is not so great. But he denies that other ideas such as laws of force are synthetic a priori. These ideas, he says, are a result of our knowledge from experience – internal as well as external – which are given order by identical or analytical statements (Mansel 1853, pp. 21*ff*).

I do not know how Mansel proposes to show that our physical theories of force can be analyzed into merely that which is presented to us and analytic statements (plus the synthetic a priori statements of mathematics). Mansel's reductionism is seemingly more liberal than that of Herschel or Mill. But it faces the same problems and with even vaguer criteria for success, since what should count as a reduction to sensations is, perhaps, clearer than what would count as a reduction to "presentations." Mansel's critique of Whewell thus seems to be based on quite vague presumptions concerning the analysis of ideas. And here Whewell would seem to have the edge since there seems no clear way of analyzing varying conceptions of force into mere presentations and laws of identity plus mathematics. Whewell asserts that and even makes ideas and facts relative: what is "presented" to us depends on ideas we impose on observation. We have certainly not yet found the answer to our puzzles concerning the negative judgment of Whewell's philosophy.

I should note here that the view which Whewell adopts is characterized by Mansel as Platonistic, that is, we have ideas not derived from "presentations" which correspond to the world or which are real (Mansel 1856, pp. 22*ff*). This characterization became important latter as Whewell was accused of illegitimately introducing Platonism into his theory or of switching from a Kantian to a Platonic view. Whether the interpretation according to which Whewell confusedly combines Kantian with Platonistic elements and switches as he needs between one point of view and the other is correct or whether it would be better to say that Whewell has a coherent view which is in some respects similar to Kant (ideas which are a product of our mind are imposed on and give order to impressions we receive from without) and in some respects similar to Plato (ideas are real and are known to be true by a kind of intuition) is an issue I will have to return to below.

Mansel's adherence to, even strengthening of, the limits set by Kant was complementary to the criticisms of Whewell made by Herschel and Mill. Mansel's critique had the same foundation in the quest to find the proper limits of knowledge by analyzing the source of all knowledge in impressions. Mansel's criticism of Whewell for allegedly introducing ideas with no proper source and thus no proper justification shared the same weaknesses with the criticism of Herschel and Mill as well. It could provide neither the necessary reductions nor the needed theory of proof. But the possibility of such reductions and thus of proof based on them were taken as dogma. Thus, the weaknesses of Whewell's theory counted against him as he struggled to develop a theory of the growth and justification of scientific knowledge, whereas the weaknesses of Mansel's analysis did not count against his critique. The only justification for such a procedure had to be a transcendental argument. But this transcendental argument was circular, unless of course one took the way out offered by Whewell and appealed to something like inconceivability. The need for such a move thus appears in Mansel's attempt just as it had in that of Herschel and of Mill. It was circular because it presumed that scientific proof had to be based on reduction and this is just what Whewell said could not be the case – whatever scientific proof or the lack of it might happen to be.

The complementarity and overlap of the analyses of the empiricists Herschel and Mill and the Kantian Mansel throw some light on the way in which the established opinion of Whewell's philosophy was built, since it shows the standards which were used to judge this philosophy and the principles which served as the foundations of attempts to construct alternatives. But this clarity does not lessen the tension which one may feel between the negative judgment and power of Whewell's philosophy. Rather, it seems to show that the deeper problems which Whewell sought to discuss concerning the nature of scientific knowledge and scientific proof where never put on the agenda. We must look further.

### 1.6. The Whewell Heritage: No One to Carry the Torch

The reaction to Whewell was not only severe at the top of the intellectual community. It went deep as well. There was virtually no one who could carry on Whewell's work. He had no students, no followers, no defenders. It is no wonder. Whewell himself had great prestige but his opponents were in fashion and powerful and also had prestige. It would have taken great courage to attempt to carry on a fight which everyone had deemed settled. Todhunter, as we shall see, wished that there had been a reconciliation between Herschel and Whewell. It was, I think, not a wish of his alone. Such a reconciliation would have made it easy for those who appreciated Whewell to do so openly and defend the value of his work. So long as it was lacking it was difficult for anyone to carry on the torch.

The difficulties were both intellectual and social. The social difficulties were posed by the fact that Whewell had argued so well for his theory on so many points that one could not hope to do much better. Yet Whewell had failed to convince. Should one, then, merely repeat the arguments? How could one defend Whewell against Mill with any new and interesting arguments, when the excellent arguments already brought forth were simply ignored?

The intellectual problems of developing Whewell's program were also enormous. And they were never explicitly formulated. In order to understand the fact that Whewell had no disciples one can ask what they might have done. From today's perspective it is no doubt easier to see. But then it was no doubt very hard. It was comparatively easy, for example, to discuss the so-called problem of induction. One seemed to have a general idea of what was called for and even a program: explain how theories are shown to be true on the basis of empirical evidence. If one followed Whewell and rejected induction, however, one either was led to his view that the truth was intuited – was seen to be necessary – by educated minds or one needed an alternative. Either option was no doubt very difficult to follow.

Rather than face these problems it was by far easier to endorse something like a Herschelian or Millian view. Yet the view that there were principles of science which could be shown to be true was in the air. How, for example, could one justify the principle of the conservation of force. Faraday used it as a metaphysical framework for his studies without proving it. Mayer proposed it but his proposal was deemed metaphysics and rejected. Helmholtz proposed it and it was at first rejected as nature philosophy and then accepted as proven. Helmholtz's friend Emil du Bois Reymond thought it could be established a priori. Was Helmholtz in this essay Whewellian or Millian or some third position? It is, I think, quite unclear (Wettersten 1992, pp. 69-72).

Whewell had, then, no followers. No one had the interest or courage to defend his arguments. The new problems which his view raised were deep and difficult. No one faced them head on. Yet, as we shall see, they were felt all around and refused to go away. Whewell had put down a challenge and his contemporaries were not up to the task of responding to it.

# CHAPTER 2

# EMBARRASSED SILENCE

Whewell became master of Trinity and married his first wife in 1841 at the age of 47. With his two masterpieces – his *History* and his *Philosophy* – to say nothing of his numerous other publications he had achieved fame and recognition and not only in England. But during his tenure as Master at Cambridge it was Mill rather than Whewell whose work set the tone even there (Smith 1869, p. 460). He was steadily engaged in controversy, but his contemporaries found his *Of the Plurality of Worlds,* in which he argued that life could not be expected to be found on other planets, far more interesting and challenging than his deep studies in the history and philosophy of science. Herschel later set the condescending rumor in the world that Whewell was not even serious in this book but was merely seeking diversion to overcome the grief which was caused by the death of his wife (Todhunter 1876, vol. I, p. 187).

Whewell died in 1866 and was buried under a heap of laurels. The recognition which he had achieved for his reform of Cambridge, for his tenure as Master, for his generous contributions to Cambridge to say nothing of the fame which his work in the philosophy and history of science as well as in philosophical theology and even ethics (Whewell 1845) was enormous. Given such an enormous output combined with a brilliant career one would expect a clear and untroubled attitude toward the man after his death. The battles could be forgotten and his great merit emphasized. And sympathy there was. It was known not only that he had been successful but also that his views had been rejected. It was also known that the deaths of his wives had caused him great grief. But even in death he was a difficult fellow: How could one render him his due without admitting that he had not been treated very well?

## 2.1. W.G. Clark, Sir John Herschel, David Brewster: A Decent Burial?

When Whewell died there was virtually no one who deemed his work to be of sufficient value to work on it, to improve it, or to criticize it. A consensus had apparently been achieved and nearly no one chose to defend it or quarrel with it. (One exception is the poor discussion of his philosophy published by Grote; see his 1900.) Among the few notices of his person and work which followed his death are three obituaries, all of which are mentioned in the *Dictionary of National Biography*. There is no doubt leeway in what one may write in an obituary and no specific appraisal of the intellectual work of a man is needed. But all three of these obituaries try to reconcile the picture of a man of high accomplishment with their rather dim views of his philosophy. They struggled to put forth a satisfying picture. Yet what we see seems to serve their own needs more than that of such an appraisal, which they did not even need to give. Whewell's death was troubling.

### 2.1.1. *W.G. Clark: Too Hasty Work and Ultra-Platonic*

W.G. Clark (1866) takes up two of the most troubling themes for Whewell's contemporaries: his enormous productivity and his challenging, heretical philosophy. Perhaps he speaks for Cambridge. Whewell's work was so important and so voluminous and yet so thoroughly rejected. And he had there no competitor or follower as, say, Clark might have been. Clark thus says of Whewell's *History* that

> It will doubtless long continue to be the standard English book on the subject, enriched and amended by the comments of successive editors. In a book which takes a bird's-eye view of all science, numerous inaccuracies must of course be apparent to microscopic investigators, and further corrections and qualifications will be required by the growth of each branch. (Clark 1866, p. 547)

He is less kind in his evaluation of the *Philosophy* which he says, "was not so successful as its predecessor. Many thinkers were unable to accept the ultra-Platonic hypothesis on which it was based, but none could fail to find in it much that was instructive if not convincing, and many brilliant guesses at truth, if not clear discoveries of it" (Clark 1866, p. 547).

The views here expressed became quite standard as Whewell was deemed to have moved from an early Kantian to a later Platonic position – a view which was based on Mansel's criticism which I have reviewed above. Todhunter, as we will see below, quickly took issue with this view, but with little effect. Clark describes Whewell as a man of courage, but one who was not cool or self-possessed. He also claims he worked in undue haste. So, he was deemed incredibly erudite and brave and a man of high integrity. But his work and perhaps his character were flawed by his impetuousness.

Clark offers a picture of Whewell which superficially resolves the conflicts which Whewell's work produced. His productivity may be explained away as due to haste and poor character. (His colleagues who take more care – produce less – are thus better). And his philosophy suffered from the same defect of lack of care. It was excessive – not merely Platonic but ultra-platonic – and good guesses were there but scarcely discoveries of truth. The conservative reaction is defended as a return to careful work and Whewell is praised as a courageous, brilliant but impetuous man: his character and his philosophy have the same advantages and flaws; the former could not be corrected but the later needs to be.

### 2.1.2. *Sir John Herschel: Collaborator Becomes Platonist*

Sir John Herschel had known Whewell since their student days. They had collaborated and been friends. Yet Herschel had led the attack on Whewell. While always friendly in tone, and while always expressing high regard for the author, he had done much to establish a view of Whewell's contributions, a view which was very poor indeed: Whewell had left the true path of science; he had adopted excessive views; he had advertised as special what was not particularly new; and he had been too ambitious, which is to say, too quick and careless. Whewell had passed Herschel by in his understanding of science and the leading scientist of the day had protested, even against his old friend. In his obituary Herschel need not have sought to explain himself and his view once more. But he did and, indeed, added to it so that the view would become even more established even while he tried not to forget the longed-for days of co-operation in building a common view.

Herschel (1866) praises very highly the work which Whewell did along with himself in reforming the curriculum at Cambridge and Whewell's early books on dynamics and mechanics which were in part designed to serve that purpose. He asserts, however, that a great change in his views occurred between the publication of the *Bridgewater Treatise* in 1833 and the History in 1837. He refers (Herschel 1866, p. liv) – he does not quote – to a passage in the Bridgewater Treatise which reads as follows,

> we cannot but ask how we obtain these principles? (from which we reason) from what other source of knowledge we derive the original truths which we thus pursue into detail? since it is manifest that such principles cannot be derived from the proper stores of mathematics or logic. These methods can generate no new truth; and all the grounds and elements of the knowledge which, through them, we can acquire, must necessarily come from some extraneous source. . . . This process, by which we acquire our first principles, (without pretending to analyse it,) is obviously the natural exercise of human experience, and the natural exercise of the understanding; our intercourse with matter and with men, and the consequent growth in our minds of convictions and conceptions such as our reason can deal with, either by her systematic or unsystematic methods of procedure. (Whewell 1839, pp. 335-336)

This quote will hardly show what Herschel maintains, since in it Whewell makes clear that there are two sides to the process, that is, experience and understanding or alternatively our intercourse with matter and men and the growth of convictions and conceptions such as our reason can deal with. Nevertheless Herschel's suggestion seems to have had its effect. Herschel portrays Whewell as becoming increasingly Platonic and, though this view is emphasized by both Clark and Herschel, it is merely at this point a characterization. This is virtually all he has to say about Whewell's major works but the view nevertheless became standard as we will see below.

### 2.1.3. *David Brewster: As Kind as He Could Be*

A third obituary was written by David Brewster (Brewster 1866-67). Brewster was, we may suppose, called upon to say a few words due both to his prominence and the vehemence of his fights with Whewell. The essay called for some more moderate tone and Brewster does manage to say a few kind words about Whewell. Though Whewell thought, he says, that science should be its own reward, he helped younger colleagues when he himself was successful. This was some concession, since he complained about Whewell's *History* on the grounds that it left aside the defects of science and the difficulties which scientists have to cope with. In intellectual matters he can hardly find a generous word to say. He finds Whewell's *Of the Plurality of Worlds* not to be serious and gives examples. (He had written a vehement criticism of this work as well (Todhunter 1876, p. 190). He has only the following to say of Whewell's two major works:

> In 1837 he published his principal work, entitled the "History of the Inductive Sciences, from the Earliest to the Present Times." In appreciating the value of this noble record of physical truth, and admiring the capacity and vast acquirements of its author, we must at the same time regret the numerous deficiencies in the work, and its doubtful decisions in the controversies of science. As a necessary sequel to this work he published, in 1841, "The Philosophy of the Inductive Sciences," a work of able and ingenious speculation, which the public did not so gratefully receive. (Brewster 1866-67, p. 30)

## 2.2. Isaac Todhunter: Conscientiously Recording Facts for the Archives

In spite of the rather low regard in which Whewell's work was then held, it would nevertheless hardly seem possible to completely ignore someone who had played such an important role in so many of the most prominent controversies of the day and who had achieved such wide recognition and high standing at no less an institution than Cambridge. Something had to be done and two works appeared. One was produced by his niece, Mrs. Stair Douglas (1881), with whom he was quite close toward the end of his life (Smith 1860, p. 459). It deals with his personal life and is for the most part a collection of letters arranged chronologically with small notes added by the editor. It is of no importance for this discussion though it is a valuable source, of course, for anyone interested in Whewell's life. The second is the most important secondary source concerning Whewell which exists and was put together by Isaac Todhunter. It consists of two volumes. The first provides a chronological account of Whewell's intellectual activities and the second a collection of Whewell's letters. They are of some intellectual import and are also arranged chronologically.

Todhunter was not the first choice to provide a biography or something like it of Whewell and his work. He reports that "some attempts were made to obtain an editor, who would arrange the materials and write a biography of Dr. Whewell, but without success" (Todhunter 1876, p. v).

Todhunter agreed to the task, but he was neither close to Whewell personally nor did he choose to defend him intellectually. He worked conscientiously through all of Whewell's works and gives us information which is quite valuable concerning reactions and editions, etc. But there is no development of a point of view and everything handled quite uniformly. The chapter on the philosophy of the inductive sciences is, for example, twenty-two pages long while the chapter on English hexameters is nineteen. He nevertheless had influence in, for example, his view that the history was accurate in its first editions – given the state of knowledge at the time it was written – but was not adequately corrected and kept up-to-date in succeeding editions, a view which is taken over by the *Dictionary of National Biography*.

A few points concerning his discussion of Whewell's philosophy are relevant here. Contrary to most commentators on Whewell Todhunter preferred Whewell's *Philosophy* to his *History* (Todhunter 1876, p. 100). The tone of his work, however, is exemplified by the following quotation.

> I shall not venture to join here in the controversy carried on by such distinguished men, except so far as to say that I cannot personally agree with Mill and Herschel to refer my conviction of the truth of the foundations of geometry to experience. . . . Whether there may not be some compromise between the views of Sir J. Herschel and Dr. Whewell may furnish matter for conjecture. (Todhunter 1876, p. 143)

Todhunter sought to appreciate Whewell where he could and to criticize where he thought it necessary. But he offered no overall appraisal of his system or his views. He contented himself with correcting some of the most unfair of the attacks on Whewell and with noting some of the minor difficulties which Whewell's view faced. He apparently hoped that some of Whewell's insights could be of use but that the work as a whole could not be defended.

We need not nevertheless turn away entirely empty handed from our attempt to find there a considered view of Whewell's philosophy. One of the crucial issues which later became important concerned the development of Whewell's philosophy. And this has much to do with the alleged coherence or incoherence of Whewell's system. This is so since changes of points-of-view as his system developed can be used to explain incoherencies. The central issue here is whether Whewell's use of Kantian and Platonic elements in his theory leads to incoherencies. The question of his development is whether he moved from an early Kantian to a later Platonic view.

Todhunter did maintain that Whewell did not change his mind to any high degree (Todhunter 1876, vol. I, p. 144). And he further rejects the view that Whewell's view is unacceptable because it is "ultra-Platonic" (Todhunter 1876, vol. I, p. 132). The only exception which Todhunter finds to the view that Whewell did not change his view in his later work concerns the *Bridgewater Treatise* but even there he defends the general unity of Whewell's view (Todhunter 1976, vol. I, p. 70). The difference which Todhunter finds is the following: in the *Bridgewater Treatise* Whewell does not deny explicitly, as he does later in his much discussed *Of the Plurality of Worlds,* that various bodies of the sky may be inhabited (Todhunter 1876, vol. I, pp. 69-70).

There is another passage however which I will discuss below which was perhaps of more importance. Todhunter suggests that Whewell began with an attempt to explain mathematics and mechanics alone. And this might have been acceptable but the extension was not. This view reflects Mansel's criticism and it is not opposed to those of the empiricists. And it appears quite strongly in the works which follow. We will have to examine it there as the root problem of Whewell's philosophy. The view is, of course, puzzling in Todhunter since, if true, it would mean that Whewell did change his mind, at least to the extent of the application of specific views – and this was no small matter. Unfortunately, we find no answer to such questions in Todhunter.

The view of Whewell's work which became quickly established is reflected in the following quote from the *Dictionary of National Biography.*

> The book (Whewell's History) necessarily became belated in many parts. Whewell meanwhile scarcely became a philosopher. He had studied Kant, and accepts Kant's theory of space and time. For later German developments he had nothing but contempt, and his friend Hare and others could never induce him even to take an interest in Coleridge. In his controversies with Mill he seems to have the advantage in some points from his greater familiarity with science and from his knowledge of Kant, whom Mill disregarded. But his constructive theory represents the old-fashioned form of "intuitionism," against which Mill carried on a successful warfare. His theory about "ideas" and "facts" is scarcely coherent, and certainly did not obtain acceptance. (Smith 1860, p. 460)

# PART TWO

# WHEWELL'S IMPACT EMERGING

## CHAPTER 3

## DISTURBING RECOLLECTIONS FAIL TO PASS AWAY

The search for the reasons for the rejection of Whewell's philosophy in spite of its apparently great influence becomes difficult in the time after the first intense debate had ended. During Whewell's lifetime leading scientists and philosophers were intensely interested in Whewell's work even though they rejected it to a man. After Whewell's death no leading figure published any comprehensive study of Whewell's work. But nevertheless we begin to see his influence in quite disparate places and ways. He is not a philosopher or historian to whom anyone can appeal as an authority. Yet he keeps turning up and sometimes even with quite favorable judgments or, sometimes, with what appear to be reformulations of his views. Still further we find that the problems he posed for the inductivist-associationist tradition – for the followers of Mill – keep turning up; sometimes there is reference to Whewell and sometimes not. These phenomena did not, however, change the quite general critical attitude toward him: no one tried to rehabilitate him as a philosopher of the first rank. They followed the pattern set by Todhunter in that they at times defended him against this criticism or that, and found this or that idea important, but they did not explain his theoretical importance.

This interest in Whewell can be seen in America in the work of Charles Sanders Peirce, in Germany in the work E.F. Apelt, Friedrich Albert Lange and Paul Leuckfeld, and in France in the work of the scientist Claude Bernard – who never mentions him – or the philosopher Louis Liard who shortly reports on him. In England his successor in the Chair for Moral Philosophy at Cambridge, Grote, included him in a study published the year before Whewell died, that is, 1865. But that is the apparent end of the matter. The interest turned to induction and something like an interest in Whewell did not reappear in England until Karl Popper arrived on the scene from Austria via New Zealand and offered renewed defense of a theory of science without induction. Popper offered a theory of science without induction and justification. Whewell offered a theory of science without induction but with justification. The return to Whewell seemed called for to respond to Popper, to present that which was new in the 19th century as new in the 20th in order to stop that which was really new in the 20th – to borrow and apply an observation of Stein (1946).

In order to see the unease, the interest and the hostility which Whewell's theory evoked in the last quarter of the 19th century, it is useful to see how he and his views turned up, refused to go away or posed more or less repressed challenges during this period. Since there are no serious studies of Whewell made in this time, we need to turn to the role his views played and the short statements about him which were made. Let us look then at the situations in England, America, France and Germany to see what the aftermath was, to see what would not go away in spite of the universal rejection of his philosophy.

### 3.1. England: The Eerie Quiet

The 1860s and 1870s in England offer pictures of unity and of conflict which could hardly be greater. On the one hand, Mill's view of scientific methodology had won the day hands down. Science and its rules had been established and even united with a firm ethical foundation. Not only was the faith in science enormous. The faith in science as the definitive discovery of the true path to knowledge was also great. Science was on the march. There had been enormous gains in knowledge. And although Faraday had played a major role, the true Newtonian principles of science had seemed at first to win, while the speculations of Faraday were rejected as Herschel had thought they should be. But change began to come quickly. In philosophy and in science the two great threats to secure science in England had been at first successfully overcome. At the time of their deaths neither Faraday (1867) nor Whewell (1866) could have had much hope other than that provided by a great faith in the value of their work and in the intellectual community's ability to discover that value in the long run. I doubt that either had that faith though neither would have denied it. The mental collapse of Faraday as described by Agassi (1971a) indicates real despair without the solace which the confidence that the value of his work would eventually be discovered such as, for example, that possessed by Kepler. Whewell saw Mill win the day; he fought to the last but found what happiness he had in his later years in his family, in his solace in his religion, and in his dedication to Cambridge.

There were also great turmoils, which seemed to have touched religion above all. The fight for the acceptance of Darwin which raged through the 1870s was the great battle. And though leading scientists and philosophers such as Herschel and Whewell had rejected his view, Darwin's ultimate victory was deemed by the rational defenders of science to offer the hope a stunning victory for science: even the evolution of life could be explained within a scientific framework. Religion could be pushed from one further domain and the faith in science and its method rendered more secure and successful.

But there was another battle which was fought much less on the public stage and with much more subtle methods. It was nevertheless of deep concern to the lovers of science who seemed on the surface to be moving forward so spectacularly and smoothly. This battle concerned the principles of science and has been dramatically portrayed by Joseph Agassi from the point of view of Faraday. Here, I wish to supplement and modify this picture offered by Agassi somewhat in order to explain the importance of methodology and in particular of Whewell's methodology.

Officially, everyone knew what the principles of science were, that is, they were Newton's or what were then taken to be Newton's. But everyone also knew that these principles had not stood their most crucial test in the 19th century: they had not been rendered capable of explaining electro-magnetic phenomena. Not only that. There was an alternative – Faraday's – which might do better in explaining the phenomena and which had lead to startling discoveries. What principles should science then have? Must Newton's be rendered workable or science abandoned? How could one change science to incorporate new principles? What were they?

These disturbing questions were not quite questions for philosophy alone. The German nature philosophers had tried that and had earned – at least in England – ridicule for their pretentiousness. But they were not to be settled by appeal to science alone either. Even if one sought to simply change science by accepting, say, Faraday's principles for electro-magnetism at least, one would still have to explain, then, what science could accomplish, whether it was unified and could change and, if so, how. These problems were not put on the agenda for discussion. They were, along with the views of Faraday and Whewell, repressed. Yet, we can now see in retrospect, that only Faraday and Whewell together could have constructed a coherent alternative. They never did. Whewell rejected Faraday's principles and Faraday, though he enthusiastically praised Whewell's view of science, never applied it to his own case. Yet this combination had to be of great importance.

Faraday had offered a new view of science which sought a new and deeper explanation of the world by extending science to new areas and by unifying theories of electricity and magnetism and even offered the hope that this new view could account for mechanics as well. His success was partial but he showed the need for a new, deeper unifying view. He had further used a method of conjectures and refutations and he had noted that observations were theory-laden. He could show the power of his ideas experimentally but he had no coherent epistemological theory which would explain how such an approach would be good for science as a whole.

Whewell had such an explanation and considered applying it to Faraday's theory. In the end he rejected this daring move. He had put forth in his theory of the consilience of inductions a theory of how a new idea may unify previously separate fields. It might thereby change science and render it more perfect, indeed, perfectly true. All of Faraday's progressive methodological innovations were part of Whewell's theory and systematized in it. His theory, then, could explain the method of Faraday's discovery and its new status. But should it be accepted as a new, deeper and true theory?

Whewell balked at this move. He could well conceive that science progresses by changing its principles; he had explained that process. But he balked when he thought of changing Newton's principles: even for Whewell this was holy ground and had to be defended. Yet he was far ahead of his contemporaries: he considered this as a real possibility. He saw that there could be a deeper theory which would require one to view Newton's theory as a mere approximation of a special case and that this could be quite in accord with established scientific method (Whewell 1967a, vol. II, p. 137). One may presume, I think, that he balked because he was ambivalent. He had insisted that science did reach the truth. But it was also an (endless?) quest for deeper theories. Had Newton not found rock bottom? He could not decide.

Scientists and philosophers were faced with a dilemma. Faraday had challenged the scientists to show that science could proceed successfully without the changes which he proposed. They could not do so. But if science could not do that, it seemed that science as a whole was threatened and not Newton's theory alone. There was only one theory of science which could come close to explaining such a revolution as an integral part of science. And this view had been repudiated as a danger to science.

Whewell was clearly needed, since it was only he, who was deemed the greatest threat to science itself in his methodology, who offered anything like ideas which could be used to allow for the change which loomed over the horizon, which might yet be avoided but which looked ever more inevitable. In this climate of dread of success Hermann Helmholtz published his "Conservation of Force." It aroused great confusion, since it was first rejected as nature philosophy. But then it was quickly accepted. Philosophers have debated just what this essay accomplished and why ever since. Here is my conjecture which is, once more, a modification of Agassi's in that I explain the possible use of Whewell's ideas in Helmholtz's essay.

Helmholtz saw that the principles of science had to be formulated so as to be extended to other fields. He was, as he said, particularly interested in stating these principles for physiology with an eye on views of the life force but it had to apply to all forces, that is, electricity and magnetism as well. He could not appeal to Newton alone since the situation was not yet clear: Would Faraday win or lose? Yet he needed to include Newton. I conjecture that he turned to Whewell. Whewell had argued that the principles of science became clear to the educated mind by the study of empirical phenomena. These phenomena did not justify the theory but rendered it so clear that one saw that no other possibility existed. This, on my conjecture, offered Helmholtz his method. He adopted Faraday's principle – which could include Newtonian experiments – and Whewell's method of proof. Having "established" the principle in this way he could leave open the question whether Faraday's view could be accepted or not.

The importance of his paper, then, was in showing that one could accept new theories – Faraday's – without abandoning the principles of science. He here uses Whewell's idea that science proceeds by finding deeper ideas which unify – and change – more specific ones. Helmholtz had provided science with a fall-back position in case it needed one. Electricity and magnetism could be explained within Newtonian or Faradayan principles so long as the principle which both Faraday and Newton accepted was adhered to. And this was proven!

Helmholtz did not say that, of course. He hardly said anything at all. In many points he was, indeed, a follower of Mill though he was not so naive, of course, as to defend anything like the canons of induction. But his main interest here was in principles and there Mill could not help him at all. He could not even appeal to his Kantian view of causality. The only alternative view, which is sympathetic to science and informed about it is Whewell's. Helmholtz, by hedging his bets, was successful. If he had said this is a new alternative to Newton, the Newtonians would have rejected it. If he had said it was strictly Newtonian it would not have been exciting. If he had appealed to Whewell it would have been rejected as not scientific. By keeping his silence he let everyone find there what they wanted. And that is what everyone wanted. Whewell had shown the way out of the dilemma but no one could say what had happened and everyone waited with bated breath. Maxwell, by giving Faraday a sound mathematical formulation, gave the opportunity to abandon long-lost stations without losing face.

To return to my subject, which is the situation in England, we can now see that changes were silently being made to allow for Faraday, should he be vindicated, and that Whewell's views were thus accepted via Germany and Helmholtz, but surreptitiously. One could thus defend Mill as the great methodologist, accept Whewell's suggestions via Helmholtz for actual science and say serenely that everything is in order because, however the intense debate came out, one need not feel that the image of science as secure was threatened.

To the conflicts between the ideology of science – the view of Mill – and the practice of science, which Whewell had already pointed out, were thus added conflicts between the ideology of science and the interpretations of the most important elements of science. These could only be maintained by maintaining an eerie quiet about the most central issues. And that is what they did.

## 3.2. America: Charles Sanders Peirce – The Only Attempt to Build on Whewell's Results Ignored

When we turn to America to see if we may find appreciative studies of Whewell there, we think, of course, of Charles Sanders Peirce. The similarities between his view and Popper's have attracted the attention of various thinkers; more recently his appreciation for Whewell as well as his sharply critical comments on Mill have been noticed. A study of Peirce might, then, help resolve the puzzles concerning the conflict between the strength of Whewell's philosophy on the one hand and the generally poor opinion of him as a philosopher on the other. In the place of mere dismissal we might hope to find the strength of his view identified, before its weakness is explained. But nowhere in Peirce's writings do we find a serious, critical review of Whewell's philosophy. Although Peirce's various comments on Whewell do show that his debt to Whewell was considerable, the only critical comment we find is the unexplained remark that Whewell was doing some fine work in logic but it was not of a fundamental character (Peirce 1960, vol. I, §15). This remark may be important for understanding Peirce, since we may presume he did want to attack those problems he deemed fundamental and which Whewell, in his opinion, ignored. But it does not by itself bring us further in the study of the reception of Whewell. In order to understand Peirce's relation to Whewell we need to use a different procedure. We need, of course, to keep Peirce's passing remarks concerning Whewell in mind, but above all we need to study the logic of the situation. We need to ask where and how Peirce takes over Whewell's philosophy, how he changes it, and how he wants to move forward. We may then ask how Peirce's use of Whewell has been received. Was Peirce's use of Whewell anymore effective in spreading Whewell's ideas than Whewell was himself? Or was, perhaps, the reaction to Peirce largely a repetition of the reaction to Whewell, perhaps because the similarities were so great?

Peirce is now increasingly admired, as Whewell is as well, for his views on scientific method. Here we can begin our study of the relation between the two thinkers not because this is the most important aspect of Peirce's thought, but because it is here that we can expect his debt to Peirce, such as it is, will be found. Is the belated recognition of the quality of Peirce's philosophy of science above all due to the fact that Peirce recognized so early and clearly the importance of Whewell? We can easily see, indeed, how Peirce took over large portions of Whewell's analysis of scientific method, even when we also note the modifications he deemed necessary. Peirce spoke quite disparagingly of the logicians, when he noted that, after thousands of years of logical studies, it was the mineralogist Whewell, who pointed out the importance of first uniting ideas together, Whewell's process of colligation, in argumentation or proof. He states,

> As for retroduction, it is itself an experiment. A retroductive research is an experimental research; and when we look upon Induction and Deduction from the point of view of Experiment and Observation we are merely tracing the affinity to Retroduction. It begins always with colligation, of course, of a variety of separately observed facts about the subject of the hypothesis. How remarkable it is, by the way, that the entire army of logicians from Zeno to Whateley should have left it to this mineralogist (Whewell) to point out colligation as a generally essential step in reasoning. (Peirce 1960, vol. V, § 581)

and

> My own historical studies, which have been somewhat minutely critical, have, on the whole, confirmed the views of Whewell, the only man of philosophical power conjoined with scientific training who had made a comprehensive survey of the whole course of science, that progress depends on the observation of the right facts by minds furnished with the appropriate ideas. Finally my long investigation of the logical process of scientific reasoning led me many years ago to the conclusion that science is nothing but the development of our natural instincts. So much for my theory of scientific logic. (Peirce 1960, vol. VI, § 604)

and

> Almost every step in necessary reasoning depends on two premisses which come to the reasoner's knowledge from different quarters. He "puts two and two together," or, in Whewell's terminology, he colligates the facts. (Peirce 1960, vol. III, §596)

and

> It is plain that colligation is half the battle in ratiocination. (Peirce 1960, vol. IV, § 45)

Peirce, following Whewell, saw scientific method as a process in which vague ideas are put forth, tested, and improved and in this process rendered clear and precise. For Whewell this process proceeds until scientists see the necessity of the ideas, for Peirce until doubt is, to some degree provisionally, removed. With some justification Peirce gave the method he largely took over from Whewell his own name: abduction. Peirce took over completely from Whewell the view that we must start with hypotheses, test them, correct them and refine them until they are clear and definite. And he also agreed with Whewell that they then pass into the body of established knowledge. But he nevertheless did modify Whewell's theory of the determination of truth or the justification of theories. Whewell said that we improve a view until we see intuitively that it must be true. Peirce also said that we improve a view to make it clear and distinct, until we can no longer doubt it. But it is not thereby shown by intuition to be true. We make it as clear as we can and when we do so, doubt will be removed. When doubt is removed, we deem it knowledge. This is not quite proof, though the difference between Peirce and Whewell even here is not great. Following Whewell, Peirce did not hold that there is an inductive proof in the sense of inference from observations to theories. Both thinkers were opposed to Mill. Peirce states,

> Whewell described the reasoning just as it appeared to a man deeply conversant with several branches of science as only a genuine researcher can know them, and adding to that knowledge a full acquaintance with the history of science. These results, as might be expected, are of the highest value, although there are important distinctions and reasons which he overlooked. John Stuart Mill endeavored to explain the reasonings of science by the nominalistic metaphysics of his father. The superficial perspicuity of that kind of metaphysics rendered his logic extremely popular with those who think, but do not think profoundly; who know something of science, but more from the outside than the inside, and how for one reason or another delight in the simplest of theories even if they fail to cover the facts. (Pierce 1960, vol. I, §70)

In the following paragraphs Peirce criticizes Mill's interpretation of Kepler and sides with Whewell, albeit without mentioning him further. Elsewhere he states,

> I am very far from holding that experience is our only light; Whewell's views of scientific method seem to me truer than Mill's; so much so that I should pronounce the known principles of physics to be but a development of original instinctive beliefs. (Peirce 1960, vol. I, §404; for a detailed critique of Mill, see Peirce 1960, vol. II, §761*ff*)

Peirce proceeds, as Whewell did, to explain that, though ideas (or instinctive belief) are a source of knowledge, this source is so intermixed with error that it must be corrected by experience and that the result – here contrary to Whewell – can only be approximate. Over the long run the process of eliminating doubt by the severe testing of theories, by analyzing their consequences, will lead us up to the truth.

Although Peirce's view of scientific method has quite recently attracted the most attention and admiration, we can already see that his theory of scientific method was by no means very original. We can admire him enormously for his capacity to sort out the wheat from the chaff in 19th century philosophy, and all the more so because he was isolated even in America, yet knew quite well just which of the work being done (or work which had been done) in Germany, England and France was the most important. He was in this respect, for example, often better than Russell. But the claims for his originality as a methodologist, deeming him, indeed, a forerunner of Popper, fail to note the degree to which he is a follower of Whewell. It also fails to note the difference between his fallibilism and

Popper's. On Peirce's view, as well as on Whewell's, we gradually remove doubt from an increasing number of theories which then become part of established knowledge. On Popper's view we continually substitute newer and better theories for older ones.

Peirce's debt to Whewell is by no means restricted to his theory of method, but his further debt has not, so far as I know, been noticed, since the views taken over by Peirce from Whewell are not so fashionable. Peirce hoped to show that abduction is that method by which science progresses by studying science itself. Since abduction is pragmatism, he is using his scientific method, abduction or pragmatism, to study science. We show, then, how knowledge has in fact developed as well as how it quite generally develops, by studying the history of science. And the history of science, Peirce tells us further, has been correctly described by Whewell. But Peirce does not think this method alone sufficient as Whewell did. He states,

> Appeal is made by logical writers to the history of science as supporting their views of right reasoning. Whewell placed his Novum Organon Renovatum (1858), a work of high logical value, almost exclusively upon the basis of this method. As a source of secondary evidence upon such general doctrines of logic as may need such help, and even of primary evidence as to the value of special modes of reasoning, this method has been (it would be disgraceful to deny it) occasionally of no little service. (Peirce 1960, vol. II, p. 38)

Peirce never developed this approach to any significant degree (see Peirce 1960, vol. I, §43*ff*). He no doubt wanted to integrate this method with others. But I have been unable to discern a clear theory about how he wished to proceed. This is, unfortunately, typical of all his efforts and shows clearly enough how heavily his isolation, the general lack of interest in his work, and the need to make his living outside of his researches must have weighed on him. In order to carry out his historical project, all that he had to do at first was to recast Whewell's work with his modifications, substituting his pragmatic removal of doubt for Whewell's attainment of the intuition of the necessary truth of scientific theories. But he no doubt wanted to achieve more, and here it becomes difficult to see just what he had in mind.

Peirce develops his view of scientific progress further by seeking to explain the fact that we do find the truth. If we merely began with random hypotheses, he observes, it would seem quite unlikely that we could ever have the degree of success we do. This problem corresponds to that problem which Whewell sought to solve with his theology (Whewell 1971, pp. 154*ff*). Given that we are capable of knowing the truth by intuition, how can we explain this? Whereas Whewell explains this success as the result of something divine in the mind of man, Peirce explains our capability of finding the truth as a consequence of similarity between the ideas we form and the world as it is, and this similarity as a product of evolution. He finds this quite plausible, since he deems the evolution of man and the world as one process.

We thus find Peirce maintaining virtually all of Whewell's methodology and even parts of his method of doing the philosophy of science. At the same time he sought to remove Whewell's Platonic theory of justification and to substitute for it a new more empirical method. This new method was his pragmatism (or pragmaticism) and that aspect of his system for which he has rightly become famous. He explains further that pragmatism is the same as abduction. Since abduction is Whewell's methodology with modifications, pragmatism is, in fact, a modification of Whewell's theory of scientific method.

Peirce wanted this new method to be much more than an improvement of Whewell's theory of scientific method however. He hoped to construct a philosophical method which he could use to develop a new comprehensive and systematic metaphysics, or, perhaps, a method which would render metaphysics scientific. Influenced by Kant and German philosophy he sought to construct a complete system of logic, metaphysics and method. He wanted to go beyond science and to develop his metaphysics by studying the phenomenology of common sense or instinctive thought.

Peirce followed Whewell in his rejection of the theory that there are individual sensations or facts to which theories may be reduced. All observation is for Peirce as well as for Whewell, theory-impregnated. And this made his theory of the rejection of non-empirical features far more subtle, complicated or difficult – I am not sure just how to describe his view here. His pragmatism should have solved this problem of reduction without sensationalism. This (pragmatic) theory of reduction is, of course, closely connected with Peirce's guiding metaphysical idea of continuity: continuity forbids the existence of distinct entities such as sensations. Although in his rejection of sensationalism he followed Whewell, he was far ahead of the profession. In spite of this his metaphysical theory, his analysis of non-scientific or pre-scientific thought, of common sense, all becomes quite difficult. I will not attempt to go further into this difficult area.

Peirce's rejection of sensationalism also no doubt led to the rejection of his work in the early 20th century since, under the influence of Russell, sensationalism came into vogue and the development of theories which rejected this assumption had to wait more than a half-century to be taken seriously once again. Quine begins to return a position closer to Peirce, but quite intuitively speaking, seems to remain somewhere between the two thinkers, defending a holism which depends for meaning on individual stimuli. His pragmatism is also more Jamesian than Peirceian even though

Peirce's pragmatism is much closer to Russell's analysis than James' is. Peirce wanted pragmatism to be a method of finding the one set of true categories which would be identified, that is to say rendered clear and distinct, by investigations of their consequences; Russell wanted, at least early in his career, to determine the meaning of concepts by reducing them to that which we know by acquaintance and relations. James was not so interested in the analytical task of determining the (proper) meaning of concepts but used pragmatism rather as a liberal theory of truth, making famous and infamous the motto: whatever works is true.

On the interpretation of Peirce I suggest here he hoped to offer a modification of Whewell which would be somewhat more empiricist because of his rejection of intuitive proof and Whewell's Platonic theory of ideas. The core of this modification was his pragmatic method which was designed to remove doubt and establish correct scientific and metaphysical ideas. This method should have provided a new phenomenological theory of ideas. The method should be used to develop a new (metaphysical) system which explains both the Whewellian methodology and new developments in logic.

Peirce's contemporaries hardly took his theory of science seriously. This theory was for Peirce himself merely the background knowledge he laid out in the hopes of using it to build his philosophical system. Peirce was not satisfied with Whewell's philosophy, since it was not "fundamental." He here seems to endorse the criticism of Whewell according to which Whewell had created no system. But Peirce saw that Whewell's work was the best available, that his theory of intuition and of proof should be replaced, but also, he thought, that his theory of ideas had to be changed. And here he saw the fundamental problems of metaphysics and logic, areas which are by no means sharply demarcated in Peirce's philosophy. For, he wanted as Russell also did to find integrated solutions to problems of them. We might say, he wanted to render metaphysics scientific. He did not see the theory of abduction or of method as his prime result, even though he identified these with pragmatism. The theory of abduction was still not the result he was seeking. It was the method he developed to solve more fundamental metaphysical problems. He never did solve these problems and the method he developed was watered down by his followers and put to uses he never dreamed of and which he looked down on.

The fact that even by his own lights Peirce never succeeded in solving those problems which he deemed most important is one significant factor in the confused reaction to Peirce. His theory of scientific methodology was at first ignored. His modification of Whewell's method – his pragmatic method of making ideas clear – was taken over by James, Dewey and others and used for purposes which were quite foreign to either Whewell or Peirce. Peirce did not want to offer a method according to which anything that works can be deemed true, a view which was dismissed by Russell and which excused him from a serious study of Peirce. Peirce deemed his views relevant for practical affairs to be sure and even in spite of noting that he himself was not a practical man. But his main interest was metaphysical. We should, he said following Whewell, look at the consequences of an idea in order to make it clear and distinct, and this is something different from looking at what works and certainly different from looking at what works for you or me. But the consequence of the treatment of Peirce by those most friendly to him led all thinkers to completely ignore the best he had to offer his contemporaries. This was his critical appraisal of the state of knowledge at the time and his attempt to deal with those methodological and metaphysical problems which arose out of this appraisal.

Late in this century Peirce's analysis of the state of knowledge of his contemporaries has been viewed as a sign of his foresight and originality. The profession had more or less caught up with him, but only to use him to show that they had no need for even newer ideas. We see once more the phenomena, described by Gertrude Stein, of the acceptance of an idea which is somewhat new, or new for some, to block the introduction of a still newer idea. This modern praise of Peirce has been deemed by the profession as a recompense of sorts for the abominable way he was treated when he was alive. In fact it is a mere continuation of the same treatment he received then: in place of a serious study of his philosophy and his problems, those parts of his view which fit the current fashion are used to further that fashion.

We can see in Peirce, then, one attempt to take Whewell's view further, whether we agree with the direction he sought to take it or not. He saw that the work of Whewell, Boole and de Morgan had rendered obsolete the foundations of logic and scientific method. He sought to build new foundations for the new structures. He sought to carry out a task whose futility Whewell had warned about as he discussed German philosophy (Whewell 1971, p. 223). Where Whewell had warned against too much ambition, Peirce wanted to construct a complete system, a system which would overcome the antithesis of philosophy, explain how the absolute truth could be achieved, and provide a framework for philosophy and science as a substitute for Aristotle.

Peirce thought he had discovered the key to a more powerful empiricist view. Instead of deeming scientific ideas to consist of mere combinations of entities observed independently, the empirical nature of ideas had to be seen in theoretical contexts. The reduction of ideas to distinct empirical observations was, he thought, impossible: we cannot find independent clear sensations. But we can render the empirical interpretation of ideas explicit by making ideas clear, by looking at how the world appears through the light of these ideas. He here seeks to modify Whewell's central idea that the purpose of empirical research is to render ideas so clear and distinct that we see their necessity. Whewell wished to render them so clear that their empirical quality would leave open no room for doubt. It is, indeed, quite difficult to distinguish Peirce's quest for clear ideas and Whewell's for the clear and distinct perception of necessary ideas.

Peirce did not present his philosophy as an attempt to re-do Whewell's philosophy as a coherent, complete system. Until now no one has analyzed Peirce's system in this way, even though observations of similarities between, say, Whewell, Peirce and Popper are by now commonplace. Peirce refused to come to terms explicitly with the criticisms of Whewell of such a system as that which he hoped to construct. And this may, indeed, have been his downfall. He did not explain how and why he thought Whewell's reservations about the quest for a system could be overcome.

Our problem concerning our evaluation of Whewell is nevertheless rendered more severe by Peirce's work. For, we see how the most influential American philosophy was built by taking Whewell seriously, by seeking to improve on his ideas instead of rejecting them. Our impression remains mixed. Peirce never tried explicitly enough to improve on Whewell. He failed to systematically reappraise Whewell in order to lay a foundation for his own work, and it seems he was simply all too grudging in his recognition of Whewell. As a consequence the failure of his system left the traditional appraisal of Whewell were it was. His appreciation for Whewell had no impact on the general appraisal of him even when some of Whewell's most important ideas did have impact because they had been reformulated in Peirce's philosophy.

## 3.3. France: Science without Induction – The Theory of Science with Induction

In the nineteenth century France found itself situated between the severe empiricism of the English represented by Mill and the excessive apriorism of the Germans represented by the nature philosophers. They were in a position to be more open to alternatives to Bacon, while still requiring adherence to a scientific point of view. But who could give impetus to real change? And how could it come about? It proved to be, as we shall see, quite difficult for the French to take advantage of the excellent position in which they found themselves. But take advantage they did.

### 3.3.1. *Claude Bernard: Quiet Acceptance by Scientists*

Whewell's rejection in England had its parallel in French philosophy of science. Nobody defended Whewell there and there is little evidence that he was at first taken notice of. In this climate of opinion we find a very wondrous thing. The physiologist-biologist Claude Bernard published a introductory text in experimental medicine for his fellow scientists which agrees with Whewell on virtually every major methodological point. And not only that. It was hailed as a great book and quickly became not only a standard text for biology but also a standard outside of this field.

We find here, as in the case of Peirce, that our problem concerning the rejection of Whewell is intensified. How can we explain his rejection, if his methodology – or one which on virtually all important methodological points is identical to it – can be so quickly and widely accepted in France? How did this come about? Such events as this are commonly taken as evidence of two things. The first is that philosophers think about science rather poorly and the second is that scientists – who know science first hand – do so better, when they take the time to write down what they know. They usually do not do that because it is all too trivial. (This is the point of view expressed in the preface to the English edition of Bernard's text (Bernard 1957, pp. v-vi). Such explanations as these will simply not do, however. As we have seen, it was the scientists who had played the most crucial role in the rejection of Whewell. If they had laughed at Mill for his ignorance and praised Whewell for his knowledge, the situation in philosophy of science would have been entirely different. In fact, they praised Mill for his circumspection and rejected Whewell as excessive.

The rejection of Whewell as a philosopher, however, was combined with grudging respect for his knowledge and some of his methodological observations. So, it was quite in accord with the standard view of Whewell, first developed by Mill, to present Whewell's methodology without his philosophy. And this is just what Bernard did. He avoided philosophical interpretations of the path to knowledge as much as he could. Above all he gave guidance as to how progress was made. He could, it would seem, take over what was so good in Whewell and reject the rest so as not to get into trouble. In this way he could make Whewell's progress acceptable to scientists who want to avoid philosophical traps. What is significant about Bernard's work, however, is that he not only writes to avoid those traps which Whewell got into, but that he also avoided the standard attempt to give a Millian interpretation to Whewell's methodology.

But Bernard's interest were not only methodological. There is good reason to believe that he was quite sympathetic to a more Whewellian, more metaphysical point of view, but suppressed so as not to get into trouble (Chevalier 1954, pp. 45*ff*). Such a view of the rational side of science is prominent in his book. And it fits nicely with his own suggestions about how physiology should be conducted, especially his suggestion that the inner environment of animals should be studied.

The generally accepted view of Bernard's work is that it is quite independent of Whewell's. On the basis of circumstantial evidence it has been noted that Bernard appears to be a follower of Whewell. Blanché raised this question of Bernard's debt to Whewell but, after failing to find any direct evidence of Whewell's influence, deemed Bernard's work independent of Whewell's (Blanché 1969, p. 176). But to suppose that Bernard was ignorant of such a prominent figure as Whewell, whose books were reviewed by men such as Sir John Herschel and translated into German, is implausible. It is also implausible to suppose that an author of a text on scientific method had not looked at the reception of such books as those of Herschel, Whewell and Mill. And any serious reader of Mill had to be familiar with Whewell to some degree. In particular, the methodological and psychological insights of Whewell which Mill had built into his system had to be known. Given all that, I find Bernard's silence about Whewell no evidence one way or the other. He certainly had good reason not to point to the similarity, if not identity, between his methodological theory and Whewell's even if his debt had been substantial: it would have made the reception of his book very difficult indeed. There is, perhaps, only one thing certain: he wanted to avoid philosophical controversy. He did this skillfully, presenting a highly controversial methodology – even among scientists – as if it were common sense, the experience of a scientist. What he accomplished, then, was to give the (French) scientists the opportunity to adopt Whewell's methodology without his philosophy: given the opportunity, they did so in droves.

The excellence of Bernard's view is sometimes deemed a mere product of his acquaintance with science. This view is not only naive, it underestimates his accomplishment. If it were true, any scientist could have done it. But Bernard's work was new and exciting and became established quickly. Bernard's acquaintance with and thought about science may have been necessary for his innovation, but does not explain it. We must look to his intellectual heritage and philosophical problems to understand his work.

Bernard had several central concerns. They turned on the question: how should one conduct (scientific) physiology. He had his own new theory about the nature of physiological problems. According to this view the central problems of physiology concerned the internal environment, even harmony, of animals. How was this harmony maintained? How was it disturbed? What are the nature of these disturbances and what effects do they have? These were some of the questions his research program put on the table. He had a central idea for physiology. And this idea should be scientific.

Now in addition to, and in conjunction with, this theoretical framework for physiology, Bernard wanted to explain how empirical research proceeded. Here his problem arose due to his perception that mere observation was not enough. Ideas, such as his theory of that of the importance of the internal environment of living beings, were needed. But empiricism was also needed. He needed, then, to explain how ideas could be studied scientifically, or perhaps, how they could be rendered scientific. In regard to both issues his great competitor was Pasteur, who deemed all sickness a product of germs, that is, of influence from the outside rather than disturbance of the internal relations, and who was an inductivist, who believed in the gathering of facts. In Bernard's *Introduction* to *Experimental Medicine* he was primarily interested in methodological issues. But the theoretical concerns, or the concern for theoretical physiology, which he hoped to put on a secure scientific footing, should not be lost sight of.

Bernard leaves no doubt open about his rejection of inductivist, Baconian, Millian views. He criticizes theorists who attempt to gather facts or who use merely statistical methods with no proper ideas, and he shows with examples how experiments can go awry without proper ideas. One experimenter, he says, failed to distinguish newly born kittens from excrements and another hoped to study the average European urine by collecting many samples from a railroad station and mixing them (Bernard 1957, pp. 131*ff*). The examples, however astounding we now find them, illustrate his point: one has to know what one is looking for, one has to have a good idea, before one conducts good experiments, before one can know what to look for. The blind belief in facts is, he says, dangerous. He states,

> I think that blind belief in fact, which dares to silence reason, is as dangerous to experimental science as the beliefs of feeling or of faith which also force silence on reason. In a word, in the experimental method as in everything else, the only real criterion is reason.
>
> A fact is nothing in itself, it has value only through the idea connected with it or through the proof it supplies. We have said elsewhere that, when one calls a new fact a discovery, the fact itself is not the discovery, but rather the new idea derived from it; in the same way, when a fact proves anything, the fact does not itself give the proof, but only the rational relation which it establishes between the phenomenon and its cause. (Bernard 1957, p. 53)

Bacon, Bernard said, did not really understand science: he did not understand the role of hypotheses and theories. He had a vision of science and saw the sterility of scholasticism but did not invent an experimental method (Bernard 1957, p. 51). The method of differences has almost no place in physiology, he says, giving a Whewellian objection: in order to use this method, we have to know all the bodies and all the conditions combining to express a collection of phenomena. He proceeds to give examples as to how its use can mislead research (Bernard 1957, p. 127). On all these points Bernard and Whewell are at one against Mill.

But how does one find such an idea? Once again we find Bernard it close agreement with Whewell. He explains that one must start with conjectures and test them. All sciences, he says, start conjecturally and then proceed to test, refine and establish ideas. He hopes to move medicine from conjecture to science (Bernard 1957, p. 139). When a theory is found wanting we need to replace it with a better one. We must be critical of previous thinkers, even great ones. He states,

> The ideas and theories of our predecessors must be preserved only in so far as they represent the present state of science, but they are obviously destined to change, unless we admit that science is to make no further progress, and that is impossible. (Bernard 1957, p. 41)

and

> In the experimental sciences, a mistaken respect for personal authority would be superstition and would form a real obstacle to the progress of science: at the same time, it would be contrary to the examples given us by the great men of all time. Great men, indeed, are precisely those who bring with them new ideas and destroy errors. They do not, therefore, respect the authority of their own predecessors, and they do not expect us to treat them otherwise.
> This non-submission to authority, which the experimental method regards as a fundamental precept, is by no means out of harmony with the respect and admiration which we bear to the great men preceding us, to whom we owe the discoveries at the base of the science of to-day. (Bernard 1957, p. 41)

Induction is, Bernard explains, interrogative reasoning (Bernard 1957, p. 45). When induction is properly understood it is the same as deduction, but is not certain in the natural sciences as it is in mathematics, where no tests are needed to confirm an hypothesis (Bernard 1957, pp. 45*ff*). He states,

> An idea that comes to us at the sight of a phenomenon is called a priori. Now we shall later easily show that this a priori idea, which rises in us à propos of a special fact, always containing implicitly and, in some sort, without our knowledge, a principle to which we tend to refer the special fact, so that when we think that we are making an induction, we are really making a deduction; only the experimenter guides himself by an assumed or provisional principle which he alters moment by moment, because he is searching in almost total darkness. In proportion as we gather facts, our principles become more and more general and more secure; so we gain the certainty that we deduce. (Bernard 1957, p. 48)

In the pursuit of truth the scientist shows his feeling for truth. This is Bernard's explanation for success which corresponds to Whewell's theory of the participation by the human mind in the divine or Peirce's theory that evolution has led to an affinity between mind and matter. The scientist's genius, Bernard says, is revealed in this feeling which is checked by the facts, even though there is and can be no firm ground for it. He states,

> The experimental method, the free thinker's method, seeks only scientific truth. Feeling, from which everything emanates must keep its complete spontaneity and all its freedom for putting forth experimental ideas; reason also must preserve that freedom to doubt, which forces it always to submit ideas to the test of experiment. Just as, in other human actions, feeling releases an act by putting forth the idea which gives a motive to action, so in the experimental method feeling takes the initiative through the idea. Feeling alone guides the mind and constitutes the primum movens of science. Genius is revealed in a delicate feeling which correctly foresees the laws of natural phenomena; but this we must never forget, that correctness of feeling and fertility of idea can be established and proved only by experiment. (Bernard 1957, p. 43)

Bernard built into his philosophy of science or methodology the most crucial aspects of Whewell's. And he did this in order to make room for ideas and empiricism in science. He shared the spirit as well as the content of Whewell's philosophy. He did not explain himself very extensively in regard to his theory of justification, that part of Whewell's philosophy which got him into trouble. But he was no Millian here either. And his comments indicate that it is the perception of the rational connections, of the idea, which allow for justification. He simply remained silent where he could. Did he do so because his own view would have been difficult to defend? Or because he had no view? We are here left with some unsolved puzzles, puzzles which became important for French philosophy of science as they slowly realized that Bernard had institutionalized in France a Whewellian methodology, just that methodology so roundly rejected by all philosophical and scientific authorities in England.

The establishment of a Whewellian view of methodology – to whatever degree Whewell may or may not have been responsible for it – intensifies our problem once more. And, indeed, we see that, as I explain below, just because Bernard's methodological interest in Whewell is renewed. But how they should treat him was by no means clear. Should they seriously reappraise him or reaffirm the traditional view?

### 3.3.2. *Louis Liard: Hypotheses plus Induction in Philosophy*

French philosophers were faced with a much more serious problem than French scientists. They could not import Whewell's views without defending his philosophy. And this they chose not to do. They adopted the normal, English view of the matter and remained loyal to inductivism. A prominently mentioned source for information about Whewell in France is *Les Logiciens Anglais Contemporaines* by Louis Liard (1878). The book is intended as an overview of theories of science developed in England. In it

Liard discusses the views of Herschel, Mill, Whewell, de Morgan, Jevons and others. It should be merely expository. He states,

> Ceci est un livre de pure exposition. On veut simplement faciliter a ceux qui, en France, ont souci des études logiques, la connaissance des systèmes, vrais ou faux, parus en Angleterre au XIX1e1 siècle. Ces systèmes font date dans l'histoire de la science; il faut désormais compter avec eux, et il n'est plus permis de les ignorer. Les renseignements qu'on en avait dans notre littérature philosophique étaient tres-incomplets et souvent erronés. Il a paru bon de les compléter et de les rectifier. (Liard, 1878, p. 0)

Due to the modest intentions of Liard we might expect to find here a basis for an appreciation of Whewell or at least an interesting presentation of his system. But Liard devotes a mere eight pages to Whewell (Liard 1978, pp. 6-14). He does describe some essentials of Whewell's point of view quite adequately but at the end we come to this appraisal,

> On a pu voir, par ce qui précède, que les méthodes inductives décrites par Herschel et Whewell sont exclusivement des méthodes de découverte. L'un et l'autre semblent n'avoir pas vu cette vérité, qu'après eux Stuart Mill devait mettre en un relief si saissant, que toute logique, même une logique inductive, est une science, non de la découverte, mais de la preuve. Whewell en particulier, malgré une connaissance profonde des théories scientifiques anciennes et modernes, séduit et égare par l'apparente simplicité qu'introduit dan la conception de la science et des méthodes l'antithèse des faits et des idées qu'il regarde comme fondamentale, a méconnu la véritable nature de l'opération inductive. Si l'on peut décrire la découverte scientifique comme l'application idées fournies par l'esprit a des faits donnes par la sensation, il reste a savoir comment les propositions ainsi obtenues ont force de loi; des lors, on est conduit a se demander comment nous pouvons être assures de la validité de ces propositions, et quels sont les procèdes auxiliaires de la preuve expérimentale. (Liard 1878, pp. 13-14)

Liard adopts Mill's view of Whewell without any consideration of Whewell's responses or of his criticism of induction. We see that the uncritical acceptance of this point of view had already spread so widely and decisively by 1878 that it may be repeated as, as Liard says, mere exposition.

The upshot of the agreement among the philosophers that Mill was right in demanding a theory of (inductive) proof as a part of any adequate theory of science and the agreement among a wide segment of the scientific population that Bernard's (Whewell's) methodology was correct had to create great tension: the philosophers were faced anew with Mill's problems but in a new and more difficult form. Mill could argue that Whewell's description of science was incomplete, since Whewell was not deemed an authority on science. In France, however, one could not say that Bernard had left some important aspect out of his description of science. Yet Bernard had no place for induction. The scientists of England had demanded that induction be a part of science and had rejected any theory of science which did not include it. The scientists of France had quietly accepted Whewell's description of scientific method without induction as complete. I will come back to the discussion in France to which this situation led when I discuss the work by Blanché.

### 3.4. Germany: Germanizing the English Heretic

Our expectations that the treatment of Whewell in Germany might be more extensive and sympathetic than that found in England is not at first blush confirmed. There are few discussions of his view and hardly any that one might call comprehensive. It may be, of course, that Whewell would be relatively uncontroversial in Germany: the use of Kant's perspective according to which there must be some sort of a priori ideas should surely find a wide base of acceptance in Germany without any argument. Let us look, then, at some of the commentaries to see if ideas of Whewell have been taken over, whether he is passé, and thus of small relevance for the discussions of the philosophy of science in Germany, or whether there are new and interesting criticisms of his view to be found there. We know that he had received some recognition in Germany quite quickly: Littrow had translated his *History* shortly after it appeared and von Humboldt wrote a very appreciative letter to Whewell which is published in Todhunter (1876, vol. I, pp. 147-149). But we have to look very carefully, indeed, in order to find any influence or serious consideration. Mach not only mentions Whewell but obviously took him quite seriously in that he took the challenge of his historical studies and psychology as posing problems which an adequate philosophy of science needed to solve (Wettersten and Agassi 1991). Mach mentions that Beneke rejected him (Mach 1917, pp. 281-282). And in this position Mach also says that Whewell's use of the theory of the idea of cause has lead to amazing results. Yet the final statement of Beneke which Mach quotes seems quite Whewellian indeed. It is:

> "Die Formen, welche für die Erkenntnis zunächst vorliegen, sind erst in der Entwicklung der Seele entstanden, vor derselben prädeterminiert in angeborenen Anlagen und Verhältnisse, welche ganz andere Formen an sich tragen." Ich wusste diesen trefflichen allgemeinen Bermerkungen nichts erhebliches hinzuzufügen. (Mach 1917, pp. 281-282)

Wundt mentions Whewell at a crucial point in his early discussion of the psychology of perception (Wundt 1862, p. 442). He seeks there to offer an alternative theory of the logic of perception, that is, of the logic of unconscious inference, to those of Mill or Helmholtz (Wettersten 1992, pp. 83-89). Külpe seems to have benefited from Whewell's history at least and develops a philosophy of science which is in its main lines identical to Whewell's, that is, it seeks to explain how we gain knowledge of real objects through the use of the two sources of knowledge – ideas and sensations – available to us (Wettersten 1992, pp. 117*ff*). But even though such thinkers by the logic of the situation and by references to him seem to have been greatly influenced by his problems and perhaps by his view, we find no systematic recognition of his views or acknowledgment of debt to him. Mill, quite ironically, seems at first blush to have been received much more favorably in Germany that Whewell was. This is puzzling indeed. But perhaps the study of the puzzling reaction in Germany will lead to a better understanding of the rejection of Whewell's views in all parts of the intellectual world.

The traces of Whewell's influence in Germany are spread out and thin. It appears that in each case even to begin an appraisal one has to reconstruct a particular logic of the situation. It would seem extreme to attempt to be comprehensive, so I will try to follow some interesting leads here and there to see where they lead. There is, for example, a book on induction by E. F. Apelt in which Whewell is repeatedly mentioned. Apelt was a disciple of Fries who in turn sought to improve on Kant. Erdmann published a short review of four English writers – above all of Mansel – in which Whewell is mentioned as a commentator on Mansel (Erdmann 1857). I will not discuss it further. Paul Leuckfeld published discussions of various thinkers who offered theories of induction and Whewell was included. Friedrich Albert Lange, the founder of the Marburg school or new-Kantians praises Whewell highly. I ignore here the strong indications of influence in Mach – which is discussed elsewhere (Wettersten and Agassi 1991, pp. 358*ff*) – and in psychology in the works of Wundt and of Külpe, which I also discuss elsewhere (Wettersten 1992). They do not seem to give hints as to the grounds for the rejection of Whewell. Indeed, they may raise our curiosity, since they show that his problems pop up even in contexts in which he is apparently totally forgotten.

### 3.4.1. *E.F. Apelt: Traditional Kantianism and the Growth of Knowledge*

After the inadequacy of Kant's system was widely perceived, German philosophy split quite drastically. On the one hand were the idealistic speculations of Goethe, Fichte, Hegel, Schelling, *et. al.* and on the other side were the philosophically minded scientists who found a leader in the middle of nineteenth century in Hermann Helmholtz. This split left little room in the middle and there were few philosophers to be found there. The scientifically minded philosophers seem to have sought refuge in the appeal to empiricism and this meant, already shortly after the middle of the century, John Stuart Mill. The metaphysically minded nature philosophers rejected such a position quite radically. There were nevertheless some philosophers who sought to maintain their Kantian views, to account for science and to avoid the extremes which had come to be associated with the development of Kant's views after Kant. One of these Kantians was Jacob Friedrich Fries (1831). He started out with great hopes for a successful career and was even as a young philosopher a serious competitor with Hegel for the chair in Berlin. But he came into conflict with the authorities due to his political views and was not allowed to teach. He was no longer in position to gain any influence even though he continued to work. This is not to say we should now make a liberal political hero out of him: he advocated, as Walter Kaufmann has pointed out, that Jews be with "Stumpf und Stiel ausgerottet" and that they be required to wear a yellow star on their clothing – unless they chose to convert to Christianity. He survived the repressions of his views under which he suffered in that he found a few followers to defend his views. One of the most important was E.F. Apelt. I should mention that it was the Jewish social democrat Leonard Nelson who revived interest in Fries as he sought, as Hilbert's protégé, to form a new foundation for science, mathematics and society. Fries is most known as a whipping boy, however: he is taken as a prime example of psychologism, of an attempt to base knowledge on a mere description of psychological processes. This characterization is, I think, somewhat unfair since he was quite sympathetic to science and sought to develop his views in such a way as to take account of science (Wettersten 1995). He sought, whether successfully or not is another question, to keep Kantian philosophy without excesses and to unite it with science. We could thus expect to find in Fries' disciple, Apelt, a man with great sympathy for Whewell, who also tried to carry out the fundamental tasks Fries had sought to execute. We might also expect some difficulties since Mansel, who criticized Whewell so sharply, defended a brand of Kantianism which was among the Kantians the closest to the attempt of Fries to find the sources of knowledge in psychological analyses.

We find, indeed, that Apelt had much more sympathy for Whewell than for Mill. He considers three English philosophers who had then recently treated induction and these are, of course, Herschel, Mill and Whewell. It would seem, indeed, that the purpose of his book is to find a place for induction within his Friesian system. He presumes that induction is part of science and the problem, then, is to develop an adequate theory of it. We may presume that at the time he wrote his book – 1854 – induction was an important subject for philosophers sympathetic to science even in Germany. He criticizes Mill's theory of mathematics as follows:

> Wie unsicher Mill in seinen Behauptungen ist, geht schon daraus hervor, dass er bald sagt: die Geometrie schliesse aus blossen Begriffen; bald: sie schliesse nicht aus Axiomen, sondern aus Hypothesen und die Voraussetzungen dieser Wissenschaft seien so weit entfernt nothwendig zu sein, dass sie nicht einmal wahr seien; bald widerum: diese Hypothesen (d.i. dies aus der Luft gegriffenen Behauptungen) seien

> Generalisationen aus der Beobachtung und wurzelten ind dem Boden der Erfahrung. (Apelt 1854, p. 173)

Concerning induction he says,

> Ebenso schwankend und unbestimmt ist das was er über die Induction sagt. S. LI sagt er: die Induction sei eine wirkliche Folgerung von Thatsachen aus Thatsachen. Seite 5 dagegen erklärt er die Induction fuer "die Operation, durch welche man allgemeine Urteile (Saetze) entdeckt und beweist. Es ist wahr, führt er fort, das Verfahren, wodurch wir einzelne Thatsachen erforschen, ist ebenso inductiv, als das durch welches wir zu allgemeinen Wahrheiten gelangen." Seite 10 erklärt er wiederum die Induction als "das Verfahren, wonach wir schliessen, dass, was von gewissen Individuen einer Classe wahr ist, auch für die ganze Classe wahr ist, oder dass was zu gewissen Zeiten wahr ist, unter ähnlichen Umständen zu allen Zeiten wahr sein wird." Seite 11 schliesst er ganz willkürlich die volständige Induction aus und beschränkt die Beduetung dieses Wortes nur auf die unvollständige, "denn zu einer Induction gehöre, dass von bekannten Thatsachen auf unbekannte geschlossen werde." Endlich Seite 25 lehrt er: die Induction bestehe darin, dass man schliesst, eine Erscheinung, die bei einzelnen Gelegenheiten stattgefunden hat, wird bei allen Gelegenheiten einer gewissen Classe stattfinden, nämlich in allen, welche den vorhergehenden in dem, was man die wesentlichen Umstände nennt, gleichen. (Apelt 1854, pp. 174-175)

Though Apelt has much more to say on this subject I will not pause to analyze his criticism of Mill. His rejection of Mill – and of Herschel – is just as clear as his favorable attitude toward Whewell and that may suffice for pursuing the puzzles having to do with the rejection of Whewell. Concerning Whewell he says,

> Whewell ist der erste Engländer, der in die Geheimnisse der Kantischen Philosophie tiefer eingedrungen ist. Während seine Vorganger von Bacon bis auf Herschel die Gleichartigkeit aller menschlichen Erkenntnis voraussetzen und die Induction als den einzigen Weg zu den allgemeinen Wahrheiten und den Principien bezeichen, hat er die Ungleichartigkeit und Verschiedenartigkeit der menschlichen Erkenntnissweisen, den wesentlichen Unterschied zwischen den zufälligen und nothwendigen Wahrheiten, richtig erkannt und begriffen, dass die ersteren aus dem Sinn, die letzteren aus reiner Vernunft, eine jede derselben also aus einer andern Quelle entspringen. Wenn wir auch die Wahl seiner Kunstausdrücke nicht durchgängig billigen können, so müssen wir doch die Sicherheit und Selbständigkeit bewundern, womit er sich dieses Grundgedankens der Kritik der reinen Vernunft bemächtiget hat, so wie die Klarheit und Ruhe, womit er diese Lehre von dem Unterschied der Erkenntnisse a priori und a posteriori darzulegen und zu entwickeln versteht. (Apelt 1854, p. 181)

Apelt continues to praise Whewell for his theory of Fundamental Ideas as the basis of necessary knowledge, for his appreciation of the difference between definitions and axioms, for his analysis of human knowledge as uniting necessary with empirical truths as well as his appreciation for Kant's theory of pure intuition. We may surely here expect to find an appreciative analysis of Whewell's view.

Apelt abruptly puts an end to these expectations, however, as he goes over to his criticism of Whewell, which is hardly what we would expect:

he complains that Whewell has no logical theory of induction (Apelt 1854, p. 184). This is the same complaint that Mill had raised and not that of Mansel, whose views are the closest to Apelt's. Both Mansel and Apelt follow a program similar to that of Fries, that is, the existence of a priori categories such as Kant has postulated are asserted to exist, but a new method is recommended for their identification and justification. This new method, or methods in speaking of the three thinkers, is descriptive and oriented toward a type of psychology which should identify the categories of thought. (I here leave aside the question of whether these views, as has often been asserted, are psychologistic; see Wettersten 1995.) In order to understand why Apelt endorses the standard inductivist complaint, we will have to see the basis of it in his own brand of (Friesian) Kantianism.

Apelt seeks to offer a theory of the logic of scientific inference. On his view, which, he says, follows that of Fries, the basic form of argument is the syllogism. Apelt hopes to explain, then, how the syllogism functions in science. He argues from his presumption that the syllogism should be the foundation of the analysis of scientific argument to the conclusion that there are three types of propositions in science. The major proposition of a syllogism is a universal assertion. It asserts relations between concepts or pure reasons. The minor proposition asserts the relations between individuals and the concepts and is based in intuition. The conclusion determines the nature or a characteristic of the nature of an individual. This is all trivially illustrated, if it should be needed, in the standard example, all men are mortal, Socrates (here Cajus) is a man, therefore Socrates is mortal.

Apelt seeks to give a broad analysis of the roles of the various types of syllogisms as well as of modus ponens and of modus tollens. But we may overlook this as it is mere preparation for the analysis of the logic of inductive arguments. There are, he says, two forms of induction. These are categorical – what is true of a part of some sphere is true of the concept of the sphere – and hypothetical – when all of the "consequences" of some "reason" occur, then the "reason" is present. This later means, it would seem, that if some rule is confirmed by the instances, it may presumed to be operative. The latter is, I think, the only form which he considers and, if I read him rightly in the following discussion, the first is no longer deemed induction but a process – abstraction – which is to be sharply distinguished from induction. His prime example of induction is his reading of Kepler – the same as de Morgan's – according to which Kepler knew all the planets and by examining the path of each one could infer by induction to the path of all. He notes here that induction is only possible under specific assumptions. But what are these? Where do they come from? How are they justified? And how are they used? These are the questions which Apelt wishes to solve.

Apelt recognizes that it is too easy to take account only of those cases of induction in which all instances may be examined. And he realizes that when only some of the cases can be examined only probability at best can be attained. The logical form of the argument cannot in such cases provide a justification. So he makes a proposal: there are guiding maxims which justify such (incomplete) inductions. These guiding maxims are used in rational inductions which are to be distinguished from mere empirical inductions. Mere empirical inductions are based in similarity and in regularity, but they fail to bring these under necessary laws. And this is the crucial aspect of rational, that is to say, scientific induction as distinguished from day-to-day induction.

How, then, do we make rational inductions? We use the guiding maxims. And what are the guiding maxims? They are the certain laws of our knowledge. They instruct us how to connect the facts when we make inductions. When we order the facts in accord with these certain laws and we find that the cases fit the laws, then we have a justified induction. Now Whewell would find this all a bit naive. For, just as he argued against Mill, he may argue here: this theory presumes that which is most difficult, that is, to find the laws under which the cases should be ordered. He may also, of course, object that this view excludes induction from the discovery of the laws or fundamental ideas and this gives it a minor role.

Apelt has what he takes for answers to these questions. He distinguishes between abstraction and induction. In abstraction we find the necessary laws which are the foundations of science. These laws cannot be proven without falling into an infinite regress so they are seen to be true by pure thought. There are thus two sorts of scientific method corresponding to the two sources of knowledge. If the source of knowledge is pure reason, we find concepts or laws by abstraction and see them to be true by intuition. There are two types of such knowledge which are philosophical and mathematical. The fundamental laws of Newton's *Principia* are an example of natural philosophy. They are known by intuition not by empirical research. It is the job of empirical research to find out the parameters of their application. In general and in accord with the analysis given, above the most general statements of science are those of natural philosophy, the lowest are empirical observations, and it is mathematics which provides the connection between the two. Mathematics is both pure as the concepts of natural philosophy are and intuitive – as in intuition of space and time – as the empirical observations are.

Whewell's error then is that he did not see that the laws of science have an independent source in abstraction. He did not see the difference between Galileo's abstractive method and Kepler's inductive method. For this reason he fails to demarcate mere hypotheses which are given without any guiding maxims from rational inductions which occur under the direction given by these maxims.

We see here how close the views of Apelt and Whewell are: both deem it necessary that the fundamental laws of science be seen to be true by intuition: no other method, either inductive or deductive, is available. They both see the crucial problem – Apelt here obviously following Whewell – in the discovery of the law or framework in which the facts are to be placed. The chief difference is that Apelt sees science as either a priori, that is, seeking laws, or a posteriori, that is, applying laws, whereas Whewell deems empirical tests to be needed before a law can be uncovered or seen to be necessary. The only advantage which Apelt's system could have would be to avoid the introduction of guesses in science which could, then, rest on proof. But even Apelt knows that this is not true. For, he asks how do we find the right leading axioms, even if we have them already available. He notes that an accident may be needed to find the right connection between the phenomena and the laws (Apelt 1854, p. 78). We thus do not succeed in eliminating guesses from science. We have no explanation of the so-called method of abstraction which would explain how we may use it correctly: why do we abstract falsely and what is the relations between false and correct abstractions? It is thus difficult to find in Apelt's view reasons why one should prefer his theory of the growth of scientific knowledge to that of Whewell's.

Perhaps we come to an answer through the back door, so to speak. Apelt's view, though by nowhere near so powerfully developed and defended as Whewell's, shares much with Whewell's system. And much of this we find not only today quite implausible but there were powerful voices then pointing to, for example, the difficulty of the view that scientists know the truth and the necessity of natural laws by pure intuition alone. But I, for one, am still puzzled. The difficulty which Apelt faced in attempting to develop a competitor to Whewell shows the need to break away from Kant more than Apelt did. Apelt could not account for the growth of science within a more conservative Kantian view, that is, one in which scientific method and scientific proof would be one. If we break away completely from Kant – a move that would bring one back to the position of Herschel and Mill – we find that we cannot account for necessary truths, for proofs or even for the facts of scientific method. Why, then, should a serious even defective attempt to do that such as Whewell's be rejected when no serious Kantian or inductivist competitors to solve these problems had been constructed?

The problem here can be illustrated, I think, by reference to Helmholtz. He quickly gained acceptance for his brand of Millian inductivism as well as for his principle of the conservation of energy. But his essay on the conservation of energy conceals a deep problem: How are we to interpret such a principle? For Apelt and Whewell this would be no great problem. But on Helmholtz's view it must be based on induction or not at all. And if based on induction how can it be a principle for all science and not merely for those cases where it has been applied? He seems, as Agassi has pointed out, to blur the difference between a metaphysical theory and an empirical law as he blurs the difference between Newtonian mechanics and Faraday's speculations. We seem to find strong reasons to move in the direction of Whewell's speculations which would seem to be agreed on by everyone. Yet both sides persist in rejecting his view on the apparent ground that he introduces guesses into scientific method. Is this rejection defensible? We must look, perhaps, somewhat further to understand it better.

### 3.4.2. *Friedrich Albert Lange and New-Kantianism: Whewell Idealized*

In the search for sympathetic and critical appraisals of Whewell – appraisals which may not dismiss him too easily – it would seem plausible to look to theorists who later developed Kantian points of view. Of especial interest are those who share attitudes with Whewell such as the desire to develop a Kantian position which avoids speculative excesses, which remains close to science by accounting for it, and which is historically oriented. We do not have to look far for such a position in late 19th and early twentieth century. The so-called new Kantians of Marburg – Friedrich Albert Lange, Paul Natorp, Hermann Cohen and Ernst Cassirer. We have to look carefully, however, to find anything at all in these thinkers about Whewell. They made no comprehensive study of the works of Whewell. And one might ask why they should have: German philosophers would not have to turn to an Englishman to learn about how to reformulate Kant's views so as to make them acceptable in the later 19th century. But the similarity of the tasks set by Whewell to those set by the new Kantians give us reason to look at the possible relations and the possibility of a different appraisal of Whewell.

If we look carefully at the work of the founder of the Marburg school – Lange – we find, indeed, that he was quite aware of the work of Whewell. In his *Geschichte des Materialismus* Lange mentions Whewell's "berühmten Geschichte der induktiven Wissenschaften" (Lange 1873, vol. I, p. 7) in various places. He takes issue with Whewell at some points and quarrels with him at others, but he has, it would seem, read him carefully. We might hope then for a detailed study and appraisal of Whewell's view, but we are only partially satisfied. Lange makes no mention of other studies of Whewell such as those of Apelt. We find that Whewell pops up here and there, but the comments offered do not seem connected with each other or to other studies of Whewell (outside of Herschel and Mill): Although Lange is better than many other commentators on Whewell, because he takes notice of Whewell's criticism of empiricist views, he hardly discusses the relevance of Whewell's theory for his own views or their development. And one must be cautious with thinkers such as Lange and the Marburg school in attributing too much importance to the fact that a thinker is discussed by them: they were encyclopedic in their references.

We may nevertheless turn with interest to the discussion of Whewell's philosophy and his dispute with Mill which Lange offers. Lange begins with mathematics, since it is in mathematics where the existence of necessary propositions not derivable from observations is clearest and, according to Lange, crucial for a Kantian point of view, such as the one he wishes to defend (Lange 1873, vol. II, pp. 16-18). After summarizing Herschel's opinion concerning the empirical nature of even mathematical truths – the view he wishes to combat – he continues,

> was die Polemik gegen Whewell betrifft, so ist dieselbe keine ganz gerechte und vorurtheilsfreie, wiewohl der lang andaurende Streit im Ganzen in den höflichsten Formen geführt wurde. Mill, der sonst eine gegnerische Ansicht durchaus loyal zu behandeln und klar darzulegan pflegt, referiert nicht immer streng richtig and bringt manche Äusserung seines Gegners in einen Zusammenhang, in welchem sie nicht gestanden hat. Der Grund dieser auffallenden Erscheinung liegt darin, dass Mill beständig das Gespenst der alten angeborenen Ideen und der platonischen Offenbarungen aus einer uebersinnlichen Welt vor Augen hat, welches in der Metaphysik so lange Zeit sein Wesen getrieben hat und dessen Zusammenhang mit Unklarheiten der schlimmsten Art wohl geeignet ist, einen nuechternen, aller Mystik abgeneigten Gegner zu reizen . . . Wir werden sogleich zeigen, wie dem Empirismus Mills in streng Kantischen Sinne zu begegnen ist; vorher wollen wir die schwachen Punkte desselben hervorheben, wie sie sich in der Debatte zwischen Mill und Whewell herausgestelt haben. (Lange 1873, vol. II, pp. 17-18)

The last sentence in this quotation raises our expectations: we may hope to find here an appraisal of the dispute between Whewell and Mill offered by a Kantian. This appraisal should explain how far the debate between Whewell and Mill progressed and then, after specifying what is still needed from a Kantian point of view, should continue to make good on whatever defects one finds in Whewell's philosophy. In pursuing the answers to these questions, however, we find difficulties. Lange (1873, vol. II, pp. 116-117) makes some nice and appreciative points concerning Whewell's point of view – especially concerning the unity of form and content or of ideas and observation, explaining that his view is not a version of Leibniz'. (We might also say, I think, that it is not a version of Plato's.) He shows that he has followed the debate between Herschel and Mill on the one hand and Whewell on the other with interest. But when we look for the point at which Lange ends his report on the debate between Whewell and Mill and begins his own further discussion of the issue at hand we are disappointed. He simply moves from a discussion of the weak points of Mill's view as brought out by Whewell to a discussion of his own view and response to Mill without explaining what he wishes to add to or how he differs from Whewell.

We begin by finding that Lange endorses arguments offered previously by Kant and/or Whewell for the impossibility of deriving mathematical axioms and/or the necessity of these axioms from experience. We find what he regards as the crucial point: even if these axioms arise due to our capacity for knowledge, the fact of their necessity proves the truth of Kant's basic point of view. We may apply necessary axioms to wrong cases, but we must nevertheless deem them necessary, and our view of their necessity is independent of experience (Lange 1873, pp. 18-20). Now all of this is not new. Lange then says that he wants to present Kant's view "in aller Schärfe." Perhaps he wishes here to begin his own view as a supplement to Whewell's but he does not explain.

As we pursue the argument we find, to be sure, a presentation which differs from that of Whewell. But we come to problems and points of view which are very Whewellian indeed. Yet Whewell has been long forgotten. Lange says, for example,

> Kant zeigte zuvörderst an dem Beispiel der Mathematik, dass unser Denken wirklich im Besitz gewisser Erkenntnisse a priori ist, und dass selbst der gemeine Verstand niemals ohne solche ist. Von hier aus fortschreitend sucht er nachzuweisen, dass nicht nur in der Mathematik, sondern in jedem Erkenntnissakt überhaupt apriorische Elemente mitwirken, welche unsere Erfahrung durchgehend bestimmen. Wie sollen aber diese Elemente entdeckt werden? Hier ist ein dunkler Punkt im Kantischen System, den auch die sorgfältigste Forschung nach der eigentlichen Meinung des grossen Denkers schwerlich je wird beseitigen können. (Lange 1873, p. 28)

Lange continues to argue that these elements cannot be discovered by mere deduction from a priori or metaphysical principles, as Fichte sought to do, nor can they be won from experience, as Mill tried to do. But a priori principals do not need to be discovered a priori. Indeed, Lange says, "Wir sind also in der Aufsuchung und Prüfung der allgemeinen Sätze, welche nicht aus der Erfarhung stammen, lediglich auf der gewöhnlichen Mittel der Wissenschaft beschränkt" (Lange 1873, p. 31).

Lange further argues that Kant's great merit lay in the fact that he distinguished sense and understanding as sources of knowledge. But his weakness was that he allowed understanding to exist quite independently of sense (Lange 1873, p. 32). And his method, then, of separating the two was dubious. Lange argues, for example, that our conception of space is not merely determined by the form but that sensations may influence the way space is perceived. This should not lead us to reject the theory of the a priori nature of space or time, however.

Lange continues to argue that the problem which the alleged existence of a "thing-in-itself" need not arise and that the suggestion that our perception of space arises out of organic conditions or of sensations as we, with our physical make-up, happen to receive them is no argument against the view of the form of space and time as a priori nor even against the conjecture that all (finite) rational beings use the same categories. He suggests not only that the a priori principles must be discovered in science but that they are unclear, that they must develop and that they may be extended.

Now all of this is found in Whewell. Why do we need more argument? Perhaps we find an answer to this question where Lange says,

> Giebt es eine Art von Hypothesen, welche sich durch die Nothwendigkeit ihres Entstehens aus unserm Geiste von allen andern unterscheidet, dann ist mit der Generalisation, dass es doch eben auch eine Hypothese sei, gar nichts gewonnen; vielmehr handelt es sich darum, den inneren Grund ihrer besondern Natur zu entdecken. (Lange 1973, p. 19)

Perhaps Lange refers here to Whewell who, by implication, has not gone deeply enough. It would have been nice to see this spelled out. What we have at best is a view which, in its essential points, is identical to Whewell's, but which needs, perhaps, to be developed more systematically – at least in the mind of Lange and the new Kantians. This is puzzling and disappointing. If Whewell had gone so far, why does he receive so little recognition? And how can we appraise the new-Kantian's additions to Whewell's perspective if we have no discussion of just what they took these to be and why they deemed them needed? (I will not go into the differences in the systems here. I might note, however, that Whewell sees himself as a realist and the new Kantians are idealists of sorts. Oswald Külpe found it necessary to reject their view of the role of concepts even though he also asserted two sources of knowledge: he wished to explain how these two sources lead to knowledge of real entities (Külpe 1912; Wettersten 1992, pp. 117-126). We are left then with a view which picks up on many ideas introduced by Whewell. But these are introduced as new. We can only be more curious rather than less as to why he was rejected before: his ideas come back to the fore in various places where we would not expect them or where they might seem unneeded.

Before leaving the new Kantians I wish to briefly mention the last well-known member of the school, Ernst Cassirer. Cassirer was one of the most erudite of philosophers. He discussed in his voluminous writings almost any philosopher one can think of. He did not discuss Whewell, so far as I know, even though he followed Lange in adopting a view with so much in common with Whewell. In 1920 he published a detailed study of post-Kantian systems, but he does not mention Whewell (Cassirer 1974). We thus find Whewell's ideas making headway at the same time his own theory is hardly discussed. (Klaus Christian Köhnke published not so long ago a book of 624 pages devoted to the new Kantians, in which Whewell is not mentioned even once; see Köhnke 1986.) Does Whewell have an unacknowledged influence? Are his ideas simply being rediscovered? And, in either one case or the other, way does he remain so obscure a figure? We might indeed note, that just that idea which has been so overlooked or deemed confused – the idea that a priori principles need to be discovered by empirical, critical research before they become clear – has been of central importance for Peirce in America and the new Kantians in Germany.

### 3.4.3. *Paul Leuckfeld: Establishing Mill at the Turn of the Century*

Due to renewed interest in Kantian points of view which were close to science at the turn of the century in Germany – the new Kantians were working on such and Oswald Külpe was beginning work on his own more realistic version, to mention two – one might expect to find there a more favorable review of Whewell at the least. But we still find little mention of him. One is by Paul Leuckfeld in his *Zur Logischen Lehre von der Induktion* (1898). We find that Leuckfeld begins anew giving a brief overview of Whewell's view with an emphasis on his view of induction. I find his summary at the beginning quite passable – there is no point here, I think, to discuss details – but we might become uneasy when he says that Whewell, with Herschel and against Mill, deems an hypothesis credible when its predictions are found to be true without noticing the sharp difference in their views of inductive argument – which is here under discussion – between Herschel and Mill on the one hand and Whewell on the other.

The failure to distinguish between the perspective of Herschel and of Whewell sharply enough – he presents their disagreement at one point as a matter of emphasis of the idea that presupposed principles should be used (Leuckfeld 1898, p. 388) – leads quite naturally to posing a Herschellian or Millian question for Whewell concerning the role of induction and finding Whewell's answer inadequate. He states, "Dabei bleibt in N.O.R. die Frage nach der Rolle, die den Ideen im inductiven Schlusse vom logischen Standpunkte aus zukommen soll, offen" (Leuckfeld 1898, p. 389).

Leuckfeld does not discuss whether such a theory is possible or needed, even though he does mention Whewell's criticism according to which Mill cannot account for actual scientific practice with his alternative of the four methods. And, he concludes his essay with the remark that,

> es handelt sich aber bei ihm eigentlich nicht um die logische Begründung, sondern darum, wie die Überzeugung, als psychologischer Vorgang, allmälig intensiver wird. Was nun im Speciellen die eingetoffenen Voraussagungen, die zusammengefallenden Inductionen, endlich die Vereinfachung der Theorien durch Hypothesenbildung,

> anbetrifft, so wird selbstverständlich durch alles dies der Character des Erprobungsverfahrens nicht im geringsten geändert. (Leuckfeld 1898, p. 390)

Leuckfeld thus picks up the common views of Whewell according to which he needs but does not have a theory of induction and that he is interested in a psychological question, but also that his theory of discovery is not necessarily touched by these defects. This leads quite naturally to a view according to which Herschel and Whewell are quite similar since Herschel's view, perhaps, also needs a theory of induction such as that supplied by Mill.

We are left then somewhat disappointed: old and erroneous ideas which were developed by Whewell's critics are presented as mere reports of the situation without even noticing Whewell's objections and the sharp difference between his view and that of Herschel. He has received here even less appreciation than that offered by Apelt and Lange before, who seem to have no influence, even though Apelt is referred to by Leuckfeld.

Our search for a solution to the puzzles caused by the rejection of Whewell in the 19th century when his ideas keep arising in the 20th have hardly been resolved by the survey of comments on him in 19th century Germany. We have indeed found reason why at least some important thinkers in Germany might have taken him seriously. But even these thinkers either repeat the old reasons we have found above for rejecting them, but which did remove our trouble, or they just ignore him as a predecessor with whom they agree to some unstated degree. We can only find our puzzlement increased.

# CHAPTER 4

## THE TWENTIETH CENTURY SNEAKS A WORRIED LOOK AT OLD JUDGMENTS

The fact that Whewell's ideas were so close to, and even seemed to have influence on, succeeding thinkers of some importance led to some unavoidable curiosity about his views in spite of the established opinion of them. The tension between this view and the excitement his writings could raise was enticing. The tension was highest and received its greatest impetus from France. This was above all due to the work of Bernard. In other countries small studies – an essay here or there – appeared and only a very few systematic studies were produced. As might be expected these systematic studies were made by relatively minor thinkers. Whewell was deemed somehow relevant, but not so important. He was also hardly studied. It would be a good topic, say, for a proper but not too important doctorate.

In spite of the small prestige which any study of Whewell in the early part of this century had, we can hope to learn from them since they had to face some of the most important issues of the philosophy of science and its history. And above all we may hope to find some systematic analyses, analyses which take into account the various attacks on Whewell's philosophy, the strength and coherency of these attacks both individually and together, and the strength of Whewell's view vis à vis his competitors. The later commentators on Whewell have had to reconstruct this discussion and explain it to their readers. We may hope, then, to find answers here to the questions which were raised at the end of our study of the results of the first major discussions, that is, how do these criticisms go together or not go together to build the basis for the established view, and how can we explain, or explain away, the apparent weakness of many of these objections, or, how can we explain their success in relegating Whewell to the archives in spite of their weakness?

Discussions of Whewell's philosophy have been produced at various times and for various reasons. Their connection may not appear close. Those essays now found in standard bibliographies may seem to have served only narrow aims; they have hardly been absorbed or appraised. A new look at them may, then, be enlightening. We may hope to show their connections and appraise the ongoing discussion. When we survey this literature, there is one fact which is most striking: not a single English study of Whewell in the first half of the 20th century is anywhere mentioned in standard bibliographies. Whewell's native land, his beloved Cambridge, had completely forgotten him. We must turn elsewhere and to relatively obscure thinkers to find any work at all.

Due to the work of Charles Sanders Peirce one might expect to find some work of interest in America. But here we again find only one study and this study has nothing to do with Peirce. It was written as a doctorate by Marion Rush Stoll at Bryn Mawr College in 1929 and is entitled *Whewell's Philosophy of Induction.* It was a product of the interest in America in the new-Kantians – the school of Friedrich Albert Lange – as well as that of the French – especially the work of André Lalande whose work I discuss below. The similarity of approaches between Whewell, on the one hand, and the new-Kantians and Lalande, on the other hand, was noted even though the debt of both to Whewell was not. A new study seemed called for – at least it might be worth a doctorate. Stoll's study is well-informed and sympathetic to Whewell, as much previous literature was not. And it seeks to offer the first composite picture of Whewell which anyone, anywhere attempted! Perhaps we may find here a new basis for finding answers to those questions concerning the strength of Whewell's system vis à vis his competitors.

The interest in Whewell in France is due, it seems clear, to the overpowering influence of Claude Bernard. The similarity between the enormously successful methodological essay of Bernard and the enormously unsuccessful methodological proposals of Whewell could not go unnoticed in France. It was André Lalande who forcefully called attention to it. A schism in France had been taken over from the 19th century between philosophers (or scientists) who had absorbed Mill and scientists (or philosophers) who had absorbed Bernard. What then could one do with Whewell, who could have been – whether he was or not – Bernard's teacher in methodology? Such questions troubled André Lalande. He sought to combine Whewell and some aspects of Mill and discovered the role of hypotheses in many predecessors such as Le Sage or Peirce. We see nowadays modern commentators seeking to follow the same path to the unification of inductivist and anti-inductivist views.

Robert Blanché saw already in the 1930s that even Lalande's sophisticated attempt to unify some aspects of Mill with deductive theories of science would not do. As a consequence a new study of Whewell was needed, which conceded the strength of his arguments against induction. The one serious study which sought to investigate the questions of why such a powerful philosophy was rejected, and how it might be reappraised, then, is *Le Rationalisme de Whewell* by Robert Blanché. It was published in 1935 and is, we may hope, the most important of the three by far, since it takes notice of the earlier work by Stoll and it makes an apparently serious attempt to set right some of the mistakes which were made in building the early interpretation of Whewell. It is, in a sense, our last hope, since no systematic attempt to review, explain and develop an interpretation of Whewell has been made since.

It would seem useful to begin the search for a deeper and more sympathetic appraisal of Whewell's work in Germany. Not only did his work make a quick impression there – his history was translated into German by Littrow almost immediately – but we might expect that the philosophical style in Germany might lead to more sympathetic treatments. This expectation was partly born out by comments here and there which appeared in the 19th century. The studies by E.F. Apelt, Paul Leuckfeld and Erdmann are mentioned in the standard literature. We also saw the impact of Whewell's work on Friedrich Albert Lange. When we look for more comprehensive studies of Whewell, however, we find only one book dedicated to him. And this book, *Die theoretische Philosophie William Whewell's und der Kantische Einfluss*, is a doctorate written in Germany by an American! I will look at it any way in spite of the forebodings that it will not help much.

The discussion of Whewell's work did not, of course, end with these studies from the early part of the 20th century. It was taken up once more in England and America. But the interests of those who participated in this discussion primarily concerned, as we will see below, induction as inference from particular to general statements, that is, exactly that problem which Whewell rejected as irrelevant to scientific practice. The return to old views and not their critical examination was then called for.

### 4.1. France: Is Science Inductive?

Research into the strengths and weaknesses of Whewell's philosophy in 20th century has stemmed above all from two factors. These are the sharp and established rejection of his philosophical explanation of knowledge as incoherent, on the one hand, and the realization of the strength of his theory and history of the growth of knowledge, on the other hand. This dual interest in the strength and the weakness of Whewell's philosophy is most evident in the best studies. The work of André Lalande was apparently of especial importance in this respect and appears to be largely responsible for the work of both Stoll in America and Robert Blanché's in France.

The study of Blanché, *Le Rationalisme de Whewell*, has a rather obvious background. The work of Claude Bernard had introduced into France a methodology, which in all essential points followed that of Whewell's. It may not seem surprising, then, that Blanché asks what the relationship between Bernard's superb methodology and Whewell's allegedly confused philosophy is (Blanché 1935, p. 2). Though the logic of the situation seems in retrospect clear, it seems to be André Lalande who first noted that the dominant methodology in France, that is to say, Claude Bernard's, and that of William Whewell were the same, and that this posed serious problems for the philosophy of science (Lalande 1929). How could the methodology of Bernard, which was established in France, be reconciled with the otherwise universal rejection of Whewell and the establishment of Mill?

There were, of course many other thinkers who had adopted positions with aspects close to points of view found in Whewell and who thus played a role in setting the background for Blanché's interest. They include Pierre Duhem, Henri Poincaré, Leon Brunschig, Emile Meyerson as well as André Lalande. The field for inquiry in France in this period is incredibly rich. And it seems to have one major cause: Claude Bernard had introduced Whewell's methodology without Whewell. This meant that there was a challenge to develop a theory of science which broke with inductivism or which reconciled inductivism with the methodology of Bernard, which had, like Whewell's, no place for induction. Most studies dispensed with both induction and with Whewell. To what degree this was based on the convention established by Bernard not to mention Whewell, on the one hand, and to what degree it was based on ignorance, on the other hand, is not easy to say. I will in any case not delve into this rich field which offers so much that is enticing and turn to some observations of Lalande who did notice Whewell and tried to take account of his views.

## 4.2. André Lalande: Reconciling Scientists and Philosophers

The conflict between the standard view of scientific methodology among scientists as represented by the methodology of Bernard and the standard view of methodology among philosophers of science as represented by Mill was intense. There was not only a wide breach between the two theories, but also a disturbing question as to how such a conflict could arise in regard to science. The standard view among philosophers was not due to them alone: it had the support of the scientists in England and many in Germany at least. And the standard view of the scientists in France was incomplete: Bernard had developed a methodology alone with no theory of knowledge to accompany it. These problems could not be avoided forever. The philosophers were given the task of sorting it all out.

Bernard had given theorists of science in France – scientists, philosophers or whomever – new freedom to develop alternative views. The old inductivist views were obviously inadequate to account for method as described by Bernard yet no new view was established. This freedom was quickly taken advantage of and new theories of enormous interest were formed. Those of Pierre Duhem, Henri Poincaré and Emile Meyerson are the most notable. For the most part these thinkers simply ignored the English inductivist views. As a result the breach between the scientists and the philosophers became a breach between France on the one hand and England on the other hand. Germany was divided, but not quite a party to one side or the other.

In 1929 André Lalande in his *Théories de l'Induction et de l'Expérimentation* produced the first major attempt to come to terms with the problem of reconciling the theories of science as induction with the deductivist views of Whewell and of Bernard. (This book was first given as a series of lectures at the Sorbonne from 1921 to 1922, and first published in the *Revue des Cours et Conférences*, 1922-24.) The delay had been long coming – some sixty years. It is by no means surprising that the first major attempt tries to save both the standard view of the scientists in France, that is, of Bernard, and indicates sympathy for the standard view of the philosophers, that is, of Mill, nor it is surprising that he praises Herschel, who was the scientist who appeared to have supported Mill most effectively, nor that he ignores more problematical cases such as Pasteur who was closer to Mill.

The surprise in Lalande's work comes from the high praise and deep appreciation he has for Whewell. Lalande saw how far Whewell had come and how close the views of Whewell – who was prior to Bernard and universally rejected – were to Bernard's – who was in France universally accepted as the authority on scientific method. But this alone would not have led to any interesting work, since there was then available a standard way of explaining this seeming paradox away: Whewell was accepted insofar as he merely described science and Bernard merely described science. Whewell was rejected just where Bernard had remained silent. This view depended, however, on finding some further view, which would be a substitute for Whewell's philosophy. Lalande knew that the attempt of the English to reconcile Whewell's methodology and Mill's (or Herschel's) philosophy would not do. And he had seen that Whewell offered hope of a deeper and more powerful philosophical approach which was complementary to that of Bernard. So, instead of ignoring the English, as Duhem, Poincaré or Myerson did, or tamely following them, as Liard did, he defended and developed the views of their leading heretic as best he could, even while accounting for any strengths the established English view might have had.

Lalande hoped to develop a theory of scientific method and of scientific proof – or something like it. He believed that in order to do that he had to solve three problems. The first and most important of these is the problem of scientific method or of technique: how, in fact, does science proceed? The second of these is the problem of the principles of science: what are the (simple) rules which guide research? This problem concerns the rules which scientists use to achieve agreement and should not concern justification. The answer to it, though, should be simple and economical; it is, as he called it, methodology. The third problem is that of the foundation of science. This problem concerns the quite general grounds for the assurance, which we obviously have, that scientific theories are true, or the quite general grounds we have for deeming scientific theories true.

The first problem would seem to be relatively straightforward for someone who not only appreciated Whewell, but deemed his work to beg and large correct, describe as a description of science at least. Following Whewell, Lalande deems this task to be the most important of the three: solutions to the other two problems should grow out of it. It is thus understandable and justified, he says, that he devotes two thirds of his book to the discussion of this one problem alone.

With this information we would expect an intensive study of the facts of the history of science indeed. But this is not what we find. Rather, we find a discussion of theories of scientific method, better, of the theories of scientific method which are forerunners of his own, that is, of Bernard's. There is nothing wrong with a study of the development of the philosophy of science. It is, indeed, a field which is sorely neglected and which this book seeks to contribute to. Whewell advocated it as part of his approach, (Whewell 1967b, vol. II, pp. 119*ff*) but it was rejected along with his other suggestions. It should now be reintroduced, and I try here and in (Wettersten 1994) to further this goal. But Lalande does not explain why he wants to do this. Why, for example, does not Lalande simply defend Whewell's *History* plus, perhaps, Bernard's methodology?

My own conjecture is that Lalande wished to show that Whewell was not an isolated figure, that his view had a long history. This conjecture is, I think, corroborated by the fact that Lalande seeks predecessors as well as followers who also defended hypotheses. In doing this he is not always so critical, since he leaves their philosophies out of the picture when they are inconvenient instead of discussing their development critically as Whewell recommended. We thus find an historical discussion of methodology which does not fit the critical methodology it is used to bolster: the power of Whewell's use of a self-reflexive method is here as well in all other studies overlooked. This study is, by the way, now repeated by those who wish to show that fallibilism is not the invention of Popper alone but has a long history; the same figures – Lesage, Peirce, etc. – come to the fore. But the critical study of the development of problems is pushed aside in favor of the quest for the discovery of views similar to the "up-to-date" views, to those in fashion today.

The solution to the first problem should give us a basis for solving the second problem, that is, the problem of finding the simple and preferably few rules which guide all scientific research. Such rules – Lalande calls them principles – are very close to what Popper shortly thereafter – five years later – called methodological rules. They are rules which guide research and enable science to achieve agreement but they are not – rather, they are not intended to be – rules of justification: they merely serve to bring about consensus. Lalande hoped, perhaps, to avoid problems of inductive inference and, if so, was a short distance away from Popper's view. But he did not, as I explain below, succeed in holding to this distinction as Popper later did.

Lalande's second problem concerns which rules or principles guide research. There are, he says, three principles which guide scientific research. (I presume here, incidentally, that he is here seeking to offer something like an alternative to Mill's canons). He calls these three principles the principle of deducibility, the principle of complementary probability and the principle of universalization.

The principle of deducibility is the principle that no true theory has false consequences. Lalande suggests that the method of seeking to refute alternative solutions in order to see which one survives is very important, even though it poses difficulties. We might, for example, have failed to consider the correct alternative – are there not an infinite number of possible hypotheses to explain phenomena? Or, we might note that a theory is refuted in a precise formulation only, but in a different, looser or vaguer formulation it may be reconcilable with the facts. Or, we might note that in any refutation a whole set of hypotheses are refuted and not one in isolation. Such problems as these, says Lalande, are not reasons to reject the principle – which is in any case indubitable as a matter of logic – or its importance in science. Rather, we need to supplement it.

Lalande offers, then, his second principle which is that of complementary probability. According to this principle the probability of an hypothesis is the complementary of the probability that the observed facts would be observed if the hypothesis were false. If the probability of the occurrence of events predicted by a theory is very low, then the probability of the truth of the theory, should they occur, is very high. Lalande offers no development or discussion of this view. In retrospect we might find it interesting to see how close at some points he comes to Popper's intuitions – in this case that a theory which predicts the most is a priori the most improbable. But by the same standard we see how he fails, since he continues to speak of the probability of the truth of hypotheses given some evidence. He thereby fails to carry out his plan of avoiding the problems of confirmation or of proof in his discussion of principles. He begins here to speak not merely of the achievement of consensus but also of proof.

Lalande's third principle of universality is even more disappointing. He claims that one should deem it to be the case that that which has always occurred in accord with some law will continue to do so. He is aware of difficulties, of course, which he thinks may be overcome by the assertion that this principle is normative as in logic or methodology and not factual. He realizes that there is and can be no guarantee, but defends its use without a guarantee. He adds, however, that it is unlikely that a false hypothesis would have a wide number of true consequences. Lalande has not kept to his division of rules of procedure and grounds for belief, since his procedures involve making estimations of the probability of hypotheses given evidence and since he appeals to a theory of the justification of theories by large numbers.

Let us turn then to the third problem Lalande discusses concerning the assurance that the foundations of science are true, a problem which corresponds to Whewell's discussion of the intuition of Fundamental Ideas. Since Lalande deems the rules of scientific research to be a mere means of gaining knowledge, but no guarantee of that knowledge – some probability excepted – he asks a third question which has nothing to do with induction or the rules of science. He offers here an explanation of the process – or of the progress – of reason. It is this development of reason which is the foundation of science. It gives sense to or informs our experience. It is our path to knowledge. As such it neither demands nor permits more than it itself can provide. We have no certainty but we may have confidence or assurance, no more.

Lalande's solution to the third problem takes the feeling or intuition of necessity, which he grants, as a datum to be explained. Both his problem and his explanation are Whewellian but, because it tends to be more skeptical and less confidently realistic, moves a little backwards, in the direction of Kant. He deems reason as that which constitutes science. It provides science with its content. We cannot do better than to use and trust our reason but we have no guarantee that we will thereby understand the world as it really is.

Lalande did explicitly what other thinkers such as Duhem, Meyerson or Poincaré had done rather implicitly: he developed a philosophy of science against the background of Whewell's deductivist view. His view of science is not so interesting as the views developed by these thinkers but contrary to them he introduced an explicit and critical discussion of Whewell into French thought. This had an important even though limited impact. So far as I know no one has tried to redo Lalande's task and to better place Whewell and his impact in French thought. It did lead to the detailed study of Blanché, however, which remains the best study there is to this day.

Lalande had sought to redo the philosophy of science in a thoroughly Whewellian way. The views of Mill and the problem of induction were spoken of in rather sympathetic ways. But they were largely removed from the system. The elements that remained were apparently there because Lalande did not see how to dispense with them: he tried to separate rules of procedure and rules of proof in a sharp way but did not see how to do that – as Popper later did. Whewell is thus much more firmly established after Lalande's work than before it. But the attempts to compromise Herschel and Whewell have left their imprint and Whewell was still deemed a secondary figure.

### 4.3. America: Marion Stoll – The Composite Picture Reconstructed

We have to wait until 1929 before we find a study of Whewell in America. And this study has nothing to do with Peirce or American philosophy. It was a doctorate and was apparently written due to the influence of such thinkers as Ernst Cassirer or the new-Kantians and André Lalande. Stoll recognized that the programs of those thinkers who sought to unify rational with empirical elements in a philosophy of science – a program which was then modern – shared a great deal with Whewell's philosophy. A study of Whewell, then, might show some of the difficulties which this program faced. We may, then, be encouraged to find answers to our questions and not only due to the source of the interest in Whewell's work, but also because we here find the first systematic study of Whewell's philosophy which was ever written!

If we begin our study of Stoll by sneaking a look at the end of her book, we find an appraisal which awakes our interest and our caution: it shows appreciation for his work even while endorsing the standard view according to which he was confused. She states,

> Whewell's conclusion, however dogmatic and out-grown it may seem to the modern reader, provided a means of harmonizing the several views which Whewell believed that the facts compelled an unprejudiced observer to admit. Unlike many of his contemporaries, Whewell avoided the extremes of realism and idealism, and he recognized the limitations of the popular empiricism championed by John Stuart Mill. His moderate position enabled him to appreciate aspects of the problems of a philosophy of science which were neglected by many of his contemporaries, but which have been recognized by recent writers. (Stoll 1929, p. 121)

and

> His final solution, unsatisfactory as it is, emphasizes the problems with which any rationalistic view of science ultimately has to contend, and also indicates the difficulties

> of solving satisfactorily any epistemological dualism which results from an endeavor to harmonize the positive contentions of realists and idealists. (Stoll 1929, p. 121)

Among the problems, she says, which Whewell posed and his contemporaries neglected was "to present a well-grounded rationalistic interpretation of the nature of science and of the processes which are involved in its development" (Stoll 1929, p. 121) and she notes quite correctly that "Since Whewell wrote, revolutionary criticism of what he regarded as perfectly certain bodies of truth have revealed the limitations and inadequacies of Whewell's conception of a priori Fundamental Ideas" (Stoll 1929, p. 121).

She herself favors a view defended by Cassirer among others. This interpretation, she states, "regards reason as normative, not constitutive" (Stoll 1929, p. 121; See also pp. 118*ff*).

It is clear, then, that it is useful to spell out the way in which Whewell's theory may be deemed inadequate in order to be clear about what if anything, we might learn from him. Our hopes may be raised by the apparent insight, sympathy and erudition of Stoll's book.

To return to the beginning of the text we find at the very outset that she repeats Todhunter's observation of the difficulty of keeping up with advances in science and the resulting decline in accuracy of Whewell's history (Stoll 1929, pp. 4*ff*). As we have just seen this is deemed by her to be of crucial importance, but the reasons seem not so clear. Was his view already in trouble in the 19th century due to changes in science and to insecurity as to exactly which ideas are fundamental? Or is it for the first time, after Einstein's revolution, that we find these difficulties? Stoll shows that Whewell changed his views on specific points concerning which ideas are fundamental (Stoll 1929, pp. 57*ff*). And she concludes:

> The difficulty, as Todhunter pointed out, was mainly that Whewell tried to carry over into the inductive sciences the notions which he had acquired from his study of the deductive sciences (Todhunter 1876, I, p. 109). There he had discovered that certain fundamental principles follow necessarily from Ideas of Space and Number which, he believed, could not conceivably be other than they are. But the inductive sciences also are grounded in principles which come to appear to the cultivators of those sciences to be as indubitable as the first principles of geometry and arithmetic. These principles, however, are not merely precise determinations of notions under which all men view the phenomena from the beginning: but they are discovered in the effort to find some means of relating phenomena into a coherent system. (Stoll 1929, p. 61)

In spite of her sympathies for Whewell she finds his system inadequate for the same reason as, say, Mansel. And, though her observation concerning the Fundamental Ideas is correct, she does not say why Whewell's system is unacceptable. She quotes Todhunter to support her position. Her reference to Todhunter reads as follows:

> the fact I believe is that Mr. Whewell first contemplated merely a History of Astronomy, or of Mechanics, and subsequently enlarged his plan to include what he considered the principal Inductive Sciences. But the unity which he supposed his work to retain when thus expanded seems more imaginary than real; it is not difficult to perceive the dissimilarity of the substances thus ideally combined – the head of gold, the feet of clay. (Todhunter 1876, vol. I, p. 109)

We still have not found the difficulty concerning the extension of the Kantian categories other than that the new categories – or even the old ones – are not quite Kantian. They are not all conditions of experience, as Whewell himself explained, and they are learned. But there is no independent criticism of this view. We need to look further.

Stoll's book is organized by reference to critical appraisals of Whewell which had been offered by her predecessors. The introduction concludes with a quote from Leslie Stephen (*The English Utilitarians*, vol. III, p. 143) to the effect that Whewell's theory was radically incoherent. This, we have seen, has already been explained by appeal to a standard view of the overambitious nature of the *History* as well as the extension of his views on mechanics to all of science. It was both too much for one person and for one idea. In the second chapter this incoherence is presumed, examined and explained. The description of his views is careful and one does not entirely forget that one is reading a dissertation as the author describes Whewell's theory of the fundamental antithesis between thoughts and things, theories and facts, necessary and experiential truths, deduction and induction, ideas and sensations, the subjective and the objective, or form and matter. Stoll appreciatively describes Whewell's view according to which both sides of the antithesis were necessary for knowledge but neither side could be found strictly free from the other. Her appraisal is critical, but mildly so. She contends that Solomon Maimon in his critique of Kant had seen the point better, he had seen that the concept of the thing-in-itself was unacceptable; this concept should be replaced by another, by the given, but the given is always a goal or limit to which one strives (Stoll 1929, pp. 31*ff*).

This analysis is quite nice in one respect, for, Maimon was ahead even of Whewell in his perception of the need for a skeptical approach as a consequence of the failure of Kant to prove the necessity of his categories – which Maimon took, nevertheless, to be correct. But Stoll is wrong to suggest that is indicative of the superior level of research in Germany and that Whewell's view was characterized by a predilection for a naively realistic point of view (Stoll 1929, p. 32). Maimon was by no means characteristic of German research. He was far ahead of all of his contemporaries but, as a wandering Polish Jew, could find little appreciation in Germany, even though Kant did say that he was his best critic, and even though the King wrote to thank him for sending him a copy of one of his publications, and even though all his serious contemporaries such as Fichte or Reinhold or Anesidemus must have known his work. Maimon was a skeptic but that does not mean, as Stoll suggests, that he was no realist, and it is certainly unfair to say that Whewell was a naive realist. The views of Maimon and of Whewell are much closer than she grasps. Stoll hints that there is confusion in Whewell's view which might be removed by more subtle analysis but does not spell it out.

The following chapter is an analysis of Whewell's apriorism. She asks to what degree his view is Kantian. Did Whewell change his mind on the nature of his apriorism in his later work to emphasize more adequately (or more confusedly) its Platonic aspects and did he, perhaps, introduce contradictions? The questions are not new. We found them especially in the discussion of Mansel. We may, perhaps, hope to find better answers to the question we left unanswered above.

Stoll's analysis follows that of the new Kantians. I have already quoted both her and Todhunter, whose view she takes over, which is, of course, that view developed by Mansel, that the extension of the ideas of space and time to all of science was problematical. She does not necessarily deem this extension a mistake, and no wonder, since the new-Kantians and Cassirer followed Whewell here. Her view is evident in her discussion of Mansel's criticism (Stoll 1929, pp. 46*ff*, p. 53). She maintains that Whewell later changed his view to a more Platonic conception (Stoll 1929, p. 55) and further that Whewell did not grasp the consequences of this change, that is, that his view leads to idealism, and therefore fails to combine empiricism with rationalism to produce a realistic view of science, as he intended and as she would like to find. Here, then, is one suggestion as to just what the difficulty is. She states,

> Whewell proceeded to develop the view that there must similarly be a priori conditions determining the possibility and manner of perceiving phenomena which are data for the inductive sciences. (Similar to those which apply to all experience). This, as the previous discussion has indicated, was the original point of view from which Whewell proceeded in the development of his apriorism. But there was another aspect of his thought which received less emphasis even though it explains Whewell's conception of the Fundamental Ideas which tended to replace that view of these Ideas which he borrowed directly from Kant. This was the notion that Ideas are the relating elements of science, for which sensations provide the disconnected data. (Stoll 1929, p. 62)

and,

> this view may appear to be essentially the same that Whewell had proposed all along. Ideas are the form-giving elements of experience, introduced by the knowing subject to make of the formless matter of sensation a comprehensible whole. Yet there is a distinction, for Ideas of Space, Time, and Number, are applied to sensory data immediately and by all men, while the ideas of the more recent sciences are consciously sought by men who would find an intelligible explanation of phenomena, and their

> success is recognized by their finding it possible to make of the unrelated data a coherent whole. Starting from the notion that things can only be apprehended under the a priori forms of perception and categories of the understanding, Whewell arrived at the conviction that events can be intelligibly related into the coherent bodies of knowledge which science represents, only by the introduction of principles of order, which must be intuitively discovered, although they are a priori in the human mind. This position amounts simply to the assertion that all principles of intelligibility are contributed by the knowing mind. The idealistic conclusion that things in themselves are incomprehensible is one which Whewell refused to accept, for he assumed that science represents the right interpretation of an external universe and that man has been endowed by a beneficent Deity with the a prior Ideas which are the necessary conditions of rightly understanding nature. (Stoll 1929, pp. 62-63)

I will return to the alleged necessity of idealism which is, I think, mistakenly maintained in regard to both Maimon and Whewell and for the same reason: the alleged need to complete their systems. They both reject this demand, but Whewell may seem to want to meet it and Stoll now moves to a discussion of Whewell's alleged need of faith in addition to, or as a substitute for, idealism to complete his system. She continues,

> Whewell's rationalism necessitated this recourse to an assertion of faith, because he had to explain how the a priori Ideas can provide means of understanding relations between things. His conviction that a Beneficent Deity must be responsible for the perfect harmony of the a priori mental order and the physical order found in nature represented the only answer which Whewell could propose in an attempt to save his rationalism from the objections of his empirically minded critics. (Stoll 1929, pp. 62-63)

The analysis of Whewell's theory which Stoll offers imposes standards of system on Whewell which he rejects, fails to note the standards of system which he himself offers, and provides as evidence for the correctness of this view the analysis of a Kantian, Solomon Maimon, whose standards of system come far closer to those of Whewell and, as a result, lead him, as well as Whewell, to a view of the incompleteness of our knowledge. The uncritical use of such a standard is typical of Whewell scholarship – as well as Maimon scholarship – and leads to the usual criticisms. Let me explain this quite general point here with Stoll's analysis as an example.

Whewell's theory of science was based on a view of the facts of science. This foundation for his theory was not collected in a Baconian fashion, independent of the ideas which were used by Whewell to bring them into a coherent pattern or, if you will, system. Stoll, however, separates the two saying, "His history of the inductive sciences provided a wealth of material which Whewell seemed to regard sometimes as data from which to draw his conclusion, sometimes as illustrations of the process which, he believed, must be involved in any scientific advance" (Stoll 1929, p. 65).

She gives a reference to a letter by Whewell reprinted in Todhunter (1876, vol. II, p. 284) to support the view. It does not do so. It merely states that Whewell used his history to give examples or illustrations of his view and, indeed, for an audience of metaphysicians. This separation of the gathering of evidence from discovery and discovery from proof is foreign to Whewell's system and allows for criticism of a straw man. It is, indeed, crucial for Whewell's theory of science that colligation of facts is only possible if we apply ideas and this colligation identifies those facts which are relevant for some field of inquiry and how some facts are related. This is evident in, for example, Whewell's criticism of Mill's four methods: they presume that the relations between facts are already identified, but this is the main problem (Whewell 1971, p. 263). The discovery of facts and the ordering of facts are only attained in any science by the application of ideas. This is true of the theory of science as well and Whewell has used his own ideas to order the facts of the history of science; it is his success in doing so which proves his theory true. And here is his "system." It is only when this "system" is taken into account that one can be fair to Whewell.

Stoll deems Whewell's later more Platonic explanations of the correspondence between ideas and the world as a necessary and inadequate addition to his theory and she is here by no means alone. It is, for example, that view of Whewell's work which is found in the obituaries of Clark and of Herschel discussed above. But Whewell clearly intended these speculations as a mere analysis of the consequences of the truth of ideas which he had independently established. He states,

> That necessary truth is progressive; – that science is the idealization of facts, and that this process goes on from age to age, and advances with the advance of scientific discovery; – these are doctrines which I have endeavored to establish and to elucidate. If these doctrines are true, they are so important that I may be excused should I return to them again and again, and trace their consequences in various directions. Especially I would examine the bearing of these doctrines upon our religious philosophy. (Whewell 1971, p. 334)

and

> Suppose the case to be as I have stated it; that in some sciences at least, laws which were at first facts of observation come to be seen as necessary truths; and let us see to what this amounts in the general sciences. (Whewell 1971, p. 355)

Some may prefer to accept them whereas those who do not believe in a God may choose not to. Whewell chooses merely to spell out the consequences of his own view (Whewell 1971, p. 358). Now Whewell does say that his system needs this addition to be complete and satisfying. But it is not a foundation from which all else is to be derived. It is an explanation which, for those who accept the view that a Supreme Being exists, is satisfying.

Can this view, according to which ideas which men have accord with the world as it really is, and do so due to the plan of a Supreme Creative-Mind which has made it so, be used as a sign of the breakdown of Whewell's system, that is, of his inability to give an adequate explanation of how and why knowledge is obtainable? I think not. In order to do this one would at least need to show that Whewell's alleged proof of the facts on which this view is based is false, that is, that his independent method of justifying his view of scientific progress by reference to the facts is unacceptable, and further, that no incomplete system can be true and justified.

On this later point we may mention Maimon for Maimon, just as Whewell, deemed the quest for a complete system to be futile and dangerous. Maimon viewed Kant's proof as inadequate because it could not be complete. We thus need to accept this incompleteness and strive for progress. This is not idealistic unless one seeks to make it complete as Rheinhold, Fichte, Shelling and Hegel did. Cassirer follows the German tradition of deeming Maimon's views an inadequate system due to incompleteness; yet Maimon viewed incompleteness and the inability to find final foundations to be a consequence of his analysis and not an incompleteness in his philosophy. Whewell also believes that there can be no satisfying solution to the fundamental antithesis: it must, however, be recognized as a fact. He deems, indeed, the central mistake of all philosophy to be the attempt to find an adequate theory of the common source of facts and ideas (Whewell 1971, pp. 300*ff*, esp. p. 309).

Stoll is thus not only wrong to introduce Maimon as a critic of Kant – and by extension of Whewell – who shows the necessity of idealism, but she chooses exactly that thinker who has been so underestimated because he, just as Whewell, rejected the traditional standards of system as unreachable. If, then, either thinker is to be rejected on this basis one must first discuss this standard of system. No one does this. Stoll appears to be influenced in her reference to Maimon by Windleband and/or Cassirer who, as I mentioned above, follows in the tradition of Lange; indeed, if this analysis is correct, Cassirer's view is the development of an idealistic interpretation of Whewell which corresponds to the idealistic extensions of Maimon by Reinhold, Fichte, Shelling and Hegel. The extension of Maimon's view by Rheinhold, et. al. and the extension of Whewell by the new Kantians are each based on the quest for completeness which is rejected by both Maimon and by Whewell as leading to excesses and incoherencies.

Stoll, we have already seen, separates the problems of discovery from those of proof. And the next two chapters separate these and evaluate Whewell in the normal way, that is, Whewell's researches in the process of discovery and the use of hypotheses are highly praised and his theory of proof is deemed inadequate due to the lack of formal proof. In regard to discovery I wish to note only point. She states,

> Whewell described the process of discovery as involving first the Colligation of Facts. By the Colligation of Facts, he meant the binding together of the facts into a connected whole through the introduction of an appropriate Idea or conception. This process cannot occur, however, until the facts have been accurately observed. (Stoll 1929, p. 65)

This last observation is misleading if not downright false. Colligation is a process in which the facts are, for the first time (more) accurately described. They cannot, then, be accurately observed before. Rather, the accurate observation depends on ideas. The inseparability of ideas and sensations expresses this dependence of ideas on sensations and sensations or observations on ideas. Since all facts depend on ideas we must use ideas to describe, better, to identify them, but we also need to "decompose facts" to make them precise and for this we need ideas (Whewell 1858, pp. 56*ff*). There is, then, no independent source of knowledge in observation which could be used in (Millian) induction.

This point is of some significance for the acceptance of the standard view according to which "Whewell attempted to analyze the process of discovery; Mill was concerned with the problem of proof" (Stoll 1929, p. 99).

This approach requires Whewell to provide a formal proof and deals him deficient when he does not provide one. Yet one of Whewell's most central points is that no such proof is possible. Rather, the method by which scientific theories are established is that process by which they are rendered clear so as to be seen to be necessary. The application of standards of proof – either Kantian or Millian – as is normally done (see Stoll 1929, pp. 91*ff*) is an inadequate method for criticizing Whewell, since it is his purpose to show that the application of these very standards is a mistake.

One point in closing my discussion of Stoll remains. She notes that Whewell in the *Bridgewater Treatise* deemed some scientific laws to be contingent which he later thought to be necessary and is, I think, the first to point out this passage (Stoll 1929, pp. 8-9; Whewell 1839, pp. 231*ff*). Todhunter, as mentioned above, deems Whewell's views to have hardly changed but on this point the change seems quite obvious and severe. This is, I think, merely of interest in regard to the question of the development of his view and not to its adequacy. The question will come up again.

Stoll has taken over the main outlines of the views developed by Herschel, Mill, Brewster and Mansel. Whewell has changed his mind and become more Platonic; his theory of ideas is confused; he has no theory of proof; his history has errors and he changes his evaluations; and his view leads to idealism. We saw above that these views were suggested but hardly developed and here we might have found the answer to our question since Stoll tried to do just that. But we find the views for the most part merely put together ad hoc and the criticism of the connection which should have been her contribution goes awry, because she ignores important aspects of Whewell's view (his view of system and his method of developing it) and simply reads him wrong on others (the proof of his own view). We still lack the proper development of the standard view.

## 4.4. Robert Blanché: Science Is Not Inductive but Confusions Remain

When we come to the study of Robert Blanché's *Le Rationalisme de Whewell* we find that in the 1930s the standard view of Whewell is finally questioned to a degree in France. Blanché offers the best hope we have of a new perspective on Whewell. Whewell, as we have seen, was clearly judged by all his interpreters to be a confused thinker, and Blanché agrees. Yet, by modern standards and especially by those then found in France, Whewell is far better a methodologist than his competitors who won the day. And Blanché, following Lalande, notes this:

> C'est en effet dans la mesure ou sa théorie de la connaissance se dégage directement d'un examen de la science, qu'elle conserve un intérêt actuel, étant beaucoup plus pres de l'épistémologie contemporaine que ne l'est la philosophie scientifique de Mill, son heureuse rivale. (Blanché 1935, p. VII)

Blanché is not merely interested in the methodology of Whewell which, he says, is so highly emphasized that one is led to neglect a second aspect of his view. This second aspect concerns his view of fundamental ideas and his theology. These never cease, he says, to harm his view. He presents his purpose so:

> Nous sommes donc attache, au risque de rendre a notre auteur un mauvais service, a fixer la signification du rationalisme de Whewell tel qu'il a été pense par Whewell. Nous n'avons pas cherche a dissimuler, nous nous sommes plutôt applique a mettre en lumière, comment la dualité de l'inspiration se reflétait dans l'ambiguïté de la doctrine. (Blanché 1935, pp. VII-VIII)

The view which he wishes to develop appears, then, to be quite traditional. He asserts that Whewell's view is inadequate, not as methodology where the tradition, following Herschel and Mill, sought to concede his points while denying him originality, but in regard to his view of truth or his alleged switch from a Kantian to a Platonic view, which is in turn alleged to have led him into incoherencies. He seeks to examine in detail the view of Whewell proposed by Herschel, developed by Mill, and made acute by Bernard according to which his methodology is excellent but the philosophy in which this methodology was presented seriously defective.

We may nevertheless look with high expectations at his book, because it offers a serious treatment of Whewell's system with a better foundation than previous attempts due to Blanché's appreciation of Whewell's criticism of Mill.

These expectations are confirmed when we see that there are two important reasons to regard the book as much more than a mere rehearsal of traditional analyses. The first reason is that Blanché recognizes the superiority of Whewell's approach over that of his opponents: Whewell has, he notes, changed the problems which the philosophy of science has to answer from those of logic or of (scientific) inference to those of method or of the history of science. The second reason is that he tries anew to develop the view proposed by Herschel in his obituary and/or by Mansel. On this view Whewell's extension of Kant's theory of a priori categories found in his theory of fundamental ideas is alleged to have led him to an incoherent view. Blanché hopes to first show what the difference between the two kinds of ideas are, that is, the difference between Kantian ideas such as space and time, which are conditions for all experience, and those ideas found in sciences, which apply to only a particular field of objects. He hopes then to further explain, as Stoll before him tried and failed to do, why Whewell cannot provide a unified interpretation of both sets of ideas with his theory of fundamental ideas: the analysis of the former set of ideas leads him to a Kantian position and the analysis of the latter set of ideas leads him to a Platonic position. His theory of fundamental ideas thus has both Kantian and Platonic aspects which he cannot reconcile.

Blanché's book seems important because it, one, is far more appreciative of what Whewell accomplished, two, seeks to develop in detail one part of the standard critique, so as to identify the point (or points) at which Whewell's system breaks down, and, three, seeks to lay the foundation for a new set of problems. He asks how one can incorporate those aspects of Whewell's views which were clearly advances – his historical approach and his view of the role of hypotheses in the growth of science – into a new and more adequate view of rationality. On the basis of Whewell's research rationality is perceived as something other than transcendental or inductive proof. We are free of the blind adherence to Mill and the doctrine of inductive proof. Blanché takes up the challenge created by Bernard's success to develop a new interpretation of the methodology, which Bernard took over from Whewell with so little interpretation as he could allow himself.

In the early part of the 20th century the standard view of Whewell in France was not peculiarly different from that found in England. Whewell was hardly noticed at all, and when he was, the standard views were taken over. Blanché points out, for example, that although Lalande praised both

Herschel and Whewell for their recognition of the importance of hypotheses, his analysis is quite inadequate due to the facts that, one, Herschel was on the side of Mill and not of Whewell, and two, their views of hypotheses and thus of knowledge were quite different (Blanché 1935, p. 78). The criticisms of Whewell have been apparently used not merely for the stated purpose, that is, to show that Whewell's system is inadequate, but also for an unstated purpose, that is, to show that Whewell's researches give no reason to change the traditional analysis of problems in the philosophy of science. But we have found no arguments which would justify the use of the traditional criticisms for this latter purpose. Let us return then to the development of Blanché's argument to see not merely if the traditional view of the incoherence of Whewell's combination of Kantian and Platonic elements can be vindicated, but also how the analysis of the weaknesses of Whewell's system leads, or does not lead, to new problems and, if there be such, what they are. This second question concerning the formation of new problems is the more important of the two, but I will, unfortunately, be able here to handle it only quite briefly, since further discussion would take me too far afield from the inquiry into the reasons and motivations of the rejection of Whewell.

Blanché has one central criticism of Whewell's philosophy: it incorporates two incompatible views of the a priori elements of thought. One view is Kantian and asserts that the a priori in thought is a pure form and a rule of thought. The second view is Platonic and asserts that the a priori in thought consists of ideas which have content – they are an object of thought – and correspond to the external world. This central contradiction afflicts, according to Blanché, each aspect of Whewell's philosophy – his theory of the fundamental antithesis of philosophy, his philosophy of mathematics, his interpretation of the fundamental ideas of science, his theory of the growth of scientific knowledge or of induction, and his theory of rationality. In each aspect he – Whewell – is forced to appeal to first one and then the other interpretation in order to complete his analysis. This central contradiction leads to specific problems in each central area of his philosophy. He cannot solve his problems in each area within his system. It must be redone with a recognition of the difference between a rule of thought or form, on the one hand, and an object of thought, an idea describing the world, on the other.

In the case of the fundamental antithesis of philosophy, the central problem is how knowledge can be both the form which we impose on the world and the idea which we find by empirical research. In the case of mathematics the central problem is how mathematics can be a form and fully necessary, when it must always have content which is empirical, that is, which is presented to or constructed by intuition. In the case of "inductive" science the central problem is how a theory can be fully necessary and yet be divisible into empirical and a priori parts. In the case of scientific method the central problem is how theories which are discovered by inventing, testing and improving hypotheses with empirical research can, in the end, be deemed ideas which are the true object of thought and which truly describe the world. In regard to the progress of reason the central problem is how empirical evidence may become matters of pure intuition of necessary truth. In the background of the discussion of each issue lies the question of the development of Whewell's philosophy, that is, the question of the truth of the standard interpretation of Whewell's later work as a Platonic addition to his earlier Kantian views. Was this addition rendered necessary by his extension of Kantian views to include all of the so-called fundamental ideas of science? Is this a source of incoherency in his views?

### 4.4.1. *Fundamental Ideas*

Blanché begins his argument with a discussion of differences in Whewell's table of fundamental ideas or of consiliences of inductions (Blanché 1935, pp. 8*ff*). The place of the central ideas of space, time, cause in this table are clear. But the places of extensions to it are not. Whewell, Blanché suggests, here becomes uncertain about which ideas are fundamental (Blanché here follows Stoll 1929, pp. 57*ff*). Blanché denies that these changes can be explained as developments or improvements of his view; Whewell, he says, does not even mind giving different views within the same text (Blanché 1935, p. 10). Rather, the indecision concerning the final list is due to the heterogeneity of the theory of fundamental ideas. It is both the final stage in the process and the development, which is never finished.

> La difficulté essentielle est celle qui tient a l'heterogeneite des idées. Whewell les met toute sur le même plan, leur attribue une nature et une fonction identiques, pace que'elles ont toutes en commune deux caractères, d'une part d'être distinctes des données sensibles, d'autre part de servir de fondement a la nécessite scientifique. Pourtant, il semble bien qu'elles tendent comme d'elles-mêmes a se repartir en deux groupes différents. Lorsqu'il est question d'espace, de temps, de nombre, de cause, de ressemblance, on songe a des conditions universelles de pensée; au contraire, dan des idées comme celles de milieu de propagation, de polarité, d'affinité chimique, de forces vitales, il est malaise de voir autre chose que des objets particuliers de pensée. C'est pourquoi l'on se demande si Whewell n'a pa méconnue, suivant l'heureuse formule de Mansel, la distinction qui sépare "les lois nécessaire selon lesquelles tous les hommes pensent, et les lois contingentes selon lesquelles certains hommes pensent a certaines choses". (Blanché 1935, p. 12)

According to Blanché in the case of laws, according to which all men think, ideas may be necessary and unchanging, but, in the case of laws according to which some men think of some objects, Whewell needs not only to allow for growth and thus change, he can not account for necessity. So, when he comes to the latter he is forced into uncertainties and has to change his view to account for scientific growth.

The difference between the kinds of ideas is, Blanché (1935, p. 13) says, characterized by three sets of opposing traits which may be posed as three questions: (1) Are ideas constitutive of a universal mental structure or the crystallization of ancient inductions, which are today more or less forgotten? (2) Are they the conditions for the thought of any object whatsoever or are they merely a determinate manner of thinking of this or that object? (3) Are they form or content, are they rules of thought after the fashion of Kant's categories or are they objects of thought after the fashion of Cartesian or Platonic ideas? Blanché asserts that in each of the three cases Whewell attributes both of the relevant characteristics to ideas, but cannot do so and remain coherent.

Whewell does, of course, deem some fundamental ideas to be universal in the sense that they are used by all men as a condition of experience and others to be applicable to some set of phenomena. Both kinds of ideas however are learned in the sense that we become clear about the axioms. But he thinks these views can be reconciled. All ideas are universal to all men in a weak sense: they can be developed and seen by intuition to be both true and necessary by intuition when men reason scientifically. This capacity to perceive the necessity of ideas applies both to those which are the conditions of experience such as space or the axioms of geometry as well as those which apply to only some objects. Whewell contends, therefore, that he has no need to choose between the alternatives of ideas being a universal, necessary mental structure on the one hand or a theory applying to only some set of objects learned by the scientific method on the other hand. All fundamental ideas are universal structures of thought in the sense that they can be developed and proven to be necessary by the scientific method, as we have seen above in the discussion of Mansel and of Whewell's response to it. They are also theories about the world. They describe the world the way it really is. Blanché has not shown here any contradiction within Whewell's theory. Rather, he has insisted that Whewell follow his identifications of those theories which are conditions for experience with those that are necessary and those theories which are necessary with those which provide form but not content. But Whewell rejects these (Kantian) identifications. His theory, should it be false, would need to be shown to be so with a different argument than that provided by Blanché.

Whewell has a simply reply to the second question as to whether fundamental ideas are conditions of thought or ideas about this or that set of objects. It is clear that some ideas – space and time – are universal in that they are applicable to all experience and others are only applicable to certain objects. It is further clear that Whewell cannot say of one idea that it is both a universal condition of thought and specific to a particular field. Whewell notes this difference and deems the former ideas – space and time – to be mathematical insofar as they are universal and not part of inductive science. He still explains that they are the foundation of science and are learned by becoming clear and distinct, as are all fundamental ideas (Whewell 1852, pp. 8-14).

Blanché suggests further that Whewell has trouble with the notion of causality since, as Herschel noted, it should be universal, but in his formulation applies only to mechanics (Blanché 1935, p. 15). In his *Philosophy,* Whewell moves from the discussion of mathematics to mechanics. And causality is the first concept discussed in the introduction to mechanics. But, although, causality is partly defined by one axiom of mechanics – reaction is equal and opposite to reaction – it is, as Whewell notes, also broader: the first axiom says that nothing can take place without a cause. Now he notes that the third axiom is borrowed from mechanics and asks whether it is universal. His answer is, in so far as cause and effect are referred to quantities, yes. It is, of course, a further question – quite in accord with his theory – to specify what these are in various cases. He shifts, then, the problem to science itself. As to the third question – are ideas rules of thought or objects of thought? – I am afraid Whewell might have difficulty understanding what the author intended. He would, I think, have deemed these expressions to refer in his theory to the same phenomena as he did the expressions "Laws of mental activity" and "Fundamental Ideas" in his response to Herschel. He there states,

> . . . if, as I suggested, instead of "Fundamental Ideas," we use the phrase "Laws of Mental Activity," then our primary intellectual Code – the constitution of our minds – will consist of a Body of Laws of with the Titles correspond with the Fundamental Ideas of the "Philosophy." (Whewell 1967b, p. 678)

Whewell deemed the so-called rules of thought to be describable as ideas: the ideas are the content of the rules we follow when we think (properly) and can thus be studied as objects of thought, though they would, of course, appear in the way men think as rules of thinking. Whewell hoped to analyze and describe these rules and thus far he is a Kantian. But his theory of these rules and his methods for analyzing them are quite different. Indeed, it is the task of mathematics and of science, not of philosophy, to discover them. The task of philosophy is to describe the process by which science succeeds in identifying fundamental ideas, that is, rules of thought. It would seem that Blanché has not taken this into account.

### 4.4.2. *Mathematics*

Blanché next attempts to show the inadequacy of Whewell's basic conception in regard to the mathematics. He argues that the foundations of mathematics cannot be found in ideas, which give mathematics it's content, but rather in axioms which are purely formal, which have no content. With the aid of then relatively recent advances in the philosophy of mathematics Blanché, hoped to show that Whewell's views were contradictory. Whewell, Blanché says, confused the notion of axiom, deeming it at the same time to be a formal principle which regulates thought, on the one hand, and a foundation which gives content – content which is brought to light in the deductive system for which it serves as a foundation – on the other hand (Blanché 1935, p. 22). We now know, he says, how to separate axioms from interpretations of them. This knowledge enables us to see Whewell's confusion.

Whewell has trouble, Blanché maintains, exactly where he expects him to, that is, in order to argue against the view of Herschel and of Mill, Whewell has to treat the fundamental axioms of mathematics as rules of thought which are found in their application to experience. In response to Herschel he emphasizes that they are developed in the course of experience, but he still cannot avoid a Kantian view of the nature of axioms in his attempt to establish their necessity. He needs to maintain that they are universal, that they are applicable to all experience.

We can allegedly see this problem in Whewell's criticism of Stewart as well. Whewell hoped to show that there is no sharp distinction between definitions and axioms, and that both are necessary. Definitions, he says, are not merely hypothetical, as Stewart claimed. But in order for Whewell to make this point, Blanché asserts, the ideas which are expressed in some combination of definitions and axioms must be something more than mere rules of thought. They must have content and be the object of thought (Blanché 1935, p. 31). We thus find just that wavering of point of view in Whewell's theory of mathematics which is typical for his basic theory and which renders it confused.

As we have seen above Whewell certainly does consider ideas from two points of view, that is, on the one hand, as rules of thought or, better, in order to use his terminology, as necessary laws of the activity of the mind, and, on the other hand, as ideas which have content and which may be studied and rendered explicit with an axiomatic method. Let us look, then, to see if the attribution of these characteristics to axioms leads Whewell into difficulties in his theory of mathematics.

Blanché suggests that because Whewell maintains that all ideas have an empirical element, ideas cannot, as on Kant's view, be simple regulatory rules or, the form and content cannot be separated (Blanché 1935, pp. 32-33). But if we accept a view of ideas as the necessary form of perception – as Whewell also does – we are led, according to a Blanché, to a purely formal conception of mathematics.

At first Blanché merely asserts this thesis and suggests that Whewell himself was not indisposed in his earlier more Kantian phase – in the *Bridgewater Treatise* – to drawing this conclusion (Blanché 1935, p. 34). The passage Blanché cites, however (Whewell 1839, pp. 335-336), to support this contention does not even discuss the nature of the axioms of mathematics. He is there concerned with the nature of inference or, whether new knowledge can be obtained by deduction, which, he claims, it cannot. Since deduction is the method of both mathematics and logic, the axioms, which give us new knowledge, must have a different source. (This passage, incidentally, is the same passage, which Herschel tried unsuccessfully to use to justify his view that Whewell changed his view concerning the sources of knowledge between the publication of the *Bridgewater Treatise* and the *Philosophy*.)

In any case, Blanché also does not believe that one can attribute to Whewell the view that mathematics is purely formal. Rather, he suggests that Whewell is led to defend a Kantian view when arguing against Herschel. He there maintains mathematical knowledge is not obtainable from observation, but is never obtained independently of observation. It must be, then, rules of thought which we apply to experience. When he answers Mansel, however, he says that he is not strictly Kantian. He maintains there that the difference between space and time, on the one hand, and other fundamental ideas, on the other hand, is merely a matter of degree: they are no longer mere rules of thought but are mixed with content (Blanché 1935, pp. 37-38). And in his discussion of Stewart, Whewell maintains that they are the objects of thought and represent real objects.

Blanché's strategy, then, is to first show that Whewell moves from one expression to the other, from one use of the word 'form' to another. And, Blanché says, this indecision of language is not merely carelessness of expression. Rather it is a sign of fundamental confusion: he cannot decide if the ideas of space and time are merely the necessary conditions for all experience or whether they also have characteristics of other fundamental ideas such as being always united with facts or sensations and/or being objects of thought which describe real objects. Whewell never saw that he had to choose between the options (Blanché 1935, p. 39). Blanché seeks to show the confusions to which this leads.

Blanché seeks to make his case that Whewell was forced into two contradictory views in his philosophy of mathematics in his discussion of Whewell's response to the view and criticism of Dugald Stewart. The analysis of this debate seems to him appropriate for the task of explaining Whewell's mistake, because he thinks Stewart defended a theory of mathematics which avoided Whewell's mistakes. According to Stewart there are within mathematics both definitions and axioms: definitions provide the content and the axioms provide the rules of inference. Blanché deems this sharp distinction to be crucial for a consistent theory of mathematics. He thus deems Stewart's view of mathematics to be by far the superior and to have incorporated a clear understanding of that point which Whewell simply did not comprehend.

The most important points in the discussion concern Kant, for mistakes in Whewell's criticism of Stewart must be shown to be more than mere weakness in his critique but based in a fundamental error (Blanché 1935, p. 48).[1] According to Blanché the supposed error in Whewell's critique of Stewart is based on a failure to properly appreciate Kant. The refusal by Whewell to accept the division of definitions and axioms, suggested by Stewart, is a result of his refusal to accept the Kantian separation of knowledge or understanding of (empty) principles from intuition of content. Whewell has failed to grasp, according to Blanché, the need for Kant's separation of the intuitions of space and time from the categories of understanding.

The problem, Blanché maintains, which Kant saw and tried, however inadequately, to solve and which Whewell allegedly cannot solve is: How can one unite the truth value of mathematics and its necessity? Or, how can we unite or reconcile traditional rational and empiricist views? In order solve this problem it may seem desirable to separate the conditions for its truth from those for its necessity and then to reunite them as, according to Blanché's analysis, Kant in fact did in his separation of the forms of intuition which supplied content from the categories which supplied mere form. This allowed Kant to treat space and time both as forms of intuition and as objects of thought without confusing them (Blanché 1935, p. 51). Whewell, on the other hand, rejected this separation and with it an adequate theory of necessity. He apparently did not grasp the problem. Ignoring it, Whewell developed a realist view of mathematics, which, Blanché suggests, became increasingly clear as he developed his view. This view, then, must be deemed a product of his confusion. Blanché suggests that Whewell had kept the worst part of Kant in retaining a vague idea of the activity of the mind and rejected the best part by rejecting his clear distinctions (Blanché 1935, pp. 49-50). This was apparently due to his failure to appreciate Kant.

Whewell maintained that no complete separation of form and content was possible in either science or mathematics. No theory built on such a distinction could account for mathematics (or science). In regard to

---

[1] Ellis defended Stewart in his review of *Mechanical Euclid*. Whewell responded in the second edition of his *Philosophy* (Whewell 1967b, pp. 101*ff*). Stewart died in 1828.

mathematics and Stewart's theory of it, Whewell argued that no attempt to carry out such a program was successful and it had been attempted in geometry. The facts of knowledge in mathematics, then, required a different view of necessity, if their foundations were to be accounted for.

To appreciate Whewell's perspective, one must note that, although Whewell understood Kant's view, he saw that it was not capable of accounting for the facts. De Morgan had complained that Whewell's view of necessity was inadequate, since it was not that concept which he deemed appropriate for logic and mathematics. And Whewell was of the opinion that, although other views of necessity were possible, no other view of necessity was adequate to account for the necessity of scientific or even mathematical statements. If one starts with a specific view of necessity and imposes it on Whewell, one may find him at fault. But Blanché needs more if he is to show that Whewell's view should have been rejected when he proposed it. He needs to show that the sacrifices of the view of necessity he favors were not then needed in order to devise a theory adequate to the facts and to his problems (Whewell 1971, p. 343) and or that it lead to contradictions discoverable with the methods of his day. His discussion of Stewart's view cannot do this job.

Whewell's problems were to explain the growth and the foundations of mathematics and of science. Whewell maintained that the facts of mathematics included the facts that mathematics is applicable to all of experience or space and time and that it is never found developed independent of experience, but does not depend on particular experiences for its development as do the inductive sciences. Mathematics must have some foundation due to its universality and this cannot come from experience. Whewell proposes that this foundation lies in the fundamental ideas of space and time, which are axiomatically developed and thereby given precise expression. The question Whewell posed for his competitors – and here his major competitor was Stewart – was whether they could account for mathematics without such a hypothesis. He says they cannot.

Whewell argued in his *Mechanical Euclid* (Whewell 1837) that Stewart's view of mathematics – according to which mathematics consisted of axioms and definitions, where axioms had no content but supplied a foundation for various systems whose content was supplied by definitions – could not account for mathematics. The prime example was (Euclidean) geometry. On Stewart's view, so Whewell, geometry should be reducible to definitions and formal axioms alone. If it cannot be so reduced, Stewart's theory cannot account for mathematics. And, he argues, it cannot. Attempts to carry out such a reduction always fail because the weakest axioms and the definitions are always too weak. If one seeks to make them stronger with definitions alone they become far too complex to serve as a foundation for mathematics. So, Whewell argues, we need both definitions and axioms. Just what we express as a definition and what as an axiom is open but both together must express some fundamental idea which is the true foundation of that branch of mathematics.

Stewart died too soon to reply but Ellis did in his stead in a review of *Mechanical Euclid.* Ellis argued that Stewart had only deemed those statements axioms which were bare truisms or the first seven of Euclid's axioms or such statements as when equals are added to equals the sums are equal and additions of other sorts may be possible. And Whewell responds quite simply: such axioms are, as he said, too weak to serve as a basis for mathematics but not so weak as they should be on Stewart's view, since they too allow for the deduction of consequences with more than trivial content. It seems to me clear that Whewell's challenge to his opponents to account for mathematics without his hypothesis had not been met.

What Blanché in fact does is to demand of a theory of science that it conform with specific views of logic or of axiomatics without asking whether any view (or any view then available) which so conforms is (or was) adequate. Whewell had argued that they were not. But this foundation of his view is here ignored. Whewell may or may not sacrifice rigor. But surely his argument that whether the opposing view is more rigorous or not it cannot account for mathematics, must be met. Or, if one has to choose between rigor on the one hand and adequacy on the other, which should one choose? Since Whewell seeks a theory of the foundation of mathematics, it is clear that he would sacrifice some rigor and admit the need for a new view of necessity.

Blanché emphasizes that Whewell becomes more and more of a realist rather than an idealist. He would, presumably, have been an idealist had he seen the consequences of his Kantian position. But from his perspective Whewell's realism was welcome and an improvement on Kant's view as it gave a better account of mathematics. In order to appraise this aspect of his view we would need an analysis of what can be gained and not merely lost by rejecting Kant's or Stewart's separation of content and form. Blanché does not consider this question. He endorses the oft-repeated observation that Whewell's conception lacks the clarity of Kant's (Blanché 1935, p. 49). But he does not explain what would have been then gained or lost by the competing points of view even where Whewell himself is clear.

Blanché finds Whewell moving in his view of mathematics toward the ambivalences between form and content which have been listed above. He endorses Ellis' argument that to treat axioms as statements which supply that which is lacking in definitions sacrifices rigor in mathematics and violates normal usage. But he does not show any direct confusion. Whewell's view of necessity might be deemed inferior to other views from Blanché's point of view. But this only begins the argument. I do not want to argue here about how Whewell's view of mathematics would have to be changed. I do not defend his view of necessity as true. But Blanché certainly does not show what he hoped to show, that is, that Whewell is confused or inconsistent due to his wavering between two views: mathematics applies to all experience, it is necessary and seen to be so by intuition, but can only be learned by experience, and describes the world truly on Whewell's view.

4.4.3. *Physics*

We might anticipate that Whewell would have even less difficulty responding to objections to his theory of physics, that is, of Newtonian mechanics, than to those of mathematics. In physics the facts would probably count even more decisively in his favor. Blanché describes this problematic as it arose for Whewell and his contemporaries quite nicely. He states,

> Soit que l'on posât ses principes comme des vérités nécessaires que l'esprit était capable découvrir par ses seules ressources, soit qu'on les assimilât a des lois naturelles que l'expérience seule avait quaite pur nous ensigner, de toute façon l'on encourait le soupçon d'obéir plutôt aux exigences d'une philosophie préconçue qu'au souci de résoudre la véritable difficulté. Toute théorie de la connaissance devait désormais tenter d'expliquer comment pouvaient s'unir dans cette science exemplaire (mechanics), les deux valeurs de l'intelligibilité et de la réalité. (Blanché 1935, p. 54)

Whewell and his contemporaries, of course, conceived of this problem of accounting for both the rational and empirical aspects of Newtonian mechanics in varying ways. There were two aspects, however, which required special emphasis in all versions. One of these aspects was the explanation of the alleged necessity of the laws of mechanics and the other was the explanation of their discovery by empirical research and/or their empirical justification. One standard approach was to explain necessity as merely strong, empirical, inductive generalizations as Mill and Herschel did. The other was to separate the necessary from the empirical aspects of mechanics as Kant and Mansel did. Whewell rejected the first on the grounds that it explained away what needed to be explained: following Kant he deemed the necessity of the laws to be a fact. But he also rejected the solutions of Kant and Mansel because they failed to explain the empirical discovery of these laws and because they could not be extended to other sciences: they accounted for parts of Newtonian mechanics only at the price of sacrificing a unified theory of science.

Whewell, then, set for himself new desiderata for a solution to the problem of how to account for both the necessity and the empirical discovery of these laws by requiring that the theory of the empirical discovery of specific laws be unified with the theory of the necessity of these same laws (Blanché 1935, p. 67). This problem, Blanché argues, led him to pose two contradictory interpretations of the laws of mechanics. He states,

> nous trouvons juxtaposes deux explications différentes, et nous nous demandon s'il faut voir dan chaque loi du mouvement une interprétation expérimentale et, partant, contingente, d'un axiome de causalité conçu comme purement formel, ou au contraire l'expression la plus simple et même, au fond, la seule concevable, de cet axiome. (Blanché 1935, p. 67)

Whewell distinguishes on the one hand, in a Kantian manner, the necessary axioms of mechanics from the empirical aspects. The necessary axioms give the form and the empirical research determines the values which these axioms must be given to render them scientific laws. On the other hand, Whewell deems the scientific laws to be rendered necessary by their ability to account for empirical events in the simplest way or in such a way as to render all other accounts inconceivable. Whewell seeks to reconcile these two apparently conflicting accounts of mechanics (and thereby of all science) in the following way. Laws which we at first know only empirically, as hypotheses, which unite certain phenomena, come to be known as necessary when they are sufficiently developed, when they are rendered simple and precise so as to explain all of the events in some field exactly. It is thus no paradox that empirical views change and become necessary laws in the course of the development of our knowledge.

Now this view, as I explain below, has been criticized in recent literature as incoherent. We have seen above that de Morgan also thought that necessary truth grows but quarreled with Whewell for not distinguishing the kind of necessity he attributed to scientific theories from that necessity exhibited by logic or mathematics. In a quite similar way Blanché criticizes the unification of this theory of the growth of necessary truths with the view of the separation of the (necessary) form of scientific laws from their empirical content: the view of the growth of necessary truths is not capable of achieving the desired unification of the philosophy of ideas and the philosophy of induction. Blanché's criticism, however, seems once more to be merely external, appealing to the necessity of the separation of form and content for a theory of necessity. As such it once again fails to offer the promised contradiction.

We may ask whether Blanché offers as a consolation prize some other criticism which would help us appreciate the reaction to Whewell or the difficulties his theory faces. We may suspect that his analysis offers such because, according to him, Whewell developed a very modern methodology but due to his mistaken understanding of logic and/or mathematics, that is, of axioms and of necessary truth, provided a poor epistemological-cum-metaphysical theory of this methodology. We could then improve Whewell's view and render it up-to-date, if we could substitute for Whewell's epistemological-cum-metaphysical theory one with a proper understanding of formal procedures and axiomatic methods. This was apparently his own program in his later work on scientific method and axiomatic methods (Blanché 1955; 1966; 1969). And this approach, were it to be successful, could be interpreted as a vindication of sorts of the insistence of Whewell's critics on not varying so far from Kant's view. The fact that this success is still questionable indicates that we are still struggling with some of those problems Whewell posed and in particular what we may need to sacrifice in epistemology to account for science.

### 4.4.4. *The Methods of Science*

The fourth theme which Blanché takes up is the theory of induction or of the method of science. Blanché correctly saw that Whewell had attempted to give a unified account of knowledge or of the status of then current theory and method. His account of knowledge, according to Blanché, has specific weaknesses which are especially due to his attempt to unify this account with his theory of the growth of knowledge. And Blanché seeks to show that it works the other way as well, that his theory of the growth of knowledge or of method or of induction has difficulties as well. But Blanché's criticism of Whewell's methodology does not concern his theory of method proper, but rather his presentation of it and the difficulties of interpreting this theory so as to find there an adequate explanation of how knowledge, that is, of how realism, is possible.

The strength of Whewell's methodology is clear and for one primary reason which Blanché emphasizes: Whewell changed the problem situation. This achievement of Whewell is his most important contribution and, if Blanché is right, we may find once again our central question: were the criticisms of Whewell used to unwisely reject this change? Blanché's portrayal of Whewell's methodology is quite straightforward and does not reflect any especially new insights. Yet it is remarkable, since all other commentators before and after him seek to explain away the central point: a theory of inductive inference has nothing to do with scientific method. They take neither Whewell's protestations nor Blanché's exposition seriously, even though the latter is referred to by virtually all succeeding commentators.

The central problem which divided Whewell from Herschel, Mill and even Mansel was whether a theory of science needed to provide a logic of scientific inference or whether a theory of method alone, a theory of how truth was discovered, even in the absence of such a logic, could be adequate. Blanché puts this so.

> A les prendre en gros, les théories de l'induction se laissent assez bien repartir en deux groupes, selon qu'elles se referant a une logique de l'inclusion ou a une logique de l'implication, selon qu'elles considèrent la loi a laquelle aboutit l'induction comme une proposition générale contenant les cas individuels, ou comme un principe permettant a l'esprit de les construire. (Blanché 1935, p. 73)

On the first view the central problem is one of proof or of indicating when one is justified in making a generalization from particulars to some general statement. On the second view the primary problem is how true ideas, known to be so, are discovered; the primary subject is not logic but how this has been done. This new approach was invented by Whewell. Blanché states,

> S'il fallait indiquer le moment ou, dans l'histoire des doctrines, apprait nettement l'opposition de ces deux manières de concevoir l'induction, c'est sans doute a l'apparition de la Philosophie des Sciences inductives qu'on devrait songer, et aux critiques qu'a suscitées, notamment de la part de Mill, la théorie de l'induction qui y est exposée. (Blanché 1935, p. 74)

The formulation of this difference led, according to Blanché, to considerable difficulties. Even though Whewell had rejected the idea of developing a logic of induction he continued to use terminology which gave the impression that he wished to do so. This was so due to the fact that at times he spoke as if a theory merely included the facts as instances. It would then seem that science would fit an Aristotelian schema and induction would be the inference from these cases to the general statements. But this was not the case. He asserted that induction always involved a new idea and this new idea did not merely include the instances which were previously known but changed them, it added something to them. He saw, then, the need to reject the idea of a logic of induction or of inference from facts to theories, since the facts depended on the theories. And this marks his originality and his deep and profound difference with Herschel who clung to the idea of a logic of induction (Blanché 1935, p. 76).

Blanché proceeds to demonstrate the gap which separates the programs of Whewell and Mill and does so just where this division begins, that is, in the theory of colligation. Following his analysis of the disagreement between Whewell and Mill according to which the real differences between them were not always clear due to the similarity of language which hid the difference in ideas, Blanché shows that, even though Whewell did at times use misleading language, the ideas which he put forth are unmistakable. In particular, when Whewell speaks of colligation he always means that a new idea is added, a new connection between facts is achieved which changes the way we perceive them. For Mill it is a mere summation of facts. And this is clear even though Whewell at times speaks as if mere inclusion were intended. This is crucial, since induction must be conceived of quite differently in the two cases (Blanché 1935, p. 105). Yet Mill never came to terms with this view. Blanché states:

> Ces critiques – Mill's arguments that Kepler's ideas, all ideas, are mere abstraction based on resemblance – révèlent avec une clarté parfaite pourquoi Mill n'a pas compris la théorie de l'induction de Whewell: d'abord parce qu'il prend pour la donnée brute des sens une expérience déjà informée par l'esprit, ensuite parce qu'il ne réussit pas a se représenter la conception, principe d'intelligibilité, autrement que sur le modele des concepts abstraits et généraux. En premier lieu, il a laisse totalement échapper l'idée essentielle de l'incorporation progressive des conception aux faits eux-mêmes. (Blanché 1935, p. 107)

Whether Mill grasped Whewell's theory of induction or not is, perhaps, not clear. That he failed to appreciate its strength and to come to terms with that and for the reasons Blanché gives, is clear: Mill never conceded Whewell's starting point concerning colligation. Yet he could not answer it either, for, as explained above, he could only offer promissory notes concerning the possibility of reduction and his view of Kepler's research could not be rendered consistent, since all observation, he had to concede, included (inductive?) inferences, yet Kepler's description of the motion of the planets should be mere description without induction.

This discussion of colligation is crucial for the discussion of induction proper, since it shows that Whewell's problem of induction is radically different from the problem which Mill hoped to solve. And Blanché shows this quite nicely. He begins by stressing that Whewell saw the formation of bold hypotheses as a crucial part of scientific research. It is not a means of last resort but bold hypotheses are necessary to begin any line of research and to make progress in it. The formation and rejection of false hypotheses is crucial for the discovery of the truth. And verification is only then possible when a theory has been clearly and distinctly formulated and this is only possible when one makes conjectures and tests them against the facts. Though conjectures themselves, then, are needed and desirable, they must be tested rigorously against the facts. The danger of hypotheses lies not in their formation and use but in their uncritical acceptance or vague formulation or loose confirmation. Success occurs, then, when we have found hypotheses which clearly, distinctly and correctly describe facts and we have especial assurance that a theory is true when we find a consilience of inductions, that is, when a new idea unites the results of two previous inductions. It is important to add that in this later kind of case, as well as in a simple colligation, a new idea is added. The new idea does not merely subsume the previous ones but adds to them and thereby changes them by incorporating them into a new and deeper theory.

As explained above Whewell has often been criticized for not offering a theory of proof. Blanché points out quite nicely that, whereas Mill sought a proof which was logically correct, Whewell sought a theory which was scientifically justified. The criticism according to which Whewell lacks and needs a theory of proof, Blanché points out, has been taken over by many commentators such as Bain, Fowler, Liard and Leuckfeld (Blanché 1935, p. 123). But this criticism misses the point, since it ignores the question of whether induction is formal proof or a method for discovery. He states:

> toute la question pendante entre lui et les logiciens est précisément de savoir si l'induction est un raisonnement ou une hypothèse, si la preuve porte sur la validité logique de la démarche inductive ou sur la vérité de son résultat, si la théorie de l'induction doit être normative ou descriptive. (Blanché 1935, p. 124)

On Whewell's view, then, although proof is never completely possible and although the distinction between probability and certainty is gradual, we may arrive at the assurance that our theories our true even though we do not have a formal logic of induction.

Whewell's view is, we should remember, based on the facts of science, that is, on the fact that colligation involves a new idea and that all progress involves conjectures with new ideas. If one seeks truth via inductive proof, Blanché points out, one sacrifices discovery.

> L'attitude qu'adopte naturellement le logicien l'invite ainsi, tout ensemble et pour la même raison, a accorder la préférence a la déduction sur l'induction et, dans celle-ci, a la démonstration sur invention, a la preuve sur la découverte. (Blanché 1935, p. 128)

Now Whewell did, as Blanché notes, attempt to meet the criticism that he lacked a theory of proof by appeal to his tables of induction. These tables exhibit the form which the development of science takes, that is, inductions which are first made are superceded by new inductions, which unite the older ones with new ideas. Whewell calls these new inductions consiliences of induction. But they should not be treated as attempts to offer a form of proof for scientific theories. And here Whewell made a mistake by presenting them as an alternative without stressing the difference between his tables and Mill's canons in an adequate way. This has led to interest in the tables as an alternative to inductive proof which is misleading. It is symptomatic, we learn from Blanché, of the difficulties which Whewell faced in clearly developing and presenting the radical new ideas which he was offering. We may now see, I think, the difficulty Whewell's contemporaries had in grasping the importance of Whewell's theories. Yet we should also not overemphasize the defects in his presentation which is in fact quite often superb.

Blanché continues his discussion of the originality of Whewell's theory and the difficulty which his contemporaries had in comprehending it with a discussion of Whewell's theory of the conditions for discovery. He points out that Whewell does not offer any canons of induction, that is, in the terminology of the day, of discovery. He emphasizes the need for intellectual education and discussion in order to develop clear ideas that are as crucial for discovery. But he stresses that he offers no rules. He also offers specific methods which are useful in the various sciences which, Blanché argues, are not new suggestions of his but descriptions of methods developed by others. Neither one nor the other offer canons for discovery.

Brewster, Blanché suggests, had a point when he complained that Whewell began by offering a new organon, a substitute for Bacon, and then proceeded to argue that no such canon was possible. He argued instead that there were conditions which promoted the growth of knowledge and that discovery followed no rules but was never entirely accidental. He proposed the quite modern view, though he did not develop it, that the "social machinery of science" was quite important. Whewell's contemporaries, then, once more had trouble understanding him. He had changed the problem once more. Instead of investigating the rules by which discoveries were made, he investigated conditions for success and these conditions were merely conducive to success: luck and creativity were also needed.

Still further, Blanché points out, this change enabled him to separate the problems of the status of science from the problem of the methods by which knowledge was obtained. On the views of induction then dominant the methods of discovery were rules which yielded proof. If these were described, there would hardly be any other problem of the status of the results. On Whewell's view the two problems had to be held apart: all methods of discovery were not proof.

One aspect of Whewell's theory of induction remains to be discussed and this is his realism. Blanché here lays some groundwork for his criticism which is to follow in his discussion of "Le Progress de la Raison." Whewell suggests that scientific research may begin by describing laws of phenomena. But it does not stop there. It seeks further answers. Thus, it is not enough to merely describe the motion of the planets. The facts thus obtained raise new and deeper problems of the causes. Newton sought to answer these. The aim of science is speculative and not practical and the importance of the application of a theory has nothing to do with its importance as theory.

Now, Blanché suggests, this view leads to a continuity with Whewell's view of the religious foundation of science. We proceed to ask ever deeper questions concerning causes and in the end we must come to the problem of the final cause or of teleology. So, the end of Whewell's theory of science, his realism and his view of science as the quest for ever deeper theories, leads him back to arguments for the existence of a God from the physical design of a universe which he had earlier presented in his

*Bridgewater Treatise.* I do not think this move is inevitable and Whewell did not think so himself as I have argued above. Yet it may have been one reason why his view was unpopular. The ideology of the Utilitarians, who sought to separate a scientific, empiricist, humanistic view from religion, and the perspective of Whewell, who sought to show how "induction" as he conceived it led to a religious view, or at any rate a view of the importance of God, would seem to make them natural opponents in spite of their common appreciation for science.

This brings us to the end of Blanché's excellent discussion of Whewell's view of induction or of method. His criticisms here are minor and concern the presentation of his views which he nicely explains were in large part due to his attempt to pour new wine – his radical new ideas in the philosophy of science – into old bottles – the language of inductive logic then current. He shows above all that the key to Whewell's originality lay in the development of a new and by far improved conception of the problem situation and that this view was never properly understood or recognized for what it was. Blanché still has criticism, however. He seeks to develop in the last section the view that Whewell's Kantianism, which is developed in his view of the nature of scientific knowledge, and his realism, which is developed in his view of induction, reveal a basic confusion in his view.

### 4.4.5. *The Progress of Reason*

Whewell developed a view of the content of mathematics and of physical science on the one hand and a view of the growth or the acquisition of knowledge on the other hand. The former is his theory of ideas and the later is his theory of induction. He separates the two in order to account for both aspects of science, but he needs just as well to unite them in order to render his philosophy coherent. Blanché presents first the theory of ideas as it applies to mathematics and mechanics or physics, second, the theory of induction and, third – in the final, major section of his book to which I now turn – Whewell's attempt to explain how one moves from empirical knowledge found in inductive research to necessary knowledge found in the examination of knowledge. Blanché argues that Whewell's attempts to bridge the gap fail.

The criticism should exemplify the basic point explained above which is, to repeat, that the analysis of ideas or, rather, of mathematics and of mechanics led Whewell to a Kantian view of the necessity of these disciplines and thus to Kantian proofs. But this view will not do for the extension of his view to other sciences. An inconsistency thus arises in his view between the adoption, on the one hand, of the theory that ideas are forms or, as Kant would have it, rules of thought and, on the other hand, of the theory that ideas are objects of thought, as Plato would have it. They cannot, said Blanché, be both.

To continue my review, this is the central argument and proof of the inconsistency of Whewell's view. Blanché accepts Mansel's contention that Whewell must adopt a Kantian view of mathematics and, he adds, of mechanics, if he is to account for their necessity without appeal to observation. And he must do that to maintain his case against Herschel (Blanché 1935, p. 168). Blanché's criticism is also in conformity with Herschel's suggestion that Whewell moved in his later years to an unacceptable Platonic position. Blanché continues, then, to argue that if Whewell fails to accept a Kantian view he cannot account for necessity, as he claims to have done. But once, or if, Whewell adopts the necessary Kantian view of mathematics and mechanics, he can no longer extend his apriorism to other sciences, since these do not have the same kind of foundation in ideas, i.e., in ideas which are presupposed in all experience, the development of which comes entirely from within, and which are seen to be true independent of empirical evidence.

Let us look than at Blanché's explanation of Whewell's allegedly inadequate attempt to bridge the gap between the two aspects of his theory in his view of the "progress of reason." In order to appraise the critique we must look once again at the alleged epistemological need for a Kantian beginning, i.e. for the necessity of pure forms which are interpreted as rules of thought rather than as descriptions of reality, and at the alleged necessity of reading Whewell in this way. We may ask whether the acceptance of a positive answer to the first of these last two questions leads to a positive answer to the second and whether, if the first answer is erroneous, the second might also be.

Blanché begins with the proposal that the gap between the two views of ideas allegedly found in Whewell may be bridged, if we distinguish in each theory between the a priori form and an empirical content. He says,

> Une première solution eut consiste a séparer idéalement, dans les principes de chaque science, une forme a priori et un contenu empirique. Les idées fondamentales qui règlent tout exercice de la pensée donneraient naissance a des axiomes, d'une évidence parfaite; mais ceux-ci resteraient indéterminés, susceptibles de recevoir plusieurs interprétations; pour savoir comment ils s'appliquent au monde réel, et pouvoir les énoncer sous forme de principes scientifiques, il faudrait donc partir de l'expérience et s'élever, par des inductions successives, a des propositions telles qu'avec elles, les axiomes fussent valables universellement. Les principes seraient alors nécessaires a leur forme, qu'ils tiendraient de la structure universelle de l'esprit, et contingents quant a leur matière, qui serait fournie par l'expérience, ou, plus, exactement, par les heureuses conjectures d'esprits individuels s'efforçant d'interpréter l'expérience. (Blanché 1935, pp. 168-169)

This interpretation of Whewell's view according to which each science has principles which are formal but which may be interpreted in various ways – the correct way to be determined by induction – must be rejected, however, because he never expressly states it in regard to the inductive sciences and because it can account neither for the absolute necessity of mathematics nor for that of mechanics.

Whewell, Blanché suggests, offers another alternative which is perhaps more promising but which is confusedly mixed with the first. On this second view there is a gradual transformation of empirical truths into necessary truths. Blanché agrees with the widely held opinion that this second view is confused; he offers a relatively long discussion of it, which seeks to show both that it is ambiguous and that it will not provide what Whewell needs in order to reconcile his views of the necessity of mathematics and of science on the one hand with the empirical nature of science on the other. To begin, Blanché points out that the theory seems ambiguous in that it can be interpreted, on the one hand, as asserting that a change in that proposition which is first empirical and contingent and then necessary takes place or, on the other hand, as asserting that a mere change in the judgment, which is made about some proposition, takes place (Blanché 1935, p. 163). This apparent contradiction, it might seem, could be resolved, he suggests, if the nature of a proposition is made relative to the person who thinks it or to the act of thought. Whether a proposition is an axiom which is necessary or an empirical truth or an empty formula would then depend on the degree of intellectual development of the thinker to which it occurred. As the thought of individuals become clearer they see the necessity of axioms.

This interpretation, though it fits some of Whewell's statements, must still be rejected, however, since, even without reference to his theology, it is clear that Whewell attributes an absolute truth value to propositions. His criterion of inconceivability of the negation of a view is meant as an objective criterion of the necessity of a view. The problem here, then, is whether Whewell can unite, one, his view of the increase in knowledge of individuals which leads to an objective truth with, two, his view of changes in propositions from contingent to necessary with, three, his view of an objective truth or his realism. All three aspects, we should note, are needed, if he is to give a unified explanation of the content of our knowledge of scientific theories as necessary and the growth of knowledge through conjectures and criticism.

Blanché poses the question of whether the necessity, which is attributed to the proposition that all matter is heavy, should be attributed to all particular laws or whether the truth of some laws may be contingent and determined by empirical research (Blanché 1935, p. 174). And, he suggests, although passages of the *Bridgewater Treatise* might suggest the latter interpretation, the former interpretation is that which must be given to his final philosophy. Whewell seeks to establish a continuity between mechanics which is necessary with all other inductive sciences. And this in turn establishes a link between the necessity of mathematics, which is the foundation for mechanics, with that necessity to be attributed to the inductive sciences. The transformation of ideas from contingent to necessary or the idealization of facts occurs in all sciences and are known in the same way as the truths of mathematics are known. Nevertheless, all scientific laws are not known to be necessary: they are not (yet) seen with the necessary clarity and distinctness.

This view enables Whewell to explain the unity of all fundamental ideas. The difference between them is, on this interpretation, merely a matter of degree. It enables him further to demarcate his view from views such as Descartes which begin with axioms prior to empirical conjectures and from Schelling who ignores them. But it has inadequacies as well. It is disappointing since we find the truth by accident (Blanché 1935, p. 181) and it raises the problem of why reason should accord with experience. Blanché states,

> Si la théorie du progresse de l'évidence assure ainsi la liaison entre la philosophie de l'induction et la philosophie des idées, entre la connaissance que l'esprit demande a l'expérience et celle qu'il tire de lui-même, en revanche, en reconnaissant aux sciences de la nature une tendance a l'intelligibilité parfaite, elle pose de façon plus pressante le problème de tout rationalisme, qui est d'expliquer cette union, dans la science, des deux valeurs de l'intelligibilité et de la réalité ou, en d'autres termes, l'accord de la raison et de l'expérience. (Blanché 1935, p. 181)

This problem of explaining why reason should accord with experience or of why the ideas we impose on nature should correspond with nature is, Blanché proposes, not adequately solved by Whewell and cannot be. The reason for this inadequacy is that if one makes the foundation of each science a synthetic a priori judgment, then one needs to justify this judgment in each case with a transcendental argument as Kant did. At this crucial point in the argument Blanché uncritically appeals to the criticism of Mansel (Blanché 1935, p. 184). He maintains not only that Whewell has not given such a demonstration, but that he could not without abandoning his historical method for the determination of the fundamental ideas. Further, such a proof is appropriate for an idealistic position but not for the realism of his later years. And still worse, his appeal to God or to theology lacks any ontological arguments.

There is much that is confused here so we need to review a bit. We may start with an appraisal of what Whewell has shown with his historical approach. On his view he has shown that in fact various sciences have at their foundation fundamental ideas which are not derived from experience but imposed on it. Further, these ideas are in fact justified by their ability to order the facts clearly and distinctly. Now Mansel and Blanché demand that there be, in addition to this justification, another transcendental deduction. But Whewell does not need this. He does, in order to be consistent, need to justify his theory of justification and he does so by applying it to itself: his theory of science can account for the history of science clearly and distinctly and no other theory can, hence it is justified. That this was his intent is clear from his discussion of methodology as a progressive science just as other special disciplines. Mansel and Blanché ignore this alternative and demand quite dogmatically that Whewell use Kant's approach.

Now it may seem that Whewell has invited this criticism and misunderstanding by offering a transcendental argument himself of the principles of mechanics but not of the principles of other sciences. Yet, as Blanché has just argued and as Whewell said, the difference is a matter of degree. Why, then, should mathematics and mechanics be special? Why should we use one kind of proof for the foundations of one science and a different sort for the foundations of other sciences, when all sciences are deemed to the same? There is an illusion here, for, the proofs are not different, though the ideas are. The foundations of mathematics are applicable to all experience. We see these ideas clearly and distinctly – when we know sufficient mathematics – as applicable to all experience. And this universality renders the argument transcendental, that is, it shows the conditions for any experience. But the nature of the argument in Whewell's system, the quality which makes it binding, is that it shows that observations are clearly and distinctly ordered by (fundamental) ideas. In the case of other sciences we have, by extending the same method of proof, proofs of the truth and necessity of other ideas which apply to some observations but not to all experience and thus may appear to be arguments of a fundamentally different sort. On Whewell's theory of scientific proof, however, both kinds are of the same sort, both kinds rely on the discovery of clear and distinct ideas which precisely order observations found in (all or only some special) experience. (See discussion of Mansel above)

If we grant Whewell that he has shown how scientific theories are proven to be true, we may ask why this is the case; why does it happen to be that clear and distinct ideas in fact describe the world as it really is? This question is raised by his study of the methods used in science. But it is not an attempt to justify this theory as Blanché mistakenly sees it. It is an attempt to complete a philosophical system. And this completion can be, as Whewell explicitly states, accepted or not (Whewell 1971, p. 358). Those thinkers who do not believe in a God or who are not convinced by Whewell's arguments for his existence may still accept and, indeed, should accept his proof of the way in which scientific theories are discovered and proven. This proof is based on the facts of science alone.

Blanché continues his argument in a quite ingenious way. And he does succeed in bringing together the theory of progressive evidence and the theology. The theory of evidence, should, he argues, show how scientists move from the discovery of contingent laws to the proof of their necessity. But there is no proof of this to be found in the historical studies. We thus need to seek it in his theology. He here appeals to de Morgan's criticism according to which there may be a certain moral necessity to the laws of science – if they were to be different the world would be inferior – but no absolute necessity, no impossibility that the world could be otherwise. This weaker sense of necessity, Blanché suggests, fits his earlier work in the *Bridgewater Treatise*, but not his later work in which absolute necessity is quite clearly portrayed as the result of scientific research. The problem Blanché poses, then, is whether Whewell's theology, that is to say the only ontology which he offers, is capable of explaining this move from contingent laws to necessary truths.

Blanché argues that on this cornerstone of his philosophy Whewell compromises his view. He cannot decide whether the laws of the universe are created because God is good and, because he is good, designed the world in the best possible way, that is, giving it a moral but no ontological necessity, or whether the world had to be the way it is, whether the developments of the world are the necessary thoughts of God. And he has found in an unpublished version of the *Plurality of Worlds* a clear statement of this pantheistic version (Blanché 1935, p. 188). In order to provide an interpretation of his theory of progress which leads to an absolute necessity he needs the pantheistic version; in order to maintain his Christian theology he needs the former view. And he cannot decide so he speaks now one way and now another.

Blanché has found here a real difficulty for Whewell. Given his theory of the progress of knowledge from contingent observations to laws to necessary truths there is an ontological problem, recognized by Whewell, which is to explain the unity of ideas and facts. Whewell asserted, of course, that such a problem could never be fully resolved and analyzed the defects of German systems as due to the attempt to do so (Whewell 1971, pp. 300*ff*). He does seek to give a satisfactory explanation of his realism in his theology, but his theory of necessity is, as Blanché shows, inadequate for the purpose of his theory of science and his theology together. His theory of science, however, contains merely one view which is that scientific theories are not merely morally necessary, but absolutely necessary. And this view, I contend, is according to Whewell justified by his historical studies. The theology is not the cornerstone of his philosophy but a further development. Blanché makes here a mistake which he elsewhere warns against, that is, of reading Whewell as a deductivist who offers the highest principles as the basis for all, rather than as a thinker who works from narrower theories to higher ones: the deeper theories are built on the lower ones and difficulties in the deeper theories do not show the shallower ones to be false: a discovery of a contradiction in Newton's theory would not show that Kepler's was false even though if Newton's theory is true, it offers a foundation for Kepler's.

Now de Morgan offered an external criticism of Whewell's theory of science which indicates that the problem of explaining the necessity of principles exists for his theory of science taken alone. But this is a different criticism. It is based on the observation of the weakness of Whewell's argument from the inconceivability of any alternative to a fundamental idea. De Morgan argues that this is, as a matter of fact, false. Scientists such as himself do not find such inconceivability and therefore a weaker sense of necessity would be appropriate for scientific theories. (See above.)

One further point should be mentioned. Blanché notes that Whewell is unclear in his view of the progress of knowledge when he speaks of the change of an empirical law to a necessary truth. Does the law change in content or do we learn something new about the law or is there some other alternative? Whewell often presents his view in a way which brings no particular problems, that is, a view which is known to be true on the basis of empirical research is later known to be necessary. Thus he says, "Now in the progress of that exact speculative knowledge which we call Science, Facts which were at a previous period merely Observed Facts, come to be known as Necessary Truths" (Whewell 1971, pp. 302, 347*ff*).

Whewell was, as explained above, forced to the view that truth changes since scientific theories are true and they change. This view was criticized by Mill since contradictory theories cannot both be true. The same view begins to find its place here in the theory of the increase in knowledge: views which are known as contingent truths are apparently deemed by Whewell to be the same as those which are later known to be necessary and also to be different, since contingent truths are different from necessary laws.

This apparent difficulty is not serious, however, since an explanation of the change is at hand which may remove it without harming Whewell's theory of the growth of necessary truth or its nature. Truths which are known empirically, that is, to truly describe observed events, are later seen to be necessary truths, that is that no other description is possible. The change in the propositions is a change – a deepening – of our knowledge of them. We now see that statements whose content we understood and which we knew were contingently true are necessarily true.

Blanché began his study with a quite clear interpretation of the difficulty which Whewell faced: he wavered between a Kantian view according to which the laws of science were mere rules of thought and a

Platonic view according to which laws of science are objects of study which describe the world. This interpretation of Whewell's work was developed quite quickly, above all by Mansel and endorsed from a different perspective by Herschel. Mansel protested against the extension of a Kantian analysis of ideas beyond Kant's was improper. (One might better say beyond Mansel's view as Kant included mechanics and Mansel did not as Erdmann pointed out, arguing that Whewell was more Kantian than Mansel). Blanché takes this view further in that he recognizes Whewell's central contribution to the philosophy of science: he changed the problem which should be solved from one of the analysis of scientific inference to one of the description and improvement of scientific method. Blanché acknowledges Whewell's demonstration of the need for such a change. But he does not succeed in offering a newer and better development of the old analysis of Whewell's theory. He does not show that Whewell ever questioned realism, for example. The metaphysical dilemma, which, as Blanché shows, Whewell finds himself in, is not between Kant and Plato but between his Christian view of God as creator and his Spinozistic or Leibnizian view of the world as necessary. Still further, Blanché fails to note the nature of Whewell's own view of progress and justification as it applies to his own view and thus misinterprets the highest level view as the final success or failure of the whole when difficulties at this level should, on Whewell's view, be interpreted as difficulties of building on firm results established at a lower level and not as the test of these results.

In spite of such deficiencies Blanché's study is the best there is. This is for two reasons. The first is that he has recognized and identified the importance and the necessity of Whewell's reformulation – or one like it – of the problems of the philosophy of science. The second is that he is the first and only commentator on Whewell to take Whewell's system seriously. This systematic study, which contains interesting passages not commented on here, is the proper foundation for modern studies of Whewell; it is apparently well-known but it has not had much influence as the following discussion will show. Only through a study of Whewell's system and the facts it was designed to account for can one appreciate the depth, difficulty and interest of the problems which Whewell attempted to solve. To find him making mistakes is too easy if we do not see the problems which lead him into such difficulties. And this is necessary if we are to understand how and why problems which we now try to solve came to be formulated or misformulated as they now are.

Ironically, Blanché takes all this into account in his rejection of Mill's criticism of Whewell. He fails to see how it applies to Mansel's. For, Whewell was attacked from both sides, from the Kantians and from the empiricists. Blanché sees that Mill had failed to grasp Whewell's innovation, because he failed to see that his standards for proof were simply not applicable to science. But he fails to grasp that Whewell had similar arguments against the Kantians and in particular against Blanché's desire to separate form and content. Blanché does not wish to sacrifice the rigor which an axiomatic treatment of mathematics and science will offer any more than Mill wished to sacrifice inductive proof. But in each case the problem is the same: is it possible to account for mathematics and science without such a sacrifice? Blanché insisted that the sacrifice should not be made even though no alternative which avoided making it is available. He did, I should point out, try to develop just that in his later writings which cannot be examined here.

## 4.5. Germany: Whewell Forgotten and Ignored

Although Whewell's *History* appeared quite promptly in Germany, there are very few German commentaries on his work. He seems nevertheless to have been relatively well known, as the letter from Humboldt and references to him by Wundt and Külpe indicate. In the 20th century he is forgotten and ignored. We find very little indeed and what we do find is not encouraging. There is only one book-length study of Whewell in the literature. This is by George Curry Seward, an American writing in Germany. This may raise our doubts as to its value for a comprehension of the reaction to Whewell in Germany but, since it is often found in Whewell bibliographies, perhaps we should look at it as well.

### 4.5.1. *George Curry Seward: Vulgar German Prejudices*

The essay by Seward, *Die theoretische Philosophie William Whewell's und der Kantische Einfluss*, is the only study of Whewell in Germany which might be deemed comprehensive. It is thus hard to simply ignore it in the search for solutions to the puzzles concerning Whewell, even though we may have misgivings about it right from the start. It was a doctorate; it was published in 1938; and it was written by an American who claims, however, that he wishes to contribute to the investigation of the national basis of philosophy, which was then, he says, quite properly emphasized. His sympathy for German philosophy is extreme as he states, for example,

> Die Tatsache, dass er (Whewell) das Wort Idee (idea) in einem anderen als dem Kantischen Sinne gebraucht, beweist nichts gegen diese Behauptung (that his view of the sources of knowledge is influenced by Kant). Es zeigt nur den Mangel an genauen philosophischen Ausdrucken in der englischen Sprache. (Seward 1938, p. 20)

Seward, we find, is of the opinion that Whewell took over more from Kant than even he realized. But Whewell's incapacity to overcome his English tendency to empiricism prevented him, according to Seward, from adhering to the proper Kantian categories and this led him to contradictions. Seward's analysis, then, is in this respect, quite standard, at least in the general point of view. He states, for example, in regard to the process of induction, which as methodology he describes tolerably well,

> Kant hat gezeigt, dass diesen Verstandesbegriffen selbst Objektivität zukommt. Whewell hat dies aber nicht vermocht. Die subjektive Notwendigkeit und Allgemeinheit der Wahrheiten sind und bleiben bei ihm bloss subjektiv. Dies führt ihn zu weiteren Erwägungen, zu einem Versuch, Objektivität auf eine andere Weise zu gewinnen. (Seward 1938, p. 54)

If one goes beyond the limits set by Kant, Seward maintains, one loses the possibility of proof. He repeats the story that the earlier, allegedly Kantian view and the later, allegedly Platonic view are different (Seward 1938, p. 62). Seward finds there confirmation of his view that he was right in saying that the knowledge Whewell describes would be merely subjective. I presume that Seward means that this knowledge would be subjective without reference to Whewell's theological view. Kant has shown us, Seward claims, that laws are objective because they are imposed on the world, but that Whewell's understanding of Kant was too weak to grasp this. He was, alas, a mere Englishman. He closes his essay as follows.

> Whewell war kein Systematiker im Aufbau seiner Lehre. Er nahm nach Belieben diejenigen Gedanken, deren er bedurfte, aus dem Kantischen System und setzte sie, ohne ausseren oder inneren Zusammenhang, in seine eigene Lehre. Er fuhlte nur den Zwang zu einer ausseren Systematik, keinen Zwang aber, hier oder da einen inneren Zusammenhang zu bewirken. Das Ergebnis unserer Betrachtung lasst sich also folgendermassen ausdrucken: Whewell hat abgesonderte Gedanken Kants gebraucht und in seine Lehre ubernommen, um den Skeptizismus zu widerlegen und die Exaktheit der naturwissenschaftlichen Erkenntnis zu beweisen, aber ohne die Voraussetzungen und Folgerungen, die diesen Gedanken ihre Bedeutung im Kantischen system geben. Denn bei diesem englischen Philosophen hat der Kritizismus den einheimischen Empirismus nicht zu uberwinden vermocht. (Seward 1938, p. 68)

It might seem that it might be desirable to dwell a little longer here to seek some insight. But the commonness of the charges, the failure to find Whewell's protests to the contrary, and the nationalistic tone of the work all seem to indicate that it would be better to use our time in some other way. We might even ask how this work has become part of the literature referred to on Whewell. Of course we might conjecture that the desire to be comprehensive in references and the paucity of the literature on Whewell give even such a work as this a place, as is here the case. We must wonder all the more at the decline from the adventurous and path-breaking work of Whewell to scholarly presentations of somewhat vulgar prejudices about him.

### 4.5.2. *Modern Germany Ignores Whewell*

We have ended our study in Germany in the 1930s. We might wonder if things have changed, but there still seems to be silence in Germany about Whewell. One example of how unknown Whewell's arguments and position are, might be given, however. In a recent article Ulrich Charpa has discussed the content and to some degree the importance of Herschel's *Preliminary Discourse*; he has also defended it (Charpa 1987). He notes that Herschel has been criticized in the 20th century (Agassi 1971b) for his failure to recognize that observations are theory-impregnated. Charpa seeks to explain why this (modern) objection is true but not serious: observations have a foundation in a broader framework, which is common to scientists. And in this connection he quotes Whewell as apparently on his side without even noting the fundamental difference between Herschel and Whewell on this very point (Charpa 1987, p. 130). And later he asserts that "Mit dieser Unterscheidung von Entdeckung und Rechtfertigung schafft Herschel Raum für die Hypothese und erhellt die bis heute interessante Möglichkeit eines nachbarschaftlichen Friedens zwischen induktiver and hypothetisch-deduktiver Methoden" (Charpa 1987, p. 145).

Charpa has surely captured here one of the most powerful, if not the most powerful, points of view among philosophers of science today. But why do we still not find, even among those whose many references show that they have studied the historical record, even a mention of the difficulties that Whewell raised for exactly this program, which was put forth by Herschel and by Mill?

# PART THREE

# THE IMAGE REINSTATED
# THE REALITY COVERED OVER

$49.95

[Philosophy]. Wettersten, John (Author); Bell, James A. (Editor). **Whewells Critics: Have They Prevented Him from Doing Good? (Poznañ Studies in the Philosophy of the Sciences and the Humanities 85)**. Rodopi, 2005. ISBN: 9789042016446. Book cover has light rubbing and light sunfade to spine, otherwise looks bright and sharp. Binding a tad wiggly/separating from spine, but still solid. Pages faintly yellowed along edges, otherwise clean and crisp. Overall a largely clean and sturdy copy. Very Good. Hardcover. [1034447]

## CHAPTER 5

## THE RETURN TO OLD MISCONCEPTIONS

Beginning in the 1950s a new interest in Whewell's philosophy of science became evident in the English-speaking world. This interest has obvious roots. It was then becoming clear to some how modern Whewell's approach is. Not only do his methodology and his historical approach anticipate current interest; his aims do so as well. Whewell sought to develop a new theory of justification in science which could account for the actual practice of science. Those thinkers who work in the tradition initiated by Whewell – Duhem, Meyerson, Polanyi, Popper, Kuhn, etc. – have sought to account for the history of science and have abandoned the project of finding a theory of (inductive) scientific inference. And it is just these thinkers who in the minds of many have offered the most challenging new ideas.

In sharp contrast to the Whewellian, historical approach to the philosophy of science, adherents of the competing and dominating tradition – the tradition which seeks to build on the foundation developed in the 19th century by Mill – have sought to develop a theory of justification by constructing a theory of inductive scientific inference alone. They have largely ignored the history of science. But when they could no longer do so, they were forced to return to the roots of this tradition, to Whewell, to respond to the demand to explain the relations between their studies in the logic of scientific inference or inductive logic and the history of science.

We find several stages in the development of these studies. Since Whewell had been so thoroughly forgotten, it was necessary to once again introduce him to the philosophical public as well as to review and bring up-to-date the difficulties which his philosophy faced. Since the Whewell Renaissance was forced on the dominant tradition above all by the often despised philosophy of Popper, there was of course considerable ambivalence. Whewell was, indeed, more important than had been hitherto thought, but his confusions should not be underestimated. I discuss this first stage in Chapter 5. In the second stage attempts were made to integrate Whewell's methodology with a theory of inductive inference, thereby integrating historical with inductivist approaches to the philosophy of science. I discuss these efforts in Chapter 6. This work was complemented by reappraisals of 19th century philosophy of science, which have been primarily designed to explain away, in accord with the tradition, the deep differences between Whewell and his inductivist opponents and/or to reaffirm the rejection of Whewell's philosophy by showing the advantages of inductivist alternatives. I discuss these attempts in Chapter 7.

Whewell may seem to some to lay between these schools and especially to those influenced by the traditional interpretation of him. Although it was he, as Blanché pointed out, who introduced the great change in the philosophy of science, he remained close to the traditional view in some of his modes of expression. The dominant inductivist tradition has sought to reconcile his methodology with their program. In the 20th century it has once again seemed interesting to take up the task which Herschel and then Todhunter proposed, that is, the task of reconciling the views of the deductivist Whewell with the inductivist tradition. We may turn, then, to studies of Whewell which seek to take up this old task once again to see if they have any more success than previous attempts, to see how and if these two traditions might be unified. We may hope, thereby, to find here a better appreciation of Whewell's philosophy than was, at any rate, offered by his contemporaries, since new efforts have been made to improve the theory of induction by using some of his ideas.

We might also, of course, be skeptical and should not be surprised if these studies should be very disappointing. We need to find out, if they consider Whewell's powerful reasons for changing the problem situation or whether they consider his view systematically. Do they fall back into the very old technique of evaluating him in accord with external standards supplied by Kant, that is, by Mansel, and/or by Herschel or Mill? Are old arguments rehashed or do they take account of Blanché's study, for example, who had already shown that these arguments are based in superficial readings of Whewell? Is the standard view that Whewell does not offer an adequate theory of proof, where proof means either a transcendental argument as Kant conceived it or inductive argument as Mill conceived it, repeated here? And do they also take over the old analyses according to which Whewell has a Kantian philosophy and a Platonic philosophy and switches in the course of his career from one to the other? We may hope, of course, to find that they develop these arguments at least as good as Blanché, though even he fails to make them convincing. Finally, do they fail to grasp the central point of his methodology which Blanché explains so well and which makes the quest for inductive proof in science superfluous: each new theory is based on new ideas which change the empirical base even while making it more precise.

## 5.1. Curt J. Ducasse: The Modern Reintroduction of the Old View of Whewell's Philosophy

The first recent English language study of Whewell noted in the literature is Curt J. Ducasse's essay "William Whewell's Philosophy of Discovery" (1951). Ducasse primarily saw his task as that of providing a summary of the view of a relatively little known thinker and especially of the controversies with Mansel and Mill. The task he set himself and the nature of the discussion show how fully Whewell had been forgotten. The summary which Ducasse offered is in some respects not a bad one. He points to three central traditional problematical aspects of Whewell's theory, i.e. his extension of necessary truth to include truths other than mathematics and mechanics, his view that necessary truths may be learned from experience, and his lack of a theory of inductive proof or inference from particular statements to general ones. His discussion is somewhat weak; he seeks to report on arguments for or against Whewell's philosophy, but it is not always clear what he thinks. In the end he adopts the standard view without clearly saying that he does that.

Ducasse first explains the fundamental antithesis of philosophy and Whewell's methodology. He uses the former as a basis for his discussion of Mansel, whose view he endorses, that is, the a priori ideas need some deduction which Whewell does not give. The discussion of Whewell's methodology introduces the dispute with Mill. Here Ducasse gives the appearance of merely describing the two positions, in particular the respective interpretations which they offer of Kepler's research, taking no side in the debate. But by allowing Mill's interpretation as acceptable, he sides implicitly with him: he concedes that scientists may proceed using inductive inference of the type described by Mill. And that is just what was at stake between Whewell and Mill.

Ducasse's discussion hardly goes beyond a brief report on these old debates. In his conclusion he attempts an analysis of Whewell's importance and gives three reasons. First, he says, Whewell proposed that propositions and definitions are two handles by which knowledge grows. This assertion is, however, so vague as to be useless. Secondly, he points out quite correctly that Whewell offered the first philosophy of science which was based on wide and sophisticated studies in the history of science. This was indeed of crucial importance. The third reason is that he provided the first deductivist philosophy of science. This seems quite false, since we cannot forget Descartes, Maimon and others. Yet he was the first to develop a deductivist view and to explain how it could account for the history of science. Ducasse thus reminds his readers of important aspects of the discussion of Whewell. But it is a mere reintroduction of a forgotten view, which starts with the old misconceptions; and some of these had already been corrected by Blanché. Ducasse, unfortunately, ignored this and thereby set the stage for their repetition in the new context.

## 5.2. E.W. Strong: The Modern Reintroduction of the Old View of Whewell's History

The traditional interpretation of Whewell had recognized the importance, originality and insight of Whewell's historical observations. But on this view these historical observations were combined with a completely inadequate philosophy. E.W. Strong (1955) sought to present and develop this traditional view once more in an essay whose publication shortly followed that of Ducasse. Strong deems his essay to be complementary to Ducasse's (Strong 1955, p. 210). Whereas Ducasse was concerned with Whewell's theory of induction, Strong is concerned with his history. In keeping with the tradition Strong deems Whewell's philosophy confused – he endorses the criticism offered by Blanché (Strong 1955, p. 215) – but he deems Whewell's historical claims to be correct. He wishes to show that Whewell not only had the upper hand in the dispute with Mill, but that Mill had not even grasped Whewell's point of view. He states,

> Thus he (Mill) did not see that he had missed a new philosophy of science, an historical or dialectical rationalism. He did not see that this historical rationalism was no mere repetition of Kantian doctrine but an attempt to conjoin a doctrine of a priori relations with a dialectical progression of scientific knowledge. (Strong 1955, p. 226)

This is not, as it might appear, an introduction to a defense of Whewell's philosophy. Rather, Strong has two tasks. One is to explain how and why Whewell's history and in specific the interpretation of Kepler's research is superior to Mill's. The second is to separate this success from his confused philosophy. Strong seeks to set the record straight by presenting Whewell's philosophy of science as a philosophy of intellectual history (Strong 1955, p. 225). This philosophy of intellectual history shows how progress is dialectical,

> in any epoch, a further major step of discovery is not merely additive to extant theories, but is the offspring of a new conception coming into conflict with a prevailing point of view . . . No further step is made by any man unless he is first prepared in the method and knowledge of a science as constituted at the time and place of his study; and it is from the conditions and particular problems of his epoch that a new conception may be forthcoming to provide explanation of observations and experiments. (Strong 1955, p. 225)

This point of view is thoroughly misunderstood by Mill, according to Strong. This is evident from the fact that Mill first seeks to translate all of Whewell's terms into his own language, thereby changing the import of the theory, as well as from the fact that he ignores Whewell's claims that Mill's methods were not in fact used by scientists in order to argue merely that they are acceptable methods of proof.

Though Strong is no doubt right that Mill did not answer Whewell adequately, his argument is flawed. The central issue between Whewell and Mill was whether the translation which Mill proposed was possible and they argued about this, that is, whether generalizations were mere abstractions and/or inductions or whether new ideas were in fact added to the observations in scientific discovery. Secondly, Strong suggests that Mill's methods were in fact methods of proof which could be used after the discovery of theories to test or prove them (Strong 1955, p. 229). He fails to note that Mill said just that in his theory of ratiocination and that Whewell's criticism hit this view as well. Strong wishes to save Whewell's theory of history, on the one hand, and the standard view of inductivism, on the other hand, just as the tradition would have it, but already gets into difficulties by not noting the scope of Whewell's criticism. Whewell offered not merely a philosophy or theory of the history of science but a competitor to Mill's theory of proof as well.

The central difficulty which Strong faces, however, in his attempt to save Whewell's history or philosophy of history arises in his attempt to explain the inadequacy of his philosophy of science in a way which can be reconciled with the praise of his philosophy of history. He here merely follows Stoll's view according to which "Whewell was already sure in his philosophy that the history of science would exemplify and illustrate his rationalistic doctrine of universal, necessary relations – his belief that there is an order and connection of things." Strong continues, suggesting that

> One finds Whewell shifting from historical investigations of theories derived from facts and of controversies through which principles receive classic definitions, to demonstrations beginning with principles which the history merely serves to illustrate. Interestingly enough this subordination of history was condemned by Whewell when he found it in Mill. (Strong 1955, p. 222)

What, then, saved the day for Whewell? He quotes Stoll who says, "However, in his more detailed statements concerning the means by which scientific laws have been discovered and established, Whewell drew his conclusions from his investigations of the actual historical course of progress" (Strong 1955, pp. 222-223).

The repetition of the story that Whewell was a dogmatist seems to me really unfortunate. Todhunter suggested that Whewell unwisely responded too quickly to criticism and Blanché, also following Stoll, attributes a certain stubbornness to him as well. It is virtually impossible, however, to avoid such a reputation when one fights against a majority opinion. It seems to me hardly fair to so label someone who argued with such force, insight, originality and adventurousness. The plausibility of the charge stemmed, I think, from his defense of religion when his major opponents, above all Mill, were seeking to develop a humanistic philosophy which did not need, even if it tolerated, religion.

There is a more serious problem, however. Strong endorses Whewell's view that new ideas are needed for progress in science. This is, he thinks, a correct view of history. But in the philosophy of science, which Strong himself uses in discussing Whewell's work, Strong resorts to the view of Herschel and of Mill: theories need to be extracted from the facts. So, Whewell is, according to Strong, allegedly correct in his rejection of Mill's methodological theory of science, because, in the study of science – as Stoll had also suggested – Whewell did not follow his own correct view of science, according to which he should make daring conjectures, but the incorrect view of Herschel and Mill, according to which one should simply look at the facts. Whewell, of course, knew that he was applying his own view to science. And that he was doing so critically is without question. The absurdity of Strong's analysis, however, shows the difficulty of even coherently presenting the standard view of the success of Whewell's history and the failure of his philosophy. No one else has tried to do it. The reason, one may conjecture, is that this standard view, which was proposed by Herschel, seeks to avoid treating Whewell as a systematic thinker and one cannot develop it without taking Whewell's system seriously. Strong's attempt seems to be ignored in the succeeding literature.

## 5.3. A Methodological Interlude on the Difficulty of Coming to Terms with Whewell

The new endorsement in the 20th century of the misconceptions of Whewell's philosophy developed in the 19th century might be seen merely as an attempt to cling to an out-dated program, even though the defense of this program could only be successful if many new and difficult problems could be solved. This explanation does no doubt explain this phenomenon to a degree. But even though dogmatic clinging to out-dated positions plays a significant role in intellectual institutions, it is also useful to ask what the problems or misunderstandings are which make such clinging possible. The answer in this case lies in the theory of criticism. Whewell and his critics thought they shared common standards of rationality and criticism, which they could use to debate the difference between his view and those of his competitors. But this was not quite right. On the traditional view all rationality was proof and proof was either deductive or inductive or, perhaps, intuition as a basis for deduction. But Whewell rejects all these alternatives as incapable of accounting for the facts of justification. This was not only too much for his contemporaries. It is still too much for most of his commentators today, even though new views – especially Popper's and those who have developed so-called non-justificationist views – offer even more radical changes. These changes pose the question of how we should rationally debate changes in our standards of rationality. The debate never progressed to this level. And even Blanché, who shows a deep appreciation for Whewell's philosophy, in the end fails to note Whewell's use of those new standards of proof he discovered in his studies of science to justify his own results.

There is a further point. Traditionally criticism has been primarily directed at the study of whether a theory was true or false. It is prima facie an all-or-nothing affair and, if a theory is shown to be false, we may reject it. Given this standard it was merely necessary to find some difficulty in Whewell's new view in order to reject it. This is, of course, too easy, since a view may be rich in new ideas and highly influential, even though it is false – as Whewell's was. So, there is a place in the history of philosophy for the study of influential and/or interesting mistakes. But in order to gain such a status it is necessary to have followers and/or thinkers whose debt is obvious. This did not happen in the case of Whewell and as a result there seemed little reason for studying him: his view was false, confused and had little influence.

This seemingly simple view of the situation overlooked, however, the fact that Whewell had changed the problem situation drastically. There are at least two reasons for studying Whewell which have been overlooked. The first is his historical influence through the change in problems. The second is the study of these problems themselves. How and why did they arise? How did his formulations lead to contemporary problems? How did the suppression of his views lead to contemporary problems? Are they good or bad? These questions require us to change our interests and methods of criticism. For, the question of truth or falsity loses interest. In its place we have questions of the importance of Whewell's problems and what we can learn from Whewell's difficulties in solving them. Can we, for example, learn from his mistakes so as to pose them better? This approach requires a systematic study and an evaluation of his research against the background of his radically new problems. It has not yet been carried out, since all critics have worried about justification and used criticism primarily to refute without going any further to ask why the difficulties they found arose. This has had the effect of blocking the (re)formulation of these important problems. Now they are coming to the fore once more and the old criticisms, however, used to block them once again. They require a switch to deeper problems of rationality, of how we rationally debate changes of standards of rationality, and of the purpose and methods of criticism in the absence of justification.

## 5.4. Harold T. Walsh: Touching Up

Harold T. Walsh made the next attempt to develop a new, more modern view of Whewell. We see with some disappointment, however, that he proceeds in the normal way. He published an essay on induction (Walsh 1962b) which followed an essay on necessity (Walsh 1962a). The essay on induction made no mark. In this essay Walsh argues that Mill's criticisms do not hit their target, since they erroneously presume that, if ideas are imposed on the world, they cannot conform to it. And this is just that which Whewell denies. And, he suggests, the central issue between them concerned just what was given to us through sensations and what was supplied by the mind. Since, he continues, we have no means of deciding this issue, the question of method is the central and neglected aspect of this debate. This study certainly moves in the right direction. But it stops with the suggestion that the problem of method or of decision procedure is important. His view that we cannot decide whether the source of all knowledge is found in sensations or whether the mind is a second source is no doubt too quick. This discussion was never, so far as I know, taken up again by Walsh.

Walsh's essay on necessity was an attempt to respond to a criticism of Whewell, offered by Ducasse, according to which Whewell's notion of absolute necessity was inadequate. Ducasse had proposed that all necessity had to be relative and that Whewell had not taken this into account. Walsh agrees that the relativity of necessary statements has to be taken account of, but he argues that Whewell does do that in his later philosophy, though not in his earlier. He attributes two changes to Whewell. The first is that already mentioned in the discussion of Stoll's work, that is, in the *Bridgewater Treatise* laws of motion are not deemed necessarily true though later they are. The second change is a movement from a Kantian view in the first edition of the *Philosophy*, in which laws are absolutely necessary, to a different view starting with the second edition, in which the necessity of laws was relative to context: statements are necessary, if we must assert them in order to maintain some corpus of knowledge, if rejecting them would lead to its "dismemberment." They are then true by fiat and protected from refutation at all cost.

Walsh's essay attracted the attention of Robert E. Butts. He responded to it and quickly followed with his own essay on Whewell's theory of necessity. Since Butts' work shortly thereafter became the most prominent of the very sparse literature on Whewell, a very brief discussion of this disagreement may help us see how the debate has – I am afraid we must say – limped along. Walsh's essay is an obvious attempt to lessen the force of the old suggestion, first made by de Morgan, that a theory of science requires some notion of necessity other than that of absolute necessity or other than that notion necessity employed in mathematics. Walsh's proposal that Whewell's view of necessity developed in the course of his career, that in the end it was not a theory of absolute necessity, but rather of a progressive necessity and that this type of necessity had to be judged relative to context enables him to answer Ducasse's version of de Morgan's criticism. A sophisticated appreciation of Whewell's theory of necessity could, perhaps, enable us to see how his view had been underestimated. Ducasse's criticism is not, one should note, merely a review of de Morgan's criticism. It is also a new version of Herschel's alternative. Both Herschel and Mill argued, in a Humean fashion, that the only meaning which one could give 'necessity' was one which resulted directly from, or was reducible to, observed connections. Ducasse seeks, then, to state once again in more modern terms this view by arguing that necessity cannot be absolute as Whewell maintained, but must be based in experience (as Herschel and Mill maintained), and thus relative, as Ducasse now proposes.

Walsh's conventionalistic interpretation of Whewell's later philosophy seems to me so implausible as not to require refutation and his other comments are not new. If an answer should be needed, Butts offered it in his note on Walsh's article (Butts 1965a). As Butts points out, Whewell cannot be plausibly interpreted as advocating a view according to which the necessity of a proposition is only necessary due to the theoretical context in which it occurs (Butts 1965a, p. 179). This view fails to take account of how "our conceptual systems really do catch reality in their nets" (Butts 1965a, p. 180). Butts also shows that Whewell always held that necessary truth was progressive even though, as he concedes, there may be differences between the view expressed in the philosophy of science and the *Bridgewater Treatise*. Butts here begins to offer his own interpretation of Whewell according to which his central difficulty was to reconcile his empirical leanings with his championing of necessary truth (Butts 1965a, p. 177). And this conflict leads him in the end to a theological guarantee that what we intuit as necessary truth is indeed necessary truth. There will be ample space below to discuss this hypothesis.

## 5.5. Robert E. Butts: Old Misconceptions Defended with New Confusions

The series of essays which Robert E. Butts published from 1965 until the present is one of the most serious contemporary attempts to appraise Whewell's philosophy. Butts follows the pattern set by Ducasse and by Walsh of offering separate studies of the two problems deemed central by Whewell's commentators, that is, of the problems with Whewell's view of necessity, on the one hand, and of the problems concerning his alleged need for some theory of induction. Butts has largely ignored Whewell's innovations in methodology, perhaps because he implicitly accepts them. He has also ignored his historical approach to the philosophy of science, other than to argue that he fails to combine it with a study of scientific inference. In his early efforts Butts concentrates on the attempt to show the inadequacy of Whewell's philosophy. I see it as an attempt to show that, in spite of the new interest and relevance of Whewell's philosophy, he offers no alternative to positivist philosophy of science. In Butts' later work, which I discuss in Chapter 6, he tries to combine Whewell's methodology with a better, an inductivist view of science.

Butts explains the purpose of his essay on necessity as follows:

> My account of Whewell's theory of the Ideas, and some parts of my account of his theory of necessary truth, parallel the account in C.J. Ducasse . . .[, which] is the only reasonably accurate and detailed exposition of the fundamental elements in Whewell's theory of science written in this century. But Ducasse's purpose – to expose clearly the fundamentals of Whewell's philosophy of discovery – is more limited than mine. I wish to show both the essential features of Whewell's theory of necessary truth, and to say something about its development in the broader context of Whewell's philosophical thought. My account makes good one defect of Ducasse's work: it shows that Whewell did make some attempt to justify his otherwise purely psychologistic theory of necessary truth. Thus I hope to have gone beyond Ducasse's papers in the direction of a fuller exposition of both Whewell's theory and the context of discussion and thought in which it developed. (Butts 1965b, p. 162)

In regard to the work of Blanché he says the following:

> Robert Blanché . . . has argued that there are two fundamental problems in arriving at a fair interpretation of Whewell's philosophy. First, are ideas constitutive elements of the structure of reason, or contingent creations of individual genius? Second, are ideas forms in the Kantian sense, or substantive notions as in Descartes and Plato? Blanché thinks that Whewell never entirely answers either question, and thus that there is ". . . l'indécision fondamentale de sa philosophie, oscillant sans cesse entre l'idéalisme et le réalisme, entre l'épistémologie et l'ontologie." I hope to be able to show that Whewell does make a decision on both questions, though the attempted solution that results is hardly satisfactory. (Butts 1965b, p. 163)

This passage must surely raise our hopes that a new and more appreciative interpretation of Whewell will be provided and even if, as we note right away, Butts never explains precisely how it meets Blanché's criticism.

There are two central points in Butt's essay on necessity. The first is a new exposition of Whewell's theory and the second is a discussion of how Whewell allegedly sought to justify his realism. The first should go beyond the work of Ducasse and the second should show that Blanché's analysis is inadequate, though I should emphasize that these references to Ducasse and Blanché are found in brief footnotes. No discussion of their views whatsoever appears in the text. This is, perhaps, not so serious in the case of Ducasse, since Butts seeks to offer an exposition of Whewell's view of necessity which has little to do with his view. But in the case of Blanché it is more serious, since, contrary to the above claim, Butts comes at some points to views quite similar to Blanché – Whewell is alleged to move from a more Kantian to a more realistic position and to seek to use an inadequate theology to save his realism – as I will explain below. Butts observes that Blanché claims that Whewell never entirely answers the questions whether ideas are constitutive of the structure of reason or contingent or whether ideas are Kantian forms or substantive as in Descartes, but he ignores Blanché's observation, "A une telle question, la Philosophie de la découverte répondra par la théorie du progrès de l'évidence" (Blanché 1935, p. 2). This aspect of Whewell's philosophy is, Blanché says, the most original part of his philosophy. Butts ignores it, even though it is crucial to his own discussion, that is, he seeks to explain how necessary truths can be learned from experience.

Butts' first problem of explaining Whewell's theory of necessary truth grows out of what he regards as a perversity in Whewell's career (Butts 1965b, p. 161). This alleged perversity is his failure to be a philosophical empiricist in keeping with other thinkers such as Herschel and Mill. In spite of deeming induction his special business and the true faith, he still believes that it results in necessary truths. (He says here that Whewell believed that inductive inference resulted in necessary truths, but Whewell did not offer a theory of inference, as should by now be clear from, as one example, his reply to de Morgan.) Butts seeks, then, to make some sense out of this peculiar deviance from the philosophical empiricism of Herschel and Mill. He nowhere notes Whewell's reasons for parting company with Herschel and Mill and this leads to misunderstandings, as I explain below.

Butts' basic idea is the following. In Whewell's philosophy of science one finds two uses of the word 'experience'. On the one hand, the word is used to mean merely sensation or perception. On the other hand, it is used to refer to the use of experience in science or the imposition of ideas on observations to order them. When Whewell refers to necessary truths not being possible without experience and claims that they are learned from experience, he uses the former meaning of the term. He did this, for example, in his reply to Herschel in conceding that mathematical truths could not be known without experience even though they were not derived from experience. When he speaks of empirical truths which are not necessary but are based in experience he uses the other sense of the term. Butts states,

> Given the meaning of 'experience' as 'sensation' or 'perception', he was able to hold that no knowledge is innate, and that thus even necessary truths take some time and experience to learn; and he was able to hold that, presupposing that no Ideas are ultimate, there are still more of them to be discovered in the future as new sciences become possible, and as subsequent discussion clarifies Ideas to such an extent that they become clear and distinct, and hence becomes sources of intuitively certain necessary propositions. It seems then that Whewell made out a perfectly good case for holding both that there is a logical distinction to be made between empirical and necessary propositions, and that some truths known empirically at first are subsequently known to be necessary. (Butts 1965b, p. 167)

I wish to note that this reading does not explain the role of scientific experience in the process of learning necessary truths. Perhaps Butts denies that scientific experience has any role at in the process of learning necessary truths, since only the first kind of experience is related to the process of acquisition of necessary truths. But Whewell's theory of science is a theory of how science, that is, scientific research or, if you will, scientific experience, leads to the discovery and establishment of necessary truths. Ignoring this Butts continues,

> It is on the basis of the distinction that Whewell is able to hold that necessary truths are learned by experience (in the sense of perception) and that these truths are not established as necessary by appeal to experience (in the sense of controlled scientific experience). (Butts 1965b, p. 167)

Butts seems to be led to the view that the necessary truths which emerge in the history of science and form the basis of the various sciences are not learned through scientific experience, because scientific experience does not show them to be necessary. He argues first that necessary propositions cannot be confirmed by experience, quoting Whewell to the effect that a proposition, which is necessary, is also known to be necessary and is not even subject to tests: if a test goes wrong we reject the result and not the necessary truth. Second, he proposes that the kind of evidence which establishes a necessary truth cannot be the kind of evidence which establishes an empirical truth. Therefore, I read him to say, scientific experience is not that experience which is needed or relevant to the establishment of necessary truth.

If we presume that the purpose of scientific experience is not to offer inference from experience to theories, this all looks quite differently.

Whewell thought the purpose of scientific research was to render our ideas – whose source was within us – clear and distinct. The role of scientific experience, then, is not to provide a basis for inductive inference; it is to refine, test and improve ideas. When this process is complete, we come to ideas which are seen to be true without reference to the tests which went before, but which would not have been seen to be necessary without this process of testing and improvement or, if you will, of scientific experience. Whewell states,

> We are thus led to see that in our knowledge of mechanics, chemistry, and the like, there are involved certain necessary principles, derived from our ideas, and not from experience. But to this it may be objected, that the parts of our knowledge in which these principles are involved has, in historical fact, all been acquired from experience. The Laws of Motion, the Doctrine of Definite Proportions, and the like, have all become known by experiment and observation; and so far from being seen as necessary truths, have been discovered by long-continued labours and trials, and through innumerable vicissitudes of confusion, error and imperfect truth. This is perfectly true: but does not at all disprove what has been said. Perception of external objects and experience, experiment and observation, are needed not only, as we have said, to supply the objective element of all knowledge – to embody, limit, define, and modify our ideas; but this intercourse with objects is also requisite to unfold and fix our ideas themselves. As we have already said, ideas and facts can never be separated. Our ideas cannot be exercised and developed in any other form than in their combination with facts; and therefore the trials, corrections, controversies, by which the Matter of our knowledge is collected, is also the only way in which the Form of it can be rightly fashioned. Experience is requisite to the clearness and distinctness of our ideas, not because they are derived from experience, but because they can only be exercised upon experience. And this consideration sufficiently explains how it is that experiment and observation have been the means, and the only means, by which men have been led to a knowledge of the laws of nature. . . . In general it may be allowed, that the necessary principles which are involved in those laws become then only clearly known, when the laws of nature are discovered which thus involve the necessary ideal element. (Whewell 1967b, pp. 663-664)

It would seem clear enough that scientific experience is crucial for the discovery of necessary ideas and for the refinement of these ideas to such a degree that we may see them to be necessary. And Butts seems to try to take this aspect of Whewell's view into account when he says, "Thus we know that particular Ideas are distinctly conceived when, in the course of actual history, they initiate inductive sciences that really do explain and predict the phenomena they were introduced to deal with" (Butts 1967b, p. 168).

But Butts does not discuss the role of scientific experience in the discovery of necessary truths. Nor does he note that this role makes his own rather forced reading of Whewell unnecessary. We do not need to separate scientific experience and necessary truths, because scientific experience does not justify necessary truths; it has a different role to play.

Butts seems to say that Whewell's view of the relation of necessary truths in mathematics to experience is also Whewell's view of the relation of necessary truths in science to experience. In one sense this is true, since both kinds order experience. But mathematical ideas apply to all, rather than merely a part of experience, as scientific ideas do. They are also proven when they are rendered so clear and distinct that they are seen to be necessary. But they are not subject to experiments, and they are not part of natural science even though they serve as part of its foundation. Butts' proposal to read Whewell's statements about the necessary truths of mathematics as applying to all necessary truths ignore his explanation of the difference. It would be in effect to read him in a rather strict Kantian way. This is both implausible and unnecessary.

Butts has a second point to make as well, which seems equally implausible, but has the virtue of raising the question of the role of Whewell's important essay on the Fundamental Antithesis in Philosophy in the development of his system. Butts' problem grows out of the common analysis, according to which Whewell needs either a transcendental deduction of all fundamental ideas (Mansel) or an empirical method of proof (Mill) and can supply neither one nor the other. Butts argues that Whewell needs some method of justification for his fundamental ideas which is not found in his methodological theory but, he suggests, this cannot be a transcendental deduction, since the number of fundamental ideas is open and a transcendental deduction after the fashion of Kant would have to be complete. He does not even mention the possibility of giving an empirical justification, but it is clear that this possibility does not come into question for Butts, since, as explained above, no amount of empirical evidence could justify a necessary statement. On Butts' view, no necessary statement has empirical consequences. This is not Whewell's view of necessary statements, since, when found within scientific systems, they certainly do exclude certain events. Butts' view of their justification, that is, no empirical proof or empirical refutation of necessary truths is possible, is, of course, true and in accord with Whewell's view.

Butts generates here unnecessary problems. The seeming paradox that necessary truths are learned from experience but cannot be justified by experience, is resolved by Whewell as one can see when one remembers that for Whewell necessary truths are not analytic, that necessary truths are only seen to be necessary, and that the empirical evidence which was necessary for their development is no longer necessary to see their necessary truth. The fact that they cannot be refuted is, of course, the quite trivial point that true statements have no counter-examples. When we know that a statement is necessary and we appear to have a counter-example, we know we have made a mistake in the construction of the counter-example.

If we ignore this explanation, however, and accept the hypothesis, that Whewell needs a justification for necessary ideas not already found in his system, as well as the hypothesis, that neither transcendental deductions nor inductive inference can supply it, then one might ask if Whewell perceived this gap in his system, if he sought to solve it and how, and whether his solution is adequate. Butts will here make a new contribution by arguing that Whewell did see the problem, that he formulates it in his letter to Herschel, and that the first memoir on the Fundamental Antithesis of Philosophy was his first attempt to solve it (Butts 1965b, p. 175). He then continues to consider his theology as (another) answer to it.

The hypothesis that Whewell needs a justification for necessary ideas which is not supplied by his system as it is found in his *History* and *Philosophy* is central for Butts interpretation of Whewell, since the problem we deem Whewell to be trying to solve is crucial for our reading and appraisal of his results. And here we find difficulties. Butts begins,

> Thus it is no mystery that Whewell, holding as he did that the intuition of necessary truth, the idealization of facts, is progressive, did not provide a transcendental deduction of the Ideas. But this does not mean that his mind was free from philosophical anxiety about the question of the final justification of those propositions whose necessity has been intuited. Indeed, he tells us of his concern in a letter to Herschel. Ideas and Things are constantly opposed, yet necessarily co-existent. How they are thus opposite and yet identical, is the ultimate problem of philosophy. . . . Knowledge requires Ideas. Reality requires Things. Ideas and Things co-exist. Truth is, and is known. But the complete explanation of these points appears to be beyond our reach. In part, Whewell thought that the solution of this ultimate problem of all philosophy is beyond our reach because he would not settle for a mere epistemological justification of the fundamental Ideas. In short, Whewell's Kantianism stops being operative at exactly the point where he thinks that only a metaphysical account will give an answer to this ultimate problem of philosophy. (Butts 1965b, p. 174)

This is, indeed, a rather confused beginning. We are asked to believe that ideas, which are known to be necessary, are in need of some justification. This is prima facie absurd, but let us go a bit further. Does Whewell say that he seeks some (further) justification for necessary ideas in the passage cited by Butts? Quite clearly no. He takes it as a fact that we know necessary ideas as such and, if there should be any doubt, that we know that they are true. But most importantly, does he think his philosophy leaves questions open which he needs to solve in order to solve problems his philosophy was intended to solve or which are must be solved in order to render it defensible? Clearly not. Whewell states,

> But the complete explanation of these points is beyond our reach. At least it is not necessary for the purpose of our philosophy. The separation of Ideas and Sensations in order to discover the conditions for Knowledge is our main task. How Ideas and

> Sensations are united so as to form Things, does not so immediately concern us. (Whewell 1967b, pp. 676-677)

Butts claims that Whewell is not satisfied with "a mere epistemological justification of the Fundamental Ideas" and shows thereby that he knows that the problem he poses already has a solution. But why is this not sufficient? And what evidence do we have that Whewell did not deem it sufficient? I find none. Whewell says that he is concerned to explain a fact, the fact that we know the truth. This is surely not a call for a new justification of these truths. It is not even a call for a new justification of his system, which is justified by its capacity to clearly and distinctly account for the history of science when no other theory can.

Butts does not stick, however, to one formulation of the alleged problem so perhaps others are better. He says, for example, "Thus for Whewell the epistemological justification of the special status of the Ideas in human knowledge had to take the form of a more basic justification of his philosophy of science as realistic" (Butts 1965b, p. 174).

This is better, since it moves to a higher level. Butts argues here that Whewell sought in his letter to Herschel to respond to a criticism leveled by him according to which his theory was idealistic. This is nice, and we can see the first memoir on the fundamental antithesis as a continuation or elaboration of this response. But, as Whewell says, he is concerned with explaining his own view and the fundamental antithesis. In his own response he is concerned with explaining why Herschel has misunderstood his view as one which is, in the end, idealistic. There is no talk of justification whatsoever. Not only does Whewell take for a fact that each known Fundamental Idea is necessary, true and known to be so, but that his philosophy of science, which explains all this, is also known to be true: it is a science of science which has been successfully carried out. It raises, of course, new and deep problems just as say, Newton's theory does concerning gravity; but that it is not to say it lacks a proper justification.

Butts tries once more. He suggests that in the *Philosophy of Discovery* Whewell comes to a realization that what he formally regarded as "the true faith" needs justification and that this was of great importance (Butts 1965b, pp. 176-177). But once again the quote he offers preceding his claim asks for nothing more than an explanation of how it can be possible that humans have knowledge of necessary truths which describe the world as it really is. He claims that Whewell now returns to metaphysics to solve his new problem.

> In his youth he (Whewell) had proclaimed that the inductive philosophy is the "true faith," only to discover in his mature years that however true the faith is, however astonishing its results are, the faith nevertheless requires justification. For the method of induction and the critical philosophy to which it gives rise are neither of them self-

> critical. They must be supplemented by, paradoxically enough, a return to exactly that form of philosophy – classical modern metaphysics – that had been regarded as the arch-enemy of the "true faith." (Butts 1965b, pp. 176-177)

Whewell is apparently alleged to have called induction the "true faith" (Butts 1965b, p. 162). Do I take the quotation marks too seriously? Did he? And did Whewell regard metaphysics as the arch-enemy? Butts offers no evidence for the latter and his statement is ambiguous as to whether it was merely in general so regarded or so regarded by Whewell. The reference to the "true faith" is apparently a letter dated Aug. 21, 1834 written to Richard Jones. Butts does not say where the letter is to be found. Todhunter published a letter by Whewell to Jones which was written on this date (Todhunter 1876, vol. II, pp. 185-188) but I find no reference to "true faith" or even any expressions which could be so characterized. And, indeed, should we really be surprised that the author of the *Bridgewater Treatise,* who explained such facts as the distance of the Earth from the Sun as signs of God's plan, should now explain the possibility of knowledge with a theology?

Contrary to his claims to offer a new approach Butts now comes to the same point that Stoll and Blanché had come to previously, that is, that it was Whewell's theology and especially his theory of creation which were crucial in his explanation of knowledge (Butts 1965b, p. 177). He seeks to explain the slide from the problem of justification to that of explanation and of metaphysics as follows.

> Whewell's major problem had been one of justifying the apparently astonishing position that the intuition of necessity, the idealization of fact, was progressive, and that there are, nevertheless, necessary truths that are the foundations of the several sciences. And this problem was absorbed into the problem of explaining that our intuited forms are not empty, but really do catch reality, really do operate in the perception of real objects. In fact, these were for Whewell two forms of the same question as the passage from the Philosophy of Discovery quoted above indicates. (Butts 1965b, p. 178)

The passage quoted by Butts from the *Philosophy of Discovery* does not show either that Whewell was concerned with justifying his view of the progressiveness of necessary truth and/or that this problem was absorbed into the problem of explaining that our intuited forms are not empty, since it makes no reference to justification. In any case, Butts turns to the discussion of theology. Though he makes no reference to Herschel, Clark, Stoll or Blanché, his work follows theirs quite closely. He deems Whewell to have moved from a Kantian position to a more metaphysical or ontological view which Blanché calls Platonic. He repeats their point of view that Whewell was rather dogmatic due to his use of his religious perspective in his philosophy of science. And, just as Blanché did, he quotes from the unpublished portions of the *Plurality of Worlds* which explains the correspondence of ideas with the world as due to their common origin. He fails to note the nice point made by Blanché that Whewell is led to pantheism, which contradicts his Christian point of view.

Butts concludes that Whewell's Kantianism and his memoirs on the Fundamental Antithesis could not solve the problem of why ideas correspond to the world, so he is forced to theology. This is, in turn, deemed inadequate, since it is neither justified nor adequately spelled out. (He hints that he will later explain the theological differences between the *Bridgewater Treatise*, on the one hand, and the *Plurality of Worlds* and *Philosophy of Discovery*, on the other hand, without noting that Blanché had done that.) We are, then, to believe that the system has collapsed because the highest level is not justified. This is, of course, not necessary. There even seem solutions available, such as Whewell's more pantheistic theology, which might make the system coherent and even justifiable in its own terms, that is, by offering clear and distinct theories. But Butts merely suggests that

> Whewell's system shows the scars and open wounds of attempting to develop a body of philosophical doctrine on science out of materials partly logical and partly historical . . . The entire problem of justifying a doctrine of the progressive intuition of necessary truth might never have arisen if he could have kept separate in his thought the quite different process of giving an historical account and attempting a logical or epistemological analysis, processes which he was more likely to confuse than recognise as different. Thus he could not appreciate fully Mansel's telling criticism of his analysis of necessity, a criticism based upon a set of careful logical distinctions to which Whewell's more historically oriented mind was largely indifferent. (Butts 1965b, p. 180)

The upshot of Butts essay is, thus, one further defense of the standard interpretation of Whewell which deems him confused because he failed to meet the standards of logic, that is, of Kant and/or Mill. But this will obviously not do since, as I have shown, Butts did not comprehend how Whewell sought to develop and to justify his system nor did he even bother to take account of Whewell's reasons for seeking an historical account of the growth of knowledge. As Blanché had shown before, Whewell changed the problem of the philosophy of science in a way which cannot be ignored: any theory of science, including any theory of scientific proof, must take account of the facts of scientific research. Butt's proposal that they may be harmonized only by sharply demarcating them conforms to the tradition but is surely naive: how do we know they harmonize if we do not test such a hypothesis? And how can we test such an hypothesis if we do not follow Whewell and integrate them systematically?

Butts next effort to develop his interpretation of Whewell is found in his introduction to a collection of excerpts from Whewell's writings (Butts 1968). The essay contains two major sections, one on necessity and one on induction. Since the section on necessity is a reprint with slight modifications of the essay just discussed, I will only discuss his introduction to Whewell's theory of induction. Butts chooses to explain Whewell's theory by emphasizing his theory of the consilience of inductions, the representation of science through the table of induction, and this table as a criterion of truth. He explains that intuition is crucial. He points to the need to deduct the consequence of some induction or Idea used to order experience. And he explains the growth of science as trial and error.

Taken thus far there is no cause for criticism. But Butts makes one point which shows a serious misreading of Whewell. He complains that the kind of confirmation that Whewell seeks and needs, that is, the deduction from his theories of all the consequences, could only then guarantee the truth of a theory, if there were a finite class of particulars and each of them had been inspected (Butts 1968, p. 22). He wonders why Whewell did not notice this and explains it, in fact, quite well: "Indeed, he even holds in another work that experiments do not in any ordinary sense prove or confirm scientific laws, but rather play the role of rendering laws more clear and intelligible as visible diagrams in geometry serve to illustrate geometrical truths." (Butts 1968, p. 23)

I do not know why he does not take seriously Whewell's view, that the prime role of experiments and observation is not the proof but the development of ideas. Whewell knew that one could not infer from particular statements to necessary truths and thus rejected such inference as part of science. This is a clear position and is powerfully argued for. Still further he replied to the charge that many theories are compatible with the evidence, and thus no theory can be justified, in his response to Mill. He pointed out that as a matter of fact this does happen in science. Whewell claims he has accomplished his task when he has explained the history of science and no other theory could. And Mill certainly could not. A dogmatic assertion of the need for inductive inference will hardly do as a response to it.

Butts continues and explains further that Whewell could not distinguish psychological conviction from objective empirical truth and inferential validity. At this late date, after the work of Blanché, if Whewell's protestations are not enough, this is unworthy of reply.

From the publication of Mill's essay on logic, if not from the publication of Herschel's review, it has been standard to complain that Whewell does not offer a logic of induction. And Butts repeats the charge and uses de Morgan as his source. De Morgan, Butts tells us, "objected to two shortcomings of Whewell's view: the notion that the inductive tables display the logic of induction, and the insistence that induction involves more than the generalization of particulars" (Butts 1968, p. 25).

The first of these observations is nearly right; the second is meant to assert that this is what induction has always meant. But Butts fails to note that what de Morgan complained about in both cases is Whewell's use of the terms 'logic' and 'induction' and not his theory. De Morgan says, ". . . and though we do not quarrel with any of his conclusions, we are entirely opposed to the use which he makes of the words logic and induction" (de Morgan 1859, p. 43). Now Butts takes this criticism to mean that Whewell needed to offer a logic of induction. He reminds us of the circularity of induction, as he has described it, and the fact that consilience does not save the day: it gives us no decision procedures for determining when a "generalization over data is validly established." He further suggests that Whewell did not understand de Morgan when he said so much the worse for logic.

Did de Morgan think that Whewell's theory of science was incomplete? He does complain about his theory of justification by intuition. But this is not to say he deems it incomplete but wrong. He thinks a new theory of justification is needed. But his sympathies for Whewell's criticism of Bacon indicate that he does not expect this theory to be a new theory of induction. He also objects to Mill's putting all discovery of truth under general logic just as much as Whewell's putting all physical discovery under logic of induction. De Morgan says he is in full agreement with Whewell's appraisal of Bacon and though he does say Kepler used induction by observing the orbit of one planet after another to conclude that all planets have elliptical orbits, he nowhere explains what role inductive inference plays in science. And Whewell may rightly reply: if "inductive" logic is not part of science so much the worse for "inductive" logic. This was his view and he corrected his terminology by calling his philosophy a philosophy of discovery – which is not to say he has no theory of justification of scientific theories – and probably as a result of de Morgan's essay. The date of the preface to the *Philosophy of Discovery* in which this terminological change (1856) is announced precedes the date of the review (Jan. 8, 1859). The book, however, was first published in 1860. Blanché thinks the date of the preface is a mistake and I agree with him. Whewell there states, "and the Philosophy at which I aimed was not the Philosophy of Induction, but the Philosophy of Discovery" (Whewell 1971, p. v).

Butts has failed to grasp that Whewell's defiant response to de Morgan was a challenge to show that "induction" or "logic" were a part of science in the sense demanded by his opponents. Butts, just as the others, begs the question by deeming its mere lack already a defect. Perhaps de Morgan would have liked to see a logic of induction – whatever that might be – but he hedges his bets and leaves his options open.

Butts concludes his essay with a discussion of the criticisms of Whewell's philosophy as an inductive logic in the sense of inference from particulars to generals. But this has little to do with Whewell. Butts suggests that Mill was right in saying that Kepler really did infer from particulars to generals and that one thus needs a view of it. He offers no argument and this view has, I think, been shown above to be false.

The recent rediscovery of the importance of Whewell in the English-speaking philosophical world might have raised our hopes that new interpretations of his philosophy and his place in the history of the philosophy of science would be developed, which take account of the deep challenge he had offered to his competitors. Our skepticism that this would occur has thus far, however, been born out. The rediscovery of Whewell and his re-introduction into the philosophical discussion brought with it, as we have just seen, the re-introduction of the old attitudes toward him. Indeed, the new studies go one step backward in that they do not even appreciate the advances which Blanché had already made in France due to the influence of Bernard. How much has this uncritical acceptance of old positions damaged the discussion of Whewell? Have there been substantial improvements in spite of this weak beginning? Let us look further at some recent attempts to appreciate Whewell in spite of the explicit or implicit acceptance of old positions to see what we can find.

# CHAPTER 6

## QUIXOTIC ATTEMPTS TO REVIVE MILL'S PROGRAM

We have seen that three standard criticisms of Whewell's philosophy were quickly developed and have become a standard part of the history of the philosophy of science. These three criticisms are, one, that Whewell needs an (inductive) theory of justification in addition to his theory of scientific discovery, which is merely psychological, two, that his extension of fundamental ideas beyond those postulated by Kant require, but do not have, a transcendental justification. And three, this extension leads to confusions between Kantian and Platonic elements, and an ad hoc and incoherent philosophy of science. Later, a fourth criticism was introduced by Stoll, developed by Blanché, and taken up by Butts, which was that Whewell sought and failed to remedy the defects of his theory with his theology. One or the other of the first two but most likely both together have become the point of departure for all research on Whewell. And, as we have just seen above, recent research on Whewell's philosophy in the English language began with the naive reintroduction of these two criticisms. It is, of course, not enough to merely repeat over and over again the same points. So, recent commentators on Whewell's philosophy have been faced with the task of explaining the significance of Whewell's philosophy in spite of these criticisms. We look for someone who would question them, but we find none. We must see, then, whether contemporary commentators can accomplish the task of recognizing Whewell's accomplishments while endorsing the (erroneous) criticism developed over the years.

From what we have already learned we might be skeptical that such research should give us a genuinely new appreciation of Whewell or even come to terms with his new ideas. Whewell's strength lies in the fact that he has questioned the presumptions of these criticisms and his arguments in this regard have scarcely been taken into account. We see also, however, the need to do just that: the overwhelming portion of research in the philosophy of science since Whewell has been based on this rejection of Whewell's approach. If this research cannot be reconciled with Whewell's advances, it has been for the most part a futile digression. This is not to say that we may not have learned something from it. But it is to say that the problems which the mainstream of philosophers of science since the debate between Whewell and Mill have tried to solve ignore real difficulties of explaining scientific knowledge, such as those revealed in

Whewell's demonstration that Mill's methods were not those used in science. These philosophers pretend we know more than we do about scientific knowledge by presuming that Mill's view is approximately correct. They seek to solve problems which are irrelevant to the acquisition of knowledge in science and which are in any case as far as any one can now see insoluble, problems such as that of devising a logic of inductive inference. Can this sad appraisal of the level of rationality of the study of science be avoided? Let us look further at recent attempts to do so.

To begin with a short review, one of the four standard criticisms of Whewell's theory is Mill's: Whewell offers no theory of inductive proof. A second is, of course, Mansel's, according to which the extension of fundamental ideas beyond those allowed by Kant or Mansel requires that these be justified by some method which Whewell has not offered and cannot provide (or Whewell confusedly seeks to do so by confusing and mixing Kantian and Platonic elements). The two standard criticisms fit together, of course, since each demands a new theory of proof. They differ in that Mansel demands a transcendental proof of necessary ideas – which cannot be given – and Mill demands a theory of inductive inference. Now, if the central problem of the philosophy of science is to offer a theory of inductive proof, it would seem at first blush as if Whewell's theory must be irrelevant, since it offers no alternative to Mill. And it was, indeed, largely ignored for this reason. But Whewell's theory was nevertheless an enormous challenge for Mill, as explained above, because it forced him not merely to give a so-called logic of induction but also to show that and how this logic functioned in actual scientific research. After Mill's solutions to these problems were (as far we as we can see at this point in our argument) erroneously taken to be adequate, Whewell could be forgotten and was.

In modern discussions of the philosophy of science the problem which Whewell posed for Mill has been posed anew for inductive logicians. And this corroborates our suspicions that the problems Whewell posed were suppressed but never seriously taken into account. Under the powerful influence of the philosophy of Karl Popper philosophers of science have been forced to rediscover these problems and have turned once again to historical questions and to problems of reconciling historical studies with studies in so-called inductive logic. In these new circumstances it is not surprising that at least some philosophers should find a renewed interest in the work of Whewell: it was he who first set the problems which a logic of induction must solve in order to be reconcilable with the actual practice of science. An analysis of these problems, then, seems not only desirable from an historical perspective. Such an analysis may also help us to pose problems of inductive logic well today. From an historical perspective philosophers of science have hoped to vindicate the acceptance of the demands of Mansel and of Mill. From the point of view of contemporary research philosophers of science have hoped to show that inductive logic can meet the problems posed by Whewell's criticism. In discussing such research on Whewell's philosophy it is, of course, necessary to keep in mind that Whewell posed no such problem of reconciling inductive logic with the history of science himself. "So much the worse for logic," he said. He deemed a theory of induction or theory of discovery to be complete without a theory of inductive inference in the sense which de Morgan and the tradition have wished to reserve for that term.

The second stage in the modern revival of Whewell consists of studies of Whewell's theory of induction, in particular of his theory of the consilience of inductions. These studies seek to show how Whewell's methodological theory can be used as a foundation for a more modern theory of induction. They pursue answers to four questions: one, What is Whewell's solution to the problem of how empirical evidence proves or supports scientific theories? two, Is his solution of any contemporary interest? three, What problems remain open for us to solve now? and four, How can they be solved? There is certainly nothing wrong with Mill's program – other than its apparent futility – of presuming that Whewell has correctly described how scientific knowledge grows and within this framework to explain how inference from particular to general statements functions. It is a mistake, however, as is clear enough from the foregoing discussion, to treat Whewell as a competitor to such attempts: he does not seek to show how scientists infer from particular to general statements but, indeed, that they do not. To treat him as a competitor is thus to miss his fundamental point that an alternative is needed to the perspectives offered by Bacon, Herschel and Mill, on the one hand, and Kant (or Mansel), on the other. It leads, of course, inevitably to misinterpretations of his theory of science. Does this happen? Are those problems of induction posed by recent thinkers in the context of Whewell's analysis of the growth of knowledge taken as problems within his system? Is he interpreted as trying to solve such problems? Does this matter? Let us turn to some examples.

### 6.1. Larry Laudan: The Consilience of Inductions as Inductive Inference?

Larry Laudan has sought to interpret Whewell's theory of induction and to use his interpretation to explain problems which arise for inductive logicians (Laudan 1971a). He seeks to carry out three tasks in his essay. These are to explain Whewell's theory of the consilience of inductions, to appraise it by asking why it should be of value, which is, for Laudan, primarily a problem of whether Whewell has provided an adequate foundation for the construction of reasons for belief that some relevant theory is true, and, finally, to discuss briefly the present state of the problem, that is, the present attitude toward the importance of the consilience of inductions and the success in showing its value or its capacity to justify theories or provide reasons for belief.

Laudan offers no systematic view of Whewell's philosophy. He apparently believes that it would be superfluous to give one. On one occasion, he calls Whewell's view idealist (Laudan 1970, p. 311) and further states, "A victim of his own many-sidedness, Whewell has been largely ignored in histories of philosophy. His eclectic and intricate philosophy, half-Kantian and half-Baconian, has made him an unwieldy entity for historians" (Laudan 1970, p. 311).

Laudan then takes over uncritically the standard view of Whewell as a confused eclectic, though he deems him half-Kantian and half-Baconian instead of half-Kantian and half-Platonic as the tradition would have it. This prepares the way for his discussion of Whewell in which he restricts himself to an exposition of Whewell's theory of the consilience of inductions alone. He hopes to find there something of interest for the (Baconian) problems of inductive inference. These would have no place on the traditional view of his confusion as half-Kantian and half-Platonic.

Laudan claims that Whewell's theory of the consilience of inductions refers to three different situations which need, therefore, to be analyzed separately. These three are:

> (1) When an hypothesis is capable of explaining two (or more) known classes of facts (or laws); (2) When an hypothesis can successfully predict "cases of a kind different from those which were contemplated in the formation of our hypothesis"; (3) When an hypothesis can successfully predict or explain the occurrence of phenomena which, on the basis of our background knowledge, we would not have expected to occur. (Laudan 1971a, p. 371)

Laudan here hopes to make a contribution by pointing out that Whewell deemed all three kinds of situations consiliences of inductions. There is no question that Whewell deemed all three kinds of success in explaining facts to be important steps on the way to truth. Are they all consiliences? Laudan quotes an aphorism of Whewell which states what the consilience of inductions is and it includes only the first of Laudan's three cases (Laudan 1971a, p. 369). The quote is the following. "The Consilience of inductions takes place when an Induction, obtained from one class of facts, coincides with an Induction, obtained from another different class. This Consilience is a test of the truth of the Theory in which it occurs" (Whewell 1967b, vol. II, p. 469).

Since Laudan's discussion tends to ignore the distinction between an Induction and a Consilience of Inductions it may be useful to quote the preceding aphorism as well which states, "Induction is a term applied to describe the process of a true Colligation of Facts by means of an exact and appropriate Conception. An Induction is also employed to denote the proposition which results from this process" (Whewell 1967b, vol. II, p. 468).

In spite of the narrowness of the description of consilience of inductions given in the quote, Laudan is no doubt correct in calling the first two kinds of situations consiliences of inductions. Whewell explains consilience of inductions in terms of the second description as well as the first (Whewell 1858, p. 88) and says they are nearly the same (Whewell 1858, p. 95). Laudan gives no reference for the inclusion of the third case and I find no explicit discussion of it which is distinguishable from the second. We may, perhaps, give Laudan the benefit of the doubt. Laudan deems this analysis important, because it shows that Whewell's "demand for consilience is a much more subtle and exacting requirement than the trivial one Butts attributes to Whewell (a theory should successfully explain more than it was devised to explain)" (Laudan 1971a, pp. 371-372).

I do not quite understand this claim, however, since the only thing that Laudan adds to the view of the consilience of inductions attributed to Butts is that a theory may explain or predict unexpected occurrences and he gives no reference for this claim. Is this important? Let us look further.

Although Laudan intends to extend the normal interpretation of the consilience of inductions, he also omits a crucial feature of all inductions, including consiliences of inductions, which are a species of induction. He does not include the central feature of all induction that a new idea is added. This is made clear when he says,

> As these instances (the three kinds of "consilience") suggest, and as I shall argue in detail later, a consilience of inductions occurs when an hypothesis is shown to have achieved a certain minimum of corroborated or confirmed empirical content. More specifically, an hypothesis achieves a consilience when it has shown itself to be capable of successfully explaining either different kinds of facts or very surprising facts. (Laudan 1971a, p. 372)

Whewell has warned us, however, to take another factor into account. He states, "But when we say that the more general proposition includes the several more particular ones, we must recollect what has been said, that these particulars form the general truth not by being merely enumerated and added together but by being seen in a new light" (Whewell 1858, p. 108).

We need to recall Blanché's discussion as well in which he pointed to Whewell's misleading use of a language of inclusion when something else and something more was taken into account. Does Laudan do that?

Laudan argues that his view differs from that offered by Butts, who mistakenly emphasized Whewell's desire to maximize content. In contrast to Butts, Laudan maintains that Whewell wished to maximize confirmation. Consilience is, he says, "a criterion of acceptability which stipulates that those hypotheses are most worthy of belief and acceptance which pass empirical tests of the kind sketched above" (Laudan 1971a, p. 372).

This opinion must appear somewhat surprising, since it indicates that Whewell thought that when theories pass tests they are worthy of some degree of belief. If this is what Laudan intends, his view is, of course, not true. When a theory passes a test or makes a true prediction it shows that it has truth content, but not that it is worthy of belief. Whewell discusses the value of exploded theories such as Ptolemy's or the phlogiston theory and praises them for their success. But he says,

> That imperfect and false hypotheses, though they may thus explain some phenomena, and may be useful in the progress of science, cannot explain all phenomena; – and that we are never to rest in our labours or acquiesce in our results, till we have found some view of the subject which is consistent with all the observed facts; – will of course be understood. (Whewell 1858, p. 85)

Whewell deemed empirical success a sign of truth content but no grounds for belief until all the phenomena could be accounted for. And at this point, as has been already explained, we see the truth without any need for the empirical evidence. In what sense, then, does Whewell seek to use confirmation as a basis for belief?

Laudan complains in a footnote that even though Whewell speaks of kinds of events, he never provides criteria for determining kinds. This criticism fully misses the point: it is the business of induction to determine kinds, and when an induction has been made we have identified a kind. Furthermore, a new induction may combine this kind with another: kinds are relative to inductions. And Laudan seems later in the essay to recognize this when he explains that a consilience is historically relative, that is, it depends on which theories are taken as a basis for a new induction. If he had a theory of kinds, we may presume Laudan would want to say – he does not say it – then there would be a foundation for induction which was not historically relative. This proposal, if Laudan should have cared to make it, would once again miss the point: kinds can only be determined with conjectures within science. Independent of the critical method of science the theory of kinds Laudan seeks is impossible on Whewell's view. This impossibility is due to the fact that Laudan suppresses: according to Whewell every induction involves the imposition of a new idea – and thus a new view of "kind."

Laudan now proposes to ask why Whewell regarded consilience to be of value. "But why," he asks, "should a consilience, whether it brings about formal unification (CI-1) or an increase in empirical content (CI-2 and CI-3), be regarded as a strong indication, let alone a proof, that the theory in which it occurs is true?" (Laudan 1971a, pp. 375-376)

Laudan reports on Whewell's claims that consiliences of inductions – in his broad sense of the term – provide indications of the truth, that when such success occurs, we know we have to do with the truth. While conceding that Whewell accurately describes "patterns of belief in the scientific community," he claims as well that "his specific arguments utterly and obviously fail to establish that hypotheses which achieve a consilience are true and can be known with certainty to be true" (Laudan 1971a, pp. 380-381).

He analyzes the weakness as follows,

> There are basically two factors mitigating against the cogency of Whewell's claim that consiliences of induction guarantee truth. The one factor, which I have already mentioned, is the logical one that it is always possible for a universal theory to be refuted, no matter how successful it has been in achieving consiliences. The second factor is a development within Whewell's own philosophy of science. He goes to some pains to argue (more or less aping Kant) that experience cannot attest to the necessity of scientific truths. But if experience, no matter how extensive, cannot be used as evidence for the necessary character of scientific truths, then the consilience of inductions, which is manifestly an experiential matter, cannot justify our claim that a certain lawlike statement or hypothesis is a necessary truth. (Laudan 1971a, p. 383)

Now Laudan concedes that Whewell knew that theories which achieved consiliences in his senses were not necessarily true but could be subject to revision. He had also earlier noted that the transition from viewing an hypothesis as a possible explanation to viewing it as the only explanation is a process which defies logical characterization. He says, "As he (Whewell) says mysteriously in *On the Philosophy of Discovery*: 'Induction is inconclusive as reasoning. It is not reasoning: it is another way of getting at the truth' " (Laudan 1971a, p. 382).

In spite of this, Laudan claims that the difficulty with Whewell's theory is due to its inadequacy as inference. Unfortunately he does not explore the "mystery" that induction is not reasoning. In order to explain the "mystery" Whewell himself provides an inductive formula which says, "These particulars, and all known particulars of the same kind, are exactly expressed by adopting the Conceptions and Statement of the following Proposition," or "These facts are completely and distinctly expressed by adopting the following Definition and Proposition" (Whewell 1858, p. 111).

And, Whewell continues,

> But it must be observed that neither of the preceding Formulae expresses the full cogency of the inductive proof. They declare only that the results can be clearly explained and rigorously deduced by the employment of a certain Definition and a certain Proposition. But in order to make the conclusion demonstrative, which in perfect examples of Induction it is, we ought to be able to declare that the results can be clearly explained and rigorously declared only by the Definition and Proposition which we adopt. And in reality, the conviction of the sound inductive reasoner does reach to this point . . . But this conviction, that the inductive inference is not only consistent with the facts, but necessary, finds its place in the mind gradually, as the contemplation of the consequences of the proposition, and the various relations of the facts, becomes steady and familiar. . . . We may, therefore, without impropriety, renounce the undertaking of conveying in our formula this final conviction of the necessary truth of our inference. We may leave it to be thought, without insisting upon saying it, that in such cases what can be true, is true. But if we wish to express the ultimate significance of the Inductive Act of Thought, we may take as our Formula for the colligation of Facts by Induction, this – "The several Facts are exactly expressed as one Fact if, and only if, we adopt the Conception and the Assertion" of the inductive inference. (Whewell 1858, pp. 112-113)

It is not only clear that Whewell was aware of the difficulties concerning inductive inference which Laudan mentions as the cause of his alleged troubles. These troubles, indeed, arise only when one, as Laudan does, ignores Whewell's contentions that in induction a new idea is always added, that in successive stages theories change so that confirmation is a sign of truth content and not of truth, that justification or proof only occurs when a theory can account for all the facts and no other theory can, and that this can be seen intuitively and the "proof" may thus be complete. Whewell sees that no theory can be proven to be true by empirical evidence, since on this basis alone other theories could always be possible. So, he takes advantage of the necessity of scientific theories: if we see that they are necessary, that no other view is possible, we know that they are true. And only then. The primary purpose of consilience, then, is not, as Laudan maintains, to increase belief but to render ideas so clear and distinct that they are seen as necessary by the scientist. Empirical successes help us along the path to this goal. Whewell's theory is far from being a solution which breaks down on the problems mentioned by Laudan. It is an ingenious, though obviously false, attempt to provide a theory of justification which takes them fully into account. If they were not present, he would have had no need, as he adequately explains, to appeal to intuition as the source of certainty in science.

Laudan has one further task to accomplish which is, perhaps, the most important because it contains the moral of the story. This task is to appraise the state of Whewell's problem today. He makes two observations concerning the impact of Whewell's theory of consilience of inductions. He first maintains that many philosophers have simply taken over Whewell's position that consiliences of inductions give some type of proof and rely, just as he did, on the fully inadequate historical justification for this position which merely notes that consiliences have in the past been acceptable. He secondly maintains that some philosophers seek to solve the problems of confirmation which these problems raise, but fail to do so. In the first class comes Stanley Jevons, along with Ernst Apelt and J. Stallo, who take it over quite uncritically, Thomas Fowler, who concedes psychological importance to consilience but demands more, Charles Sanders Peirce, who often spoke of it under different names and Karl Popper, who is alleged in the fifties to have (merely?) rediscovered Whewell's view of the importance of independent tests. The major example of the second is Mary Hesse though he mentions L. J. Cohen, J. L. Mackie and William Kneale, all of whom participated in a symposium on the consilience of inductions built around an essay by Mary Hesse which I will discuss below. Laudan's contribution belongs in this discussion so we may save it for below.

In regard to his first group of philosophers, those who remain satisfied, as Whewell did, with – presumably inadequate – historical justifications of the importance of consiliences of induction, Laudan points to several philosophers who have allegedly been, or should have been, influenced by Whewell. And this is a good beginning, since the wide influence and importance of Whewell has been systematically ignored; it is high time that the record be corrected. But in his comments on these thinkers he claims that they are mere followers of Whewell and suffer from the same defects. They all either merely appeal to history to defend the importance of consiliences even though just what is defended must vary, of course, from thinker to thinker or they see the need for more but do not offer it.

Laudan's attempt to portray these various thinkers as all concerned with "the problem of the consilience of inductions" which he has presumably explained is, I think, quite unfair to all concerned. And it is not merely unfair due to the fact that Laudan got it wrong. As I have explained above, Whewell's justification is not merely historical: it is, on his view, a scientific justification of a science of science. This is, I think, not acceptable, but it is a powerful idea which Laudan seems not to appreciate. I will not discuss other examples which Laudan gives with the exception of Popper, who is undoubtedly the most important.

Laudan begins with the claim that "Popper's major 'discovery' of the 1950s was a re-formulation of the problem of consilience" (Laudan 1971a, p. 388).

It was then, he says, that Popper saw the desirability of independent tests; Popper's requirements for a scientific theory correspond, he says, to the second two requirements of consilience. "Indeed," he says, "the only significant difference between Popper and Whewell on this issue concern the degree of confidence to be accorded to an hypothesis which passes severe tests (or, in Whewell's language, which achieves a consilience of inductions)." In a footnote we learn that "Another important difference is that Popper curiously denies that consilience of type CI-1 is of any methodological significance whatever" (Laudan 1971a, p. 389).

I find much here that is misrepresented. Popper did not discover in the fifties that independent tests are important. His theory of high degree of falsifiability presented in *Logik der Forschung* in 1934 was designed to guarantee that new, independent tests of theories would be produced. Popper's view of the need for high levels of testability is also not identical to Whewell's theory of the consilience of inductions and it does not play the same role in his philosophy as consilience plays in Whewell's. And neither do Popper's statements of his revised position in the fifties.[1] Laudan mentions, indeed, the difference in the footnote quoted above.

Popper does not recognize what Laudan calls CI-1 because he seeks criticism and refutation and not, as Whewell did, clear and distinct formulations of the truth. The purpose of independent tests on each view is radically different and each thinker made a significant contribution by pointing these purposes out. Whewell pointed out that tests were the only means we had of developing ideas and that no inductive inference in science was possible given this fact. So he sought an alternative theory of justification. Popper saw that no alternative theory of justification was possible and did without it. Whewell is interested in wide empirical success – with refutations as the path to truth – and Popper in refutations as the path to theories closer to the truth.

It finally becomes clear that Laudan's essay is merely an attempt to update Mill's analysis who, he claims, quickly saw the difficulties in Whewell's view because he saw that consiliences of inductions could not justify theories (Laudan 1971a, p. 384). This assertion I find puzzling. Mill never comprehended that consiliences – and independent tests, if we include these in consiliences – were important in science and could find no place for them in his system, because they were not inductive inferences in his sense. They could not be reduced to his four methods. Whewell had stressed that, knowing as he did that Mill's methods were irrelevant to scientific research. Mill apparently did not comprehend that Whewell offered a new theory of scientific justification and that he used this theory to justify his own view. Mill came to the conclusion, which Laudan here endorses, that the history may be accurate, but inductive justification is needed. This totally ignores the logic of the situation. Whewell saw that no theory of inductive inference could explain how scientific theories were justified and sought a new one. Now, if Whewell's new alternative fails – as we all now agree it does – we should not ignore

[1] The major point of Popper's "Three Views" (Popper 1965), which Laudan cites, was Popper's assertion that his view not merely allowed realism, the position taken in *Logik der Forschung* (Popper 1982), but was a realistic position (Wettersten 1992, pp. 209*ff*).

Whewell's arguments that a new alternative is needed. Popper's view is so important, because it takes into account what Whewell knew, that is, inductive inference offers no basis for science. Now, one may argue that Popper's view also will not do. But if one wishes to take Whewell's problems seriously, one must not beg the question, as Mill and Laudan do, concerning the question of the aim of a philosophy of science and whether it makes any sense at all to pursue a logic of induction.

## 6.2. Mary B. Hesse: An Inductivist Fudges Consiliences

The new recognition that theories of science should, perhaps, take science itself into account, has lead to a few attempts by those closer to the Millian than the Whewellian tradition to do that. Such thinkers have two central problems. The first of these is to show that the practice of science really includes some method of inductive inference. The second is to show that the inductive inferences which have been described historically are valid. Such an approach might change the study of the logic of induction by encouraging the so-called inductive logicians to begin with historical studies and then move to logic, if and when these historical studies led to results which seemed to offer a suitable a foundation for studies in inductive logic. There is, so far as I know, no serious attempt to do that.

One might also take accepted descriptions of science and ask whether the methods described exemplify valid inferences from empirical evidence to the truth of theories. There is no interesting historical work I know of which one might here appeal to in order to formulate new and more realistic problems of inductive inference. The work of Whewell might seem to be the best because of his historical insight and expertise and because he did believe that theories were in fact justified in science after a process of empirical research. We might be interested, then, in such attempts to see if they can be used to revive Mill's program, since the problems which are so posed are identical to the most crucial problems Mill faced, that is, to show that his methods were used and were valid. We might be skeptical, however, when we remember that it was Whewell's descriptions which led him to the conclusion that there was no inductive inference in the sense of inference from individual empirical statements to the truth or probability of universal statements in science. Can this conclusion be avoided?

Mary B. Hesse has proposed that it can, that she can explain how a theory of inductive inference may be constructed which is adequate to account for consiliences of inductions. The theory of induction must be revised but can be saved. She states, "I shall discuss the claim that confirmation theory cannot deal with the theoretical structure of science as elaborated in deductive systems, and that the shortcomings of confirmation theory in this respect may be understood and to some extent overcome by a modification in the idea of hypothetico-deductive systems themselves" (Hesse 1968a, p. 232).

Now it appears that what she has in mind when she mentions "the theoretical structure of science" is consilience of inductions as Whewell understood them. She states, ". . . assuming what Whewell called the "consilience of inductions," scientists do regard laws as better confirmed when they are consequences of theories than when there is only non-theoretical evidence available" (Hesse 1968a, p. 233).

Her problem is to give an account of this phenomenon. We see right at the beginning that this discussion has very little indeed to do with Whewell's description of science. And Hesse was criticized twice in this regard. L. J. Cohen, in one of the three discussions of Hesse's essay which follow it, states, "Presumably she herself would therefore not wish to pay attention here to Whewell's insistence on the inventive element in inductive generalisation" (Cohen 1968, p. 249).

Hesse responds,

> I have little to quarrel with in Dr. Cohen's helpful remarks. It is true that if what Whewell means by consilience is the inductive support of a theory which has involved a new "conception," "superinduced" upon the facts, then no single confirmation system will explicate the dependence of Cohen's expression (2) on his (3). (I explain below.) I have said as much at several points of the paper, and I do not deny that this may be a correct description of certain aspects of theoretical inference. However, if this is what Whewell means by consilience (and his modes of expression are not precise enough to enable one to be confident about this), then I think Whewell is wrong to claim that "in every inductive inference" such a new conception is involved. (Hesse 1968b, p. 254)

Whewell's historical claim is thus dismissed out of hand. The passages which Hesse has trouble interpreting so as to determine if, in every consilience of inductions, a new "'conception', 'superinduced' upon the facts" read like this: "An Induction is not the mere sum of the Facts which are colligated. The Facts are not only brought together, but seen in a new point of view. A new mental element is superinduced . . ." (Whewell 1967b, vol. II, p. 469), or like this: "Although in Every Induction a new conception is superinduced upon the Facts" (Whewell 1967b, p. 469).

Does she, perhaps, doubt that all consiliences of inductions are themselves inductions? She gives no indication where she thinks the difficulties lie, but this view would certainly seem implausible, since, if consiliences of induction were not induction, all the greatest successes of science would not be inductions, as he says they are, in reference, for example, to his table of inductions (Whewell 1967b, vol. II, p. 94). We must presume, then, that Hesse's problem of consiliences has little to do with Whewell, since it ignores just that aspect which he deems crucial in any induction: the imposition of a new idea.

Larry Laudan has also objected to Hesse's formulation of a problem related to consilience on the grounds that what Hesse calls consilience of inductions is not what Whewell called consilience (Laudan 1971a, pp. 389-390). His criticism is different from that of Cohen's, however. The problem which Hesse hopes to solve is: how can we explain the (alleged) fact that a law which follows from some theory and is itself not empirically confirmed receives some degree of support when the theory from which it follows also implies some other, empirically confirmed law? Or, Laudan suggests, there may be another problem as well, which presumes that both laws are empirically confirmed and asks whether the fact that they both follow from the same theory increases the confirmation of each. Laudan protests that the problem which interests Whewell concerns the increase in our confidence in the theory which explains both laws and what is not in the laws.

Laudan is no doubt closer to Whewell than is Hesse. But Laudan, just as Hesse, ignores the crucial feature of Whewell's theory which is that in every induction a new idea is added. Not only do they both ignore this crucial feature. They also erroneously maintain that Whewell is concerned with increase in confidence and this leads them once more to ignore the fact his real interest in empirical research lies in the increase in clarity and precision which renders ideas intuitively necessary.

In her reply to Laudan's criticism Hesse does not seek to show that her problem is the same as that which concerned Whewell but rather that her problem as well as that proposed by Laudan are closely related to each other and to the consilience of inductions (Hesse 1971). That such problems as those with which Hesse and Laudan are concerned have some connections is, of course, no surprise, since any theory of confirmation of the kind each here seem to think worth pursuing would have to take account of these and other problems. We might leave it at that, with the mere observation that they have left Whewell's description and theory of science without so much as considering them. Hesse nevertheless does claim that her theory takes account of consilience as Whewell described it. She says,

> This (that a consiliative hypothesis shows that events previously thought to be of different kinds are, as a matter of fact, the same kind of event) is just what I intend in suggesting that the theory extracts and expresses the analogies between the apparently different kinds of events it is found capable of explaining. A good theory for Whewell is not one which consists of a number of disconnected pieces, each of which is confirmed by its own entailments, but which would not "unify" or "consiliate" different kinds of events. Whewell's conditions for consilience are in fact not sufficiently characterized by Laudan in CI212, CI222, CI232, nor by my suggested solution of problem (1) above (Hesse's problem described above), for any or all of these might be

> satisfied by mere conjunction of disconnected hypotheses . . . neither Laudan nor Whewell have succeeded in making clear what it is to be an adequately "consiliative" hypothesis. I claim that the concept of laws of analogous systems fills this gap. (Hesse 1971, p. 522)

Now it is not clear just what an adequate consiliative hypothesis should be, but if Hesse means, as it would seem she does, that such an hypothesis must meet the requirements of confirmation theory her argument is, of course, circular at least in regard to Whewell. Now she may simply mean that an adequate theory of consilience must explain how two different kinds of facts are unified. But here Whewell has a clear hypothesis: a new idea is imposed on the facts. Why should this be deemed inadequate? We once again find a circular argument in which the standards of Mill are appealed to without so much as bothering to notice that they differ from Whewell's, much less that Whewell has good arguments for rejecting them. In her original essay Hesse states,

> We have to conclude from this argument (an argument to the effect that, only if there is an analogy between a theory and each of two laws which follow from it, can one of these laws, which is not empirically confirmed, achieve confirmation from the theory from which it follows due to the fact that the other law is empirically confirmed) that we never rely upon a predicted law by consilience in virtue of its deducibility from a theory, nor even its analogy with a theory, unless it also has a logical dependence on or an analogy with known laws . . . It is very difficult to rid oneself of the idea of a theory as a deductive system with levels of consequences, in spite of many indications that this is a highly artificial reconstruction, of which the fate of consilience is only one example. What might very tentatively be suggested is that the function of the theory is the indication and systematic extraction (or abstraction) of analogies between a number of empirical systems, the best theory being the one which embodies the strongest and most extensive analogy. A theory would therefore be established not so much when its deductive consequences are large in number, but when the number of analogous systems realizing it in the empirical world is large . . . The whole process would be something like Mill's induction from particulars to particulars (which incidentally, if we are realistic, also takes place by analogy and not identity of instances). (Hesse 1968a, p. 244)

We see then that Hesse wishes to use the same strategy as Mill in responding to Whewell, that is, she sees all theory construction as mere abstraction of common properties just as Mill did in his theory of description. We have seen above in our discussion of Mill what a poor response this is and how little Mill took account of Whewell's arguments and history.

### 6.3. Butts Searches for a Whewellian Theory of Inductive Inference

Butts begins a second attempt to explain Whewell's theory of induction by noting that Whewell's theory of induction has not received much attention. Especially lacking, he says, is an attempt to reconstruct Whewell's logic of induction against the background of recent discussions of inductive logic (Butts 1973, p. 53). He wishes, he says, to redress the balance. Not only were Whewell's historical studies important even as a background for Mill's work – as, we might add, the tradition has it – but they also offered a theory – really two theories – of induction. He says that Herschel and Whewell were forgotten and Mill remembered because Mill offered a theory of induction and they did not. Butts wishes to show that Whewell did, indeed, offer not one but two theories of induction. That he offered two and did not distinguish them is, on Butts' view, hardly surprising. He says,

> Whewell clearly wanted his theory of method to capture this prized kind of historical event (the point at which theories are taken to be established) and to explain it. His desires may well be seen as distorting his sense of what logic is, his insistence on keeping science in historical context may well be seen as rendering inexplicable his attempt to show that induction is demonstrative, but in all the confusion something interesting emerges – his second theory of induction. (Butts 1973, p. 57)

The first of Whewell's alleged theories of induction is "the completely non-novel hypothetico-deductivist account of the justification of scientific results" (Butts 1973, p. 54).

Butts begins by explaining Whewell's view of colligation or the imposition of ideas on the facts to order them. He offers two interpretations, one worse than the other. The first is Herschel's proposal for reconciliation with Whewell – Herschel is not mentioned in this regard – that Whewell meant nothing more than that in induction one generalized beyond the observed facts. The second is that "each proffered colligation becomes a candidate for inclusion in a (hopefully standard) observation language. Each superinduced idea thus claims a place as a predicate in the observation language of science" (Butts 1973, pp. 58-59).

Butts here notices the point which Blanché discussed so nicely: that Whewell's view of scientific progress involves the introduction of new ideas at each stage and is not, even though his language may at times suggest it, a theory of mere inclusion of given facts under higher level theories. Butts claims further that this makes Whewellian conjectures not falsifiable. "If the theory introduces just the ideas (predicates) that explain the data, and the data, so to speak, cannot speak for themselves, how can a theory ever turn out to be incorrect?" (Butts 1973, p. 59)

The answer, of course, is by making predictions which turn out to be false. It is popular among contemporary followers of Herschel and Mill to claim that if observation terms are theory-laden, theories cannot be refuted. This seems to me simply naive and I will not argue against it, but I do wish to point out the importance that Whewell gives to conjectures and refutations: this factor plays no or scarcely any role in too many modern descriptions of Whewell's theory of induction or of scientific discovery. Whewell states,

> A facility in devising hypotheses, therefore, is so far from being a fault in the intellectual character of a discoverer, that it is, in truth, a faculty indispensable to his task. It is, for his purposes, much better that he should be too ready in contriving, too eager in pursuing systems which promise to introduce law and order among a mass of unarranged facts, than that he should be barren of such inventions and hopeless of such success. Accordingly, as we have already noticed, great discoverers have often invented hypotheses which would not answer to all the facts, as well as those which would; and have fancied themselves to have discovered laws, which a more careful examination of the facts have overturned . . . To try wrong guesses is, with most persons, the only way to hit upon right ones. The character of the true philosopher is, not that he never conjectures hazardously, but that his conjectures are clearly conceived, and brought into rigid contact with facts . . . It is true, that when we have rejected all the inadmissible suppositions, they are often quickly forgotten: and few think it necessary to dwell on these discarded hypotheses, and on the process by which they were condemned. But all who discover truths, must have reasoned upon many errors to obtain each truth; every accepted doctrine must have been one chosen out of many candidates. If many of the guesses of philosophers of bygone times now appear fanciful and absurd, because time and observation have refuted them, others which were at the time equally gratuitous, have been confirmed in a manner which makes them appear marvellously sagacious. To form hypotheses, and then to employ much labour and skill in refuting them, if they do not succeed in establishing them, is a part of the usual process of inventive minds. Such a proceeding belongs to the rule of the genius of discovery, rather than (as has often been taught in modern times) to the exception. (Whewell 1858, pp. 78-80)

This is all ignored in Butts' version of Whewell's theory of induction. We go straight from colligation to the consilience of inductions as test. Butts apparently deems this proper because of the impossibility of testing theories if observation terms are theory laden. Whewell's view of the matter is hardly taken into account. In the transition section Butts discusses tests only of the same kind of event with no mention of the role of refutation. And even further, he suggests that we should perhaps not take Whewell's use of such terms as 'proof' or 'evidence' etc. too seriously but merely speak of the psychological side which is also present in Whewell's description.

The key comes next however. Butts here describes consilience first as mere inclusion of previous inductions in a more comprehensive one. He deems this quite standard with the exception of the addition of tables. (He denies what he takes to be Laudan's view that consilience involves severe testing. This depends on whether one takes as consilience Laudan's extended sense in which case a consilience may be the result of passing a severe test or Butts' narrow conception in which consilience is merely the unification of two previous inductions.) Butts notices, however, that Whewell maintains that a consilience of inductions is also an induction and this means that a new idea is added. We come to the point made by

Blanché and ignored by Laudan that a new idea is added and that this changes the empirical basis. Butts continues, "Whewell ought to have turned his attention to the question of what it is that justifies this semantical tampering; instead, he merely hints at an answer to this question and goes on to give his hypothetico-deductive account of justification of induction" (Butts 1973, p. 65).

The so-called hints are in fact, we learn in a footnote, a part of the theory now to be presented, that is, the "act of attention" which is, I might add, needed to "see" induction and the process of refinement. Whewell does not, of course, try to solve any problem of "justifying" "semantic tampering," since he saw no such problem: that scientific language changes was a fact of science which had to be explained.

The interpretation of Whewell's alleged theory of the justification of induction which Butts now gives begins with an explanation of that point at which Whewell leaves inductive logic to say that the inductive reasoner reaches a point where he sees the truth of the theory. This is the beginning of the so-called second theory, that is, the theory which does not appeal to empirical confirmation but to other properties.

Butts has by no means shown that Whewell ever had a first theory and Whewell has been quoted by me to deny it above. But we may interpret the discovery by Butts, that there is more to Whewell's theory than the theory of confirmation he first attributes to Whewell, as progress and see if the second interpretation Butts proposes might be right. Butts describes this theory as "the strikingly original (in the British methodological tradition) theory that the task of inductive logic is to generate a rule or rules on the basis of which theories are accepted or rejected, where such rules receive whatever justification they have from outside hypothetico-deductive systems of science" (Butts 1973, p. 54).

How does Butts find his interpretation of Whewell according to which theories are selected by application of rules? He observes that on Whewell's view those theories which are seen to be true are those which are consiliences of inductions. He is not satisfied, however, to leave this to intuition, as Whewell did. So, he asks further, as Whewell did not, What are the criteria by which we know them to be true? The answer he finds is that these are the qualities exhibited by theories which are consiliences of inductions, that is, primarily simplicity. Consilience is not a measure, he says, of the content of a theory, because then any theory of high content, even an ad hoc theory, would qualify. (It is, one should say, not merely a measure of content, though content is also included.) The change in the observation language serves Whewell's purpose because it enables one to attain simplicity. We thus have external standards by which to judge theories.

This is almost good. It clearly recognizes three important aspects of Whewell's view, that is, (1) that in each induction a new idea is added, (2) that ideas are seen to be true without appeal to confirmation, and (3) that the imposition of new ideas changes the observation language. But it still reads something into Whewell's view which is not there: we do not use these criteria as a test of truth. We apply them along with correspondence of the facts to refine our views. But neither correspondence with the facts alone nor, I add, satisfaction of such criteria alone, can be sufficient to show that a theory is necessary. This must be intuitively seen to be the case: it must be clear that no other theory is possible. It would seem obvious that neither correspondence with the facts nor any other criteria such as simplicity could be a test for that.

The presentation of this so-called second theory is an introduction to a thesis which is perhaps more important from Butts' point of view. This is the theory that Whewell does not offer a theory of confirmation but one of exemplification, where Butts understands confirmations to occur only where refutations are possible but exemplifications to occur without the possibility of refutation. He states, "Whewellian testing is not a form of seeking for instances of a hypothesis that either confirm or refute it – indeed, his account says nothing at all about the possibility of finding the laws to be false" (Butts 1973, p. 72).

This claim that Whewell is not interested in, and says nothing about, refutation is startling. Butts quotes Whitehead in explaining his view to say "experiment is nothing else than a mode of cooking the facts for the sake of exemplifying the law," and, Butts continues, "Whewell's treatment of experiment in his discussion of the laws of motion seems to put him in the Whitehead-Feyerabend camp" (Butts 1973, p. 73).

Perhaps Butts merely meant his denial that Whewell was interested in tests and refutations to apply to one aspect of his view. But in the end it is clear that Butts thinks this view of Whewell's theory to be quite general. He states,

> That the history of inductive science should be read by Whewell as a structure of successive exemplification is important, given his notion that the distinction between facts and theories is only relative, or does not exist at all. Beginning with the most humble empirical generalization, whose terms are exemplified by "data," and arriving at unified theories exemplified by less general laws, science exhibits a closed structure that Whewell labels 'the idealization of facts'. (Butts 1973, p. 77)

and,

> Induction being (in some sense) demonstrative, laws established by induction can be refined on the basis of apparently recalcitrant experience, and on the basis of the same experience we can list exceptions to the laws, but those laws cannot be falsified. All of which again strongly argues the point that for Whewell, Popperian corroboration plays no role whatsoever in scientific induction. (Butts 1973, p. 76)

Given the above quotation concerning the importance which Whewell gave to tests, to rejecting and to refuting hypotheses, one may be puzzled by these claims. To explain the puzzle one might note that Whewell described a path of discovery which leads to, and is indispensable for, the justification – by intuition – of the truth. Now Butts wishes to keep the context of discovery separate from the context of justification. So he might say, though he does not, that for Whewell tests and refutations are part of the process of discovery but no part of the process of induction which is justification. This separation hardly fits Whewell's theory, however.

Butts' examples and the evidence he offers for his view come, indeed, from Whewell's analysis of established laws, that is, when a theory is seen to be necessary there can be no refutation. But this is trivial and has nothing to do with the process of proof but with the result of proof. I can find no plausible basis for Butts view; he at least owes us an explanation of Whewell's concern for conjectures and refutations if we are to take him seriously.

The moral of Butts' story is, he says, that Whewell has enriched our understanding of the complexities of actual science and has pointed the direction which inductive logic should take in the future. In spite of this belated recognition of Whewell, we still have hardly left Mill's program of developing an inductive logic which can be shown to be applicable in science as Whewell described it. Perhaps we have a suggestion that such a theory should take account of empirical confirmation but also something else such as simplicity, but this is at best only hinted at. Due to Butts' failure to take account of the role of refutations in the development of science we have lost an excellent description of scientific progress. This seems to be due to the desire to read Whewell as providing an alternative to inductive logics. This excuses one, of course, from questioning the project itself and thus once more Whewell's central points are pushed aside with somewhat quixotic attempts to explain them away without ever mentioning what they are. The tradition of uncritically preferring Mill to Whewell persists even in those who wish to redo the wrong which has been done in neglecting Whewell.[2]

### 6.4. Butts Revises Interpretation of the Consilience of Inductions

In a further essay, Butts (1977) seeks to find a new view of the consilience of inductions which might render it defensible and/or useful today. His

[2] Butts' view is, incidentally, further developed in an earlier article on Newton's Rules (Butts 1970).

problem begins where the last essay left off in that he deems the hypothetico-deductive method to be defective. What he apparently means by this is that due to the impossibility of sharply separating theoretical from observational terms and due to the theory-ladenness of observational terms, it is difficult to develop a theory of the confirmation of theories within the framework of this model. He also apparently believes that scientific theories cannot be put in this form. Butts takes the opposite course, then, to that taken by the best defenders of hypothetico-deductive approaches to science, e.g. Whewell and Popper: they defended hypothetico-deductive theories because of the theory-ladenness of observation terms and have abandoned inductive inference. Butts, in contrast, looks further for an inductivist alternative.

In order to find a new approach to the consilience of inductions which integrates it in some inductive theory, he turns to an essay by Wesley Salmon in which it is suggested that a proposed theory should be plausible. For Salmon this presupposition is a basis for his Bayesian approach to the philosophy of science. Butts does not wish to commit himself here. Rather, he wishes to argue that Whewell's theory of the consilience of inductions offers help in the formation of plausible theories, should we want to do that. It offers a rule that we should be conservative, that we should use successful theories.

Butts here repeats his view that Whewell's complete system could have little appeal for anyone today and that he not only made logical slips – pointed out by such thinkers as de Morgan and Mill – but that he was proud of them (Butts 1977, p. 72). But he quickly moves on to note that all induction involves the imposition of a new idea. He attributes this interpretation to Mary B. Hesse, who, as we have seen above, was uncertain and does not mention Blanché who developed it. His acceptance of Blanché's reading of Whewell leads him to note (Butts 1977, p. 74) that consilience of inductions does not add anything to standard confirmation theory: it leaves the problems as they were. This is correct and Butts even notes that Whewell had something else in mind. Butts finds the discussions of inductive logic interesting, but he does not wish to discuss whether they are Whewellian or not. (We have seen above that they are not). He hopes to develop a new view of consilience.

Butts' new view is the following. When a consilience of inductions occurs, theories or kinds of events previously deemed different are seen to have a common *vera causa*. This does not occur due to a mere combination of the views but rather due to the imposition of a new idea which changes the meaning of the terms involved. Butts proceeds further to recommend Whewell's methodology. He claims that Whewell offered a rule which should be endorsed. This rule is: bet on past winners (Butts 1977, p. 83). Butts states,

> Despite his lack of care in logical matters, and despite the excessive oddness of much of his discussion of induction, I nevertheless think that the heart of his discussion of methodology can be saved and reconstructed partly along the lines suggested in the essay by Salmon mentioned earlier. (Butts 1977, p. 82)

Whewell, Butts continues, was trying to provide a logic of discovery as well as a theory of how theories could be justified. The best way to proceed is to pick out hypotheses which are relevantly similar to those which have been successful in the past. Therefore we should seek systems of hypotheses which exhibit consilience and simplicity. This, he says, is something like Salmon's plausibilities. He hedges his bets, however, and does not claim a Bayesian approach of the sort discussed by Salmon will work. Butts' interpretation of Whewell is based on the following passage from Whewell which he quotes and which I repeat here:

> Although, as we have said, we can give few precise directions for this cardinal process, the Selection of the Idea, in speculating on phenomena, yet there is one Rule which may have its use: it is this: – The idea and the facts must be homogeneous: the elementary Conceptions, into which the facts have been decomposed, must be of the same nature as the Idea by which we attempt to collect them into laws. Thus, if facts have been observed and measured by reference to space, they must be bound together by the idea of space: if we would obtain a knowledge of mechanical forces in the solar system, we must observe mechanical phenomena. Kepler erred against this rule in his attempts at obtaining physical laws of the system; for the facts which he took were the velocities, not the changes of velocity, which are really the mechanical facts. Again there has been a transgression of this Rule committed by all chemical philosophers who have attempted to assign the relative position of the elementary particles of bodies in their component molecules. For their purpose has been to discover the relations of the particles in space; and yet they have neglected the only facts in the constitution of bodies which have a reference to space – namely, crystalline form, and optical properties. No progress can be made in the theory of the elementary structure of bodies, without making these classes of facts the main basis of our speculations. (Butts 1977, pp. 84-85)

Now, Butts says, we can interpret this rule in a quite trivial way to simply say that when we introduce a proposition about *x*'s we should not suddenly regard it as talking about *y*'s (Butts 1977, p. 85). But he says there is another more interesting interpretation which is to read it as recommending that "consistency in application, as well as adherence to already successful modes of explanation." One should, then, retain a "well tested vocabulary" at the risk of being deemed inconsistent or not even scientific. We should build success on success.

This interpretation of this rule is surprising, because it apparently contradicts the interpretation of Whewell's theory of the consilience of inductions offered above in which a new idea and thus a new vocabulary could always be imposed. Butts mentions this right away, but thinks its adoption will not get us far. So he proposes we change our thinking about consilience in order to achieve what Whewell wanted methodologically.

And this, he says, is that we may see from the history of science which types of theories are successful.

I agree with Butts' interpretation of Whewell according to which consilience of inductions involves "semantic tampering." I find this fine. Instead of seeking to interpret Whewell's proposed rule in terms of Butts' conservatism, however, why don't we try to interpret it in terms of Whewell's theory of semantical change. On this view the rule is a response to a definite problem. Since our descriptions of the world change, we need to check to see if our theory-laden descriptions and our theories are unified. If they are not, we will be blocked; if we unify them, we may proceed. This is not trivial, since it is an answer to a new problem. It does not require conservatism, since we are free to change our descriptions if we desire. It is a demand that the individual researcher be consistent and not that he be consistent with previous science.

This is not to say that the rule takes us very far. It does not tell us, as Whewell would surely point out, how we are to know which ideas and descriptions fit together. Nor does it tell us when we should change our descriptions. It is also the case that one can use any theory one wishes as a heuristic for constructing new theories, and one need thereby pay no attention to the intentions of the author. So there is no reason to complain that Butts used Whewell's theory as a mere springboard for further inquiries. But just as Butts tells us his idea does not fit with his own interpretation of Whewell's view of consilience, he reintroduces the claim that it is Whewellian. One should at least be clear about what one attributes to Whewell. I find no reason to attribute Butts' conservatism to the adventurer Whewell.

But we do not have to rest the case against Butts' proposal with a different interpretation of the above passage. Whewell explicitly rejected this proposal. He states,

> Finally, this caution (the caution against inquiring into causes) does not even gain its own single end, the escape from hypotheses. For, as we have said, those who will not seek for new and appropriate causes of newly studied phenomena, are almost inevitably led to ascribe the facts to modifications of causes already familiar. They may declare that they will not hear of such causes as vital powers, elective affinities, electric, or calorific, or luminous ethers or fluids; but they will not the less on that account assume hypotheses equally unauthorized; – for instance – universal mechanical forces; a molecular constitution of bodies; solid, hard, inert matter; – and will apply these hypotheses in a manner which is arbitrary in itself as well as quite insufficient for its purpose. (Whewell 1858, p. 127)

# CHAPTER 7

# THE REAPPRAISAL OF WHEWELL'S PLACE IN THE HISTORY OF THE PHILOSOPHY OF SCIENCE BEGINS

The first stage of Whewellian research in which his views and his alleged confusions were explained (Chapter 5) and the second stage of Whewellian research in which philosophers of science sought to find there a new foundation for a theory of inductive inference (Chapter 6) merged into a third stage of Whewellian research in which new historical appraisals of his view and/or its importance were undertaken. One approach, taken by Gerd Buchdahl, was to reinterpret the debate between Whewell and Mill once more with the apparent intention of showing that, even though we now see Whewell in a much more favorable light, we can also see how strong a point Mill had. In a second approach philosophers such as David Hull and Michael Ruse have attempted to resurrect the view of a Herschel-Whewell methodology which thereby largely explains away his importance today even while granting he was an interesting figure in the 19th century. Among all commentators it is only Joseph Agassi who sought to reappraise him so as to show how important he was and how unfortunate the neglect of him has been.

## 7.1. Gerd Buchdahl: Saving Mill by Explaining What He Really Meant to Say

Due to the prominence which the contrast between deductivist and inductivist approaches has recently attained, Gerd Buchdahl explains, he has set himself the task of analyzing these alternatives. He does not explain his remark, but the only present-day deductivists who are mentioned are Popperians (Buchdahl 1971, p. 347) and, perhaps, Kuhn (Buchdahl 1971, p. 367). One central theme of the essay – I would like to say the central theme but the essay wanders a bit from subject to subject – is the need for inductive validation which Whewell is taken to dispense with and Mill to require. Buchdahl states,

> the fundamental logic behind this scheme (Whewell's Tables of Induction) is that of displaying a set of hypotheses to which the vast variety of data has been fitted. The tables of induction do no more than indicate graphically the deductive movement from higher-level to lower-level hypotheses, and from the latter towards the data. So far,

> therefore, we never pass beyond the process of endless fitting and continual testing, however much the architectonic of the tables may represent a stable and completed monument of theoretical achievement. Any "logical" movement that exists here is entirely from hypotheses to the data which are exhibited formally as their deductive consequences. (Buchdahl 1971, p. 345)

He continues, contrasting this view with Mill's.

> As regards the methodological aspects, what we may call "the inductive process," the method of hypothesis is given by Mill only a subsidiary role, and is joined by additional inductive schemas as for instance the derivation of causal laws from instantial premisses by means of the employment of a number of "causal methods," the details of which need not concern us here. What is important for the moment – and this concerns the logical aspect of the question, i.e. inductive inference – is Mill's view that "probation" is more than mere "fitting of data to hypotheses . . . It is not sufficient to show that the death of individual men "fits" the hypothesis that all men are mortal. On the contrary, what remains to be shown, even after all methodological precautions have been taken, e.g. elimination of irrelevant circumstances, counteracting causes and the like – a task of inductive methods – is that these individual deaths do in fact permit the inductive conclusion (inference) that all men are mortal. It is not enough to show that the data would fit some given hypothesis, but you require to "prove" that the hypothesis actually holds, "is true," or at least has a finite antecedent probability. (Buchdahl 1971, p. 346)

This seems a clear and fair enough description of the difference between Whewell and Mill, provided we overlook the fact that Whewell also thought both that theories should be shown true and that mere fitting the facts was not enough. This aspect is here omitted by Buchdahl but built into the discussion later on, so we must come back to it. Buchdahl summarizes,

> In general, then, Mill insists on some independent logical requirement that has to be satisfied before an hypothesis can be accepted; and no amount of additional "consolidative" supplements such as the existence of strong consilience, of analogy, of simplicity and so on, can take the place of induction in this sense of the term, evidently very different from that used by Whewell. As Mill sees it, the deductivist – like present-day Popperians – forever remains at the level of hypothesis. (Buchdahl 1971, p. 347)

The problem to be discussed here does not seem to be all that complex. We know what inference from particulars to universals is. Do we need that or not? Whewell says no; Mill says yes. Buchdahl, however, finds many dark difficulties and the need for a sophisticated analysis to see what was at stake. Indeed, he suggests, neither Mill nor Whewell had adequately understood the other. He states, "Whewell never grasped the force of Mill's contention that he had not produced 'proof' nor genuine generalizations, transcendence beyond the particular to the universal" (Buchdahl 1971, p. 360).

And, in regard to Mill,

> Mill construes Whewell as claiming that we can arrive at statements of necessity; even, that we can know when we have arrived them. And he holds this claim to entail its own refutation of the whole conception of necessity itself. Moreover, it prevents him from even beginning to grasp that Whewell's necessitarianism was meant to provide an alternative to Mill's own theory of validation. (Buchdahl 1971, p. 354)

Buchdahl, then, will tell us what was at stake and explain some of the confusions which arose in the debate. The result is, indeed, surprising. The debate, Buchdahl explains, concerns primarily neither the possibility of actually using inductive inference in science – as I contend – nor does it concern the validity of inductive inference. Rather,

> Mill's insistence on the importance of inductive inference turns out to be tantamount to some form of realist interpretation of theory; just as in the earlier set of arguments, he had insisted on a realist interpretation of Whewell's colligating conceptions – equivalent to the first sense of "description." Further, we saw that with this goes a sharper distinction than Whewell will permit between "fact" and "theory." In Whewell, where in some sense all is theory, the conviction of a need for special "proofs" of the adequacy of one's theory easily evaporates; yet on the other hand, Mill who so much emphasizes the need for such special proofs cannot in the end be said to have provided more than a formula, rather than an adequate justification of his programme of inductive inference. (Buchdahl 1971, pp. 363-364)

But both theories are, Buchdahl continues, inadequate and for a surprising reason. Both are fallibilists, but cannot reconcile their fallibilism with their respective methods of validation. "In sum, Whewell, like Mill, finds it very difficult to reconcile the validational principles of his logic, suggesting closure, with the openness and fallibilism of his methodology" (Buchdahl 1971, p. 367).

So, we may now, I presume, see that we are all – inductivists and deductivists alike – fallibilists and we all have now the same problem which Mill but not Whewell appreciated. This is: how can our fallibilism be reconciled with validation?

This is, indeed, a startling set of results and one might ask how he came to them. Let us look at four aspects of his argument: one, his discussion of the inductivists as well as of Whewell's view of necessity; two, his discussion of Mill's views of proof or validity; and, three and four, two discussions of Mill's use of the theory of description. In the first discussion Buchdahl suggests that not only do inductivists wish to add something to mere hypotheses but that there must be inferential reasoning as well. This reasoning, however, Buchdahl suggests, is not certain but fallible. And still further, it may involve an underlying assumption of the uniformity of nature or perhaps also of laws of causation. And for still others, such as Kneale, it may involve the assumption of principles of necessitation. Buchdahl uses vague terminology in describing the relation between the demand for inference and the need for such principles. At first it is a mere underlying assumption, then a characterization of different types of inductivism and, then, finally, as quoted above, the demand for one is asserted to be equivalent to the demand for the other. The upshot is that inductivists are realists and, indeed, of different sorts.

Can that be said of Whewell? On Buchdahl's view, no. On Whewell's view the demand for what appear to be the same principles has no ontological significance. Buchdahl states,

> . . . necessitarianism is also an important aspect of some central examples of deductivism. The most prominent thinkers here are Kant and, influenced by him, Whewell. They differ however from inductivist necessitarians in disclaiming any ontological significance for nomic necessity. Paradoxically, not unlike Hume of the Treatise, necessity for them is grounded in something "subjective" or "mental," though Hume's psychologising solution is converted into something with transcendental or a priori status. (Buchdahl 1971, pp. 348-349)

This allegedly subjective aspect of Whewell's view is further emphasized. He says,

> In other words, Whewell emphasizes the "theory-ladenness" of our data. In general, then, his is a subjective and dynamic approach . . . there is not really some independent realm of "fact," and certainly no underlying ontological basis, against which to check this edifice; and the requirement of a separate inferential step does thus not arise. For this reason alone we should expect Whewell to lack the notion of inductive inference of the kind espoused by Mill. (Buchdahl 1971, p. 350)

We have here the heart of the matter. Inductive inference is only then demanded when we are realists, but then it is, indeed, necessary. Whewell was, however, a realist. He did hold that ideas really do describe the world. Buchdahl nevertheless claims that, on the one hand, Whewell is not a realist because he follows Kant and, on the other hand, that he is confused because he does not follow Kant. He allegedly mixes what he calls transcendental and metaphysical questions, that is, he presumes that fundamental ideas describe the world. One might argue that Whewell is forced into an idealist position. But this is only plausible if one recognizes that he himself strenuously fought against this view and offered his theory of how he had avoided it in, for example, his first memoir on the Fundamental Antithesis of Philosophy or his response to Herschel.

Now Buchdahl might claim that he has shown that Whewell is forced into idealism. To be sure Buchdahl never says that Whewell himself deems his theory realistic. But he does say that Whewell deemed ideas to be general truths and that a psychological interpretation does not capture what he meant. But he quickly uses Mansel's criticism to the effect that the necessary laws must be mere forms. He states, "But Whewell pretends that the resulting formulations, i.e. the interpreted axioms, are instances of the existence of necessary truths" (Buchdahl 1971, p. 353).

And this, he says, is a misreading of Kant. We find again that it is demanded of Whewell that he agree with Kant without argument. Since there is hardly any point in taking this demand seriously, we may turn to the discussions of Mill's theory of descriptions.

The first discussion of Mill's theory of descriptions presents the dispute as if it concerned a realist-idealist dispute, where Mill defends realism by maintaining that descriptions are merely copies of what is given. Buchdahl's characterization is, unfortunately, false. Mill was a naive realist and Whewell a sophisticated one. But Buchdahl recognizes no realism outside of naive realism. He states,

> In other words, observational and conceptual aspects are always (for Whewell) inextricably intertwined. This is not an answer (the view that facts and theories or facts and inductions are both needed but relative) that would satisfy a realist like Mill. He simply contends that the concept-ladenness of observations must not be confounded with induction, which is the centre-of-gravity in which is anchored his realist metaphysics of validation, the counterpart of Whewell's notion of "idea." (Buchdahl 1971, p. 362)

Buchdahl turns to a second discussion of Mill's theory of description which he introduces by saying, "Curiously, the opposition between Whewell and Mill is also argued out in terms of a second meaning of description, almost the opposite of the realist interpretation just distinguished, thus offering fruitful occasion for mutual misunderstanding" (Buchdahl 1971, p. 362).

In fact Mill introduces two discussions of description directly following one another (Mill 1973, 292*ff*). In the first he is concerned to show that descriptions of the world are possible without the imposition of any idea. They describe merely what is there. This is, of course, necessary in order to answer Whewell's criticism that descriptions are theory-laden. In these claims he follows Herschel quite closely. But Mill has a second problem as well. Whewell argued that scientific research begins with colligation of facts, where colligation of facts is the imposition of some idea on the world in order to select and to order relevant facts. If, however, we begin in this way, there is no independent basis of facts from which we might induce general statements. It was thus necessary for Mill to distinguish two processes: one consisted of finding facts from which inductive inferences were to be made and the second consisted of making them. But in order to solve this problem Mill had to concede, indeed, he emphasized, that various descriptions were possible as a basis for induction.

Now Buchdahl maintains that various descriptions were not deemed by Mill copies of reality but were considered to be more like a mere model or a fiction or possible representation of the phenomena, (I do not find such terms in Mill's exposition) and as such are neither true nor false but offering – as Buchdahl puts it – "no more than convenient understanding"

(Buchdahl 1971, p. 362). But Mill says something quite different which Buchdahl simply ignores:

> Considered as a mere description, the circular theory of the heavenly motions represents perfectly well their general features: and by adding epicycles without limit, those motions, even as now known to us, might be expressed with any degree of accuracy that might be required. The elliptical theory, as a mere description, would have a great advantage in point of simplicity, and in the consequent degree of accuracy that might be required; but it would not really be more true than the other. Different descriptions, therefore, may be all true: but not, surely, different explanations. (Mill 1971, p. 299)

Mill then, does not have two senses of description but two discussions of description with different purposes which he tries to reconcile, that is, descriptions are deemed mere copies of reality in order to show that ideas are not imposed on the world but various descriptions of the same state of affairs are possible – and therefore must be true – in order to allow for various colligations of facts to serve as a basis for induction.

Buchdahl continues by saying that Mill requires both description which requires no inductions and explanation which does. Mill does not succeed, however, in constructing both, Buchdahl informs us, and will tell us what the confusion is.

> To make his point, Mill has however to shift his argument to a different aspect of "induction," the "transcendence" involved this time relating to the putative existence of a causal mechanism. Thus, whilst the different astronomical hypotheses of Ptolemy, Tycho and Copernicus may be just so many different "descriptions," Aristotelian, Cartesian and Newtonian dynamical theories are putative explanations, and it is only these that can be true or false, because only the latter make existential claims. (Buchdahl 1971, p. 363)

This portrayal is false. Rather, many alternative descriptions can, on Mill's view, be true – and others, no doubt false – while only one hypothesis can be true. Mill says, "Can it be said of these (the latter group), as was said of the different descriptions (the former group) that they are all true as far as they go? Is it not clear that only one can be true in any degree, and the other two must be altogether false?" (Mill 1973, pp. 299-300)

And then, Buchdahl adds,

> Mill's "shift" is easy to understand, for the difference between "mere model" and "genuine explanation" is difficult to define, if the contention is that it is a difference in kind and in some form ultimate; and with that difficulty the "induction" invoked in the second case becomes problematic. However, Mill conflates the distinction with another, more clear-cut: viz. that between kinematical and kinetical theory; between accounts which restrict themselves to a "description" of motions, and those which involve forces introduced in order to explain these motions. By conflating the two contrasts, Mill makes it easy for himself to answer Whewell; for whilst the difference between "mere method of representation" and genuine "explanation" is one that is extremely difficult to define, and which moreover appears to have little relevance to the employment of

> induction, the second contrast can be dealt with in a straightforward way, if one insists (as Mill does) that the hypothetical employments of explanatory agents involves "induction." It is however no wonder that Whewell and his contemporaries, and indeed most subsequent readers of Mill have been confused by these vacillations in his argument. (Buchdahl 1971, p. 363)

Buchdahl argues, then, that Mill's demand for inductive inference was equivalent to the demand for realism. This is so because Mill, unlike Whewell, was not satisfied with merely fitting theories to the facts. This point was not seen by Whewell, however, because in his examples Mill confused the distinction between the difference between explaining and describing with the difference between a kinematical and a kinetical explanation.

What did Whewell say about Mill's example? He said he could find no difference between the alleged description (the portrayal of the motion of the planets) and the alleged explanation (the theories of Descartes or Newton) other than the first group referred the movement of the planets to motion alone and the latter group to force. Whewell analyses Mill's position as follows.

> The only meaning which I can discover in this attempted distinction of Description and Induction is, that when particular facts are bound together by their relation in space, Mr. Mill calls the discovery of the connexion Description, but when they are connected by other general relations, as time, cause and the like, Mr. Mill terms the discovery of the connexion Induction. (Whewell 1971, pp. 248-249)

I am afraid I can find no difference between Buchdahl's analysis of Mill's confusion – his explanation of Whewell's failure to grasp Mill's point – and that analysis of Mill's confusion offered by Whewell himself.

Having understood Mill's confusion quite well, could Whewell have failed where Buchdahl has succeeded in divining the real import of Mill's demand? Can Mill be vindicated in this way? Buchdahl concludes his argument so.

> Mill's insistence on the importance of inductive inference turns out to be tantamount to some form of realist interpretation of theory; just as in the earlier set of arguments, he had insisted on a "realist" interpretation of Whewell's colligating conceptions – equivalent to the first sense of description. Further, we saw that this will permit a sharper distinction than Whewell will permit between "fact" and "theory." (Buchdahl 1971, p. 365)

Mill certainly did intend to draw a sharp distinction between fact and theory and to do so with his theory of description. But one can hardly say he succeeded. The assertion by Buchdahl that Mill's insistence on induction is an insistence on realism rests on mistaken portrayals of both Mill's theory and Whewell's theory. He mistakenly portrays Mill's theory as involving two views of description and he mistakenly portrays

Whewell's theory as not realistic. The preference of the tradition for Whewell cannot be explained or retroactively made plausible in this way.

### 7.2. The Myth of a Herschel-Whewell Methodology

The view that Whewell added a rather exaggerated and confused philosophical interpretation to his excellent historical studies and methodological observations has been in the air ever since Herschel reviewed Whewell's two major works. A part of this view has been that the methodology of Herschel and Whewell (or of Brewster and Whewell) were for all intents and purposes the same. This view is not only still with us. It has been recently given even more prominence. There has obviously been a great deal of confusion about the relations of the three most memorable actors on the stage of nineteenth century philosophy of science – Herschel, Whewell and Mill. Most theorists such as Blanché, Lange, Whewell and Mill himself saw Mill's *Logic* as a continuation of the views of Herschel, which were opposed to Whewell. Other thinkers – especially more recent analysts – have sought to revise this view to assert the similarity of the views of Herschel and of Whewell and to separate Herschel from Mill. They might even have chosen to appeal to Herschel himself as authority for this interpretation. Prima facie the facts of the case give some comfort to both moves: Mill developed philosophically the views proposed by Herschel in his description of science and Whewell used some aspects of the description of methodology offered by Herschel in his description of science.

One aspect, then, of the question of whether Whewell's critics have prevented him from doing good is whether his methodological theory offered any advances, which were wrongly rejected, perhaps only much later to be accepted. This question was answered in the negative until Blanché. Either Whewell was deemed to have had no unique methodology, since his methodological view – as distinct from the philosophical interpretation of it – was for all intents and purposes identical with Herschel's, or his suggestions were built into other views. This view is especially popular now, since Whewell's theory has become more interesting and Mill's much less. Perhaps philosophers now want to suggest that it was Herschel, or Herschel and Whewell together, but not Mill, and certainly not Whewell alone who made the needed innovations. Perhaps philosophers do not wish to admit that they and their predecessors have simply selected the poorer of the various alternatives offered as a foundation for their work. So, if Whewell is to be mentioned, one can mention Herschel-Whewell and thereby suggest that a mere question of recognition is at stake; no question of the revision of our heritage, as we define it, need be raised.

It would seem interesting, then, to review the logic of the situation once more. Were there really significant methodological as well as philosophical differences between Herschel and Whewell? If so, why were they not discussed? And what problems arose due to this failure, if any? We may seek once more to understand the rejection and neglect of Whewell to see what methodological advances, if any, he offered. In order to conduct such a review, we have to go back to see what Herschel offered before the publication of Whewell's essays and what methodological changes he provided.

### 7.2.1. *Herschel's Methodology: Hedging One's Bets*

Herschel's *Preliminary Discourse* of 1830 was no doubt an important book. Herschel was one of the most influential scientists of the day and thus his book had great authority. It was no doubt taken seriously by his two major successors, that is, by Whewell and by Mill. For some time, that is, from 1830 until the publication of Whewell's *Philosophy* in 1840 or Mill's *Logic* in 1843, the only serious competitor it had was Brewster's *Life of Newton*, which was published in 1831. Herschel's text thus served to set the stage for the two great efforts which followed it and which became the center of discussion in the 1850s. This later debate has been seen by many thinkers as a debate between Whewell on the one hand and Mill as a defender of Herschel's point of view on the other. These include Whewell, Mill, Lange, and Blanché. But Mill not only accepted many of Whewell's methodological observations, which might have been in accord with Herschel's point of view, he also quarreled with Whewell about the use of descriptions, the nature of observation and the use of the four methods of induction, where Herschel's position is not so clear. Is it possible then to deem the methodologies of Whewell and Herschel, who argued primarily about the interpretation of scientific method, but somewhat less about its methods, as good as the same, even though one would not say this of the methodologies of Whewell and Mill?

Herschel's *Preliminary Discourse* is an apology for and introduction to science. The book has three parts. The first part of the book is a portrayal of the benefits which science brings mankind; the second is a description of the methods by which the results of science are obtained; and the third section is a description of some of the central theoretical results of the sciences. The popular defense of science in the first part is based on its ability to give power to men. It is of no particular interest here. Joseph Agassi has commented on Herschel's advocacy of a success philosophy and his use of such a philosophy in his view of science: science is that which enables mankind to gain power and ascendancy in the world where, given man's other characteristics alone, he would be in a very difficult situation indeed (Agassi 1971b). The third part need also not concern us, even though Herschel's division of the study of the methods of science in one part and the central ideas of science in another was, it would seem, taken over by Whewell in his own later philosophical work.

We need, then, to look at Herschel's discussion of the methods of science. In order to discuss Herschel's portrayal it would be useful to know his problem situation. Agassi claims that the most dramatic development of the early nineteenth century and the prime source of methodological problems was the overthrow of Newton's optics by Young and Fresnel. The methodological problems arose due to the overthrow of theories which, according to the standard methodological interpretations, had to be true. How, then, was error in science possible? How should we explain it?

The overthrow of Newton's optics was not the only factor, however. The standard methodology was Baconian and Bacon had come under attack. So there was a new problem as well – which Agassi says was due to Brewster – which was to explain the importance – or lack of importance – of Bacon. Of considerable importance for Brewster and Herschel was the fact that Bacon's theory could not explain the successes of science. It could not explain the successes of Copernicus, Galileo, Kepler or Newton. How, then, could it be the true theory of scientific method?

Herschel's essay is an attempt to explain how Baconian methodology can be the true methodology of science. He combines an endorsement of Baconian methods with a description of those methods in fact found in science, methods which apparently are given no or too little a place in Bacon's theory. The whole is an attempt to present these as a unified theory. It comes apart at the seams and becomes a mere endorsement of Baconian methods along side of the recognition of the need for hypotheses and independent tests, as Agassi has pointed out. Let us look a little more closely at the text.

Any attempt such as Herschel's to reconcile Baconian theory with scientific practice had at least four central problems. The first of these four problems was due to the fact that observations are in fact not given in a simple form but result from the application of theory or interpretation of events whereas Bacon's theory says that observations are the independent foundation of science. The second problem is that Bacon's theory is given credit because it represented the discovery of how to do science, yet it followed the most stunning events of the scientific revolution, i.e. Copernicus, Galileo and Kepler, to say nothing of other figures such as Gilbert. The third problem is that in practice the collection of facts proceeds with theory and even the identification of events depends on them, but the Baconian methodology asserts that facts are collected first as a basis for induction. The fourth problem is that as a matter of fact conjectures and hypotheses are important factors in scientific research, but on Bacon's theory all discovery is made by induction from facts to theories.

Before discussing Herschel's treatment of these four problems and the methodology which is a product of this discussion I wish to make a proviso about Bacon. I make no attempt here to correct or criticize Herschel's reading of Bacon. It had been suggested then and is still suggested that standard portrayals of Bacon's system are oversimplifications or misrepresentations. He was, it is said, more subtle in his proposals or, as others would have it, simply more confused than the standard representations of his view allow. This will not concern me here, since it was Herschel's task to reconcile the oversimplified version of Bacon with the practice of science and to do so because this was the version which, it was then widely agreed, was the correct description of scientific method. Such reinterpretations in any case could at best reduce the gap between Bacon and Herschel, perhaps showing they were confused in the same ways, that is, in their combination of deductive with inductive methods.

In the first two chapters of the second part of his essay Herschel discusses experience as the foundation of all knowledge. In the first of these two chapters he proposes that experience is the basis of all knowledge of the world – this is contrasted with mathematics, since we can there come to truths, say, the properties of triangles, without appeal to experience (Herschel 1966, pp. 75-76). But experience, he is quick to point out, is not entirely unproblematic. Though experience never is wrong, our judgments about it may be.

> These, and innumerable instances we might cite, will convince us, that though we are never deceived in the sensible impression made by external objects on us, yet in forming our judgments of them we are greatly at the mercy of circumstances, which either modify the impressions actually received, or combine them with adjuncts which have become habitually associated with different judgments; and, therefore, that, in estimating the degree of confidence we are to place in our conclusions, we must, of necessity, take into account these modifying or accompanying circumstances, whatever they may be. (Herschel 1966, p. 83)

Herschel had already explained above what had to be done.

> But before experience itself can be used with advantage, there is one preliminary step to make, which depends on ourselves; it is the absolute dismissal and clearing of the mind of all prejudice, from whatever source arising, and the determination to stand and fall by the result of a direct appeal to facts in the first instance, and of strict logical deduction from them afterwards. (Herschel 1966, p. 80)

Herschel explains what he regards as the sources of false judgments. These are false opinion and the senses. The major difference between them is that errors due to false opinion are to be removed by a skeptical attitude toward hypotheses, which he calls prejudices, whereas the errors due to the senses may be removed by noting that other sense impressions contradict them, as when we see that an object takes the color of the rays of light which fall upon it rather than some definite color which we would be inclined to call its own.

Herschel thus retains the Baconian formula that we must clear our minds of all prejudices and recommends that we do so. But he recognizes as well that to do so is a task for science itself and not merely a task for those who wish to begin to do science. In the second chapter of the second part, then, he explains the progress of science as due to the analysis of phenomena. We do not, he notes, start with the simplest phenomena but aim toward the discovery of them and their relations to each other by analysis. There are however, no rules for such analysis. He states,

> but, it will now be asked, how are we to proceed to analyse a composite phenomenon into simpler ones, and whether any general rules can be given for this important process? We answer, None; any more than (to pursue the illustration we have already had recourse to) general rules can be laid down by the chemist for the analysis of substances of which all the ingredients are unknown. Such rules, could they be discovered, would include the whole of natural science; but we are very far, indeed, from being able to propound them. However, we are to recollect that the analysis of phenomena, philosophically speaking, is principally useful, as it enables us to recognize, and mark for special investigation, those which appear to us as simple; to set methodically about determining their laws, and thus to facilitate the work of raising up general axioms, or forms of words, which shall include the whole of them; which shall, as it were, transplant them out of the external into the intellectual world, render them creatures of pure thought, and enable us to reason them out a priori. (Herschel 1966, pp. 96-97)

Herschel thus recognizes that we cannot start with the mere observation of facts. In order to determine what the facts are, analysis is needed. But at least in some passages Herschel presents this analytical process as merely an addendum to Bacon's methodology, since it is the analysis which enables us then to "raise axioms" as the above quotation shows. This process of arriving at axioms on the basis of facts is, he says, what we mean by induction (Herschel 1966, p. 102). We thus may proceed from facts to theories, according to Herschel, even though analysis of observations is required before induction may take place, and even though there are no rules for analysis and even though, if there were, they would constitute – as above quoted – the whole of science. It is not clear, then, whether analysis is the whole of science or merely preparatory to induction. Before judging on this matter, however, we need to discuss his views of the collection of facts and of induction which are developed in subsequent chapters.

Before continuing the discussion of Herschel's methodological views, it is useful here to turn to his discussion of the role of Bacon. This is useful not only because the role of Bacon is the subject of the third chapter of the second part; it is also of some independent interest, since Herschel deems Bacon to be the true discoverer of the proper view of science. Why? He states his general view as follows.

> It is to our immortal countryman Bacon that we owe the broad announcement of this grand and fertile principle; and the development of the idea, that the whole of natural philosophy consists entirely of a series of inductive generalizations, commencing with the most circumstantially stated particulars, and carried up to universal laws, or axioms, which comprehend in their statements every subordinate degree of generality, and of a corresponding series of inverted reasoning from generals to particulars, by which these axioms are traced back into their remotest consequences, and all particular propositions deduced from them; as well those by whose immediate consideration we rose to their discovery, as those of which we had no previous knowledge. In the course of this descent to particulars, we must of necessity encounter all those facts on which the arts and works that tend to the accommodation of human life depend, and acquire thereby the command of an unlimited practice, and a disposal of the powers of nature co-extensive with those powers themselves. A noble promise, indeed, and one which ought, surely, to animate us to the highest exertion of our faculties; especially since we have already such convincing proof that it is neither vain nor rash, but, on the contrary, has been, and continues to be, fulfilled, with a promptness and liberality which even its illustrious author in his most sanguine mood would have hardly ventured to anticipate. Previous to the publication of the Novum Organum of Bacon, natural philosophy, in any legitimate and extensive sense of the word, could hardly be said to exist. (Herschel 1966, pp. 104-105)

Herschel deems, then, Bacon's description of the general path of science not only to be true but also to be largely responsible for the success of science and this very success is the foundation of his apology for it. The failure to follow Bacon's philosophy is given as the grounds, why the thinkers of Ancient Greece and the Middle Ages produced so little knowledge. But Galileo at least cannot be explained away in this fashion and the success of him and other scientists cannot be explained by their adherence to Baconian philosophy. Or can it? Their success was, according to Herschel, due to their adherence to the facts. They thus practiced Baconian philosophy without having articulated it. Herschel states,

> It has been attempted by some to lessen the merit of this great achievement, by showing that the inductive method had been practised in many instances, both ancient and modern, by the mere instinct of mankind; but it is not the introduction of inductive reasoning, as a new and hitherto untried process, which characterises the Baconian philosophy, but his keen perception, and his broad and spirit-stirring, almost enthusiastic, announcement of its paramount importance, as the alpha and omega of science, as the grand and only chain for the linking together of physical truths, and the eventual key to every discovery and every application . . . An immense impulse was now given to science, and it seemed as if the genius of mankind, long pent up, had at

> length rushed eagerly upon Nature, and commenced, with one accord, the great work of turning up her hitherto unbroken soil, and exposing the treasures so long concealed. A general sense now prevailed of the poverty and insufficiency of existing knowledge in matters of fact; and, as information flowed fast in, an era of excitement and wonder commenced, to which the annals of mankind had furnished nothing similar. (Herschel 1966, pp. 114-115)

Bacon, then, was the key for the development of science since, what Galileo had done merely by instinct, Bacon had correctly described. He thus provided the key to the further development of science. It is this Baconian heritage, then, that must be maintained. And Herschel will, he says, describe the Baconian methods in the following chapters.

The claim Herschel makes for the importance of Bacon is quite problematical as may be rendered clear by asking: did Newton's success depend on Bacon? It would seem fantastic to say that it did. I do not believe anyone does so explicitly, but it is done implicitly and with a vague claim: it was the fact that Bacon made the methods of science clear and that he was so enthusiastic, which led to science. But this will not do, since, as Brewster pointed out, Newton did not use Bacon's methods. And science was followed just as intensely in France before Bacon gained influence as afterwards. Science was no doubt different because of the prestige of Bacon. Much pointless gathering of facts, for example, might have been avoided, if it had not been for his prestige. But we lack any plausible explanation of why thinkers such as Newton or Hooke or Boyle needed his methodological advice, of why they could not learn how to do science from Copernicus, Brahe, Kepler and Galileo. What role, then, does the ideology of Bacon play in the history of science? We need, as 19th century thinkers began to realize, better answers than we then, or now, have. Herschel, at any rate, was committed to the view that scientific method could be identified with Bacon's and sought to explain that.

We need to consider two problems which Herschel discusses more closely to see if he can present a plausible version of his Baconian methodology. These are the problems of how scientists gather facts and what is the role of fact-gathering in science, on the one hand, and the nature and role of induction, on the other hand. In the first of his two chapters on the problem of the how scientists gather facts and of their role in science (Herschel 1966, Chapter IV, Part II). Herschel discusses the problem of how observations are to be made. He argues that they must be observations of those facts which occur uniformly and under the same conditions. And thus it is of extreme importance that one observe all the relevant conditions under which some event occurs. It is important that descriptions of events be exact. It is equally important that one report only what one observes and that one avoids all mixing of the views and language of erroneous theory with that of simple fact.

Up to this point we have a pretty good Baconian description of the problem of observation. We gather facts precisely, without using prejudices or false theory. We may even, Herschel suggests, make accidental observations of great importance. But at the end of his chapter he strikes a quite different tone. For, here he states that

> To make a perfect observer, however, either in astronomy or in any other department of science, an extensive acquaintance is requisite, not only with the particular science to which his observations relate, but with every branch of knowledge which may enable him to appreciate and neutralize the effect of extraneous disturbing causes. Thus furnished, he will be prepared to seize on any of those minute indications, which (such is the subtlety of nature) often connect phenomena which seem quite remote from each other. (Herschel 1966, p. 132)

Herschel thus moves from a Baconian view, in which facts should be gathered quite free of any prejudice, to one in which facts can only be gathered by someone who is quite well-informed. On the one hand theoretical elements in the descriptions of events are to be avoided; on the other hand true theoretical elements in the description of events is necessary.

The following chapter (Herschel 1966, Chapter V, Part II) continues this discussion of the nature and role of observations by discussing how events are to be classified. He begins by noting how necessary some nomenclature is, in order to simplify the discussion of classes of events. But he warns against too elaborate systems of classification which achieve order but do so arbitrarily: the classification should be natural. And this can be achieved by noting analogies or resemblances between events. We must refrain from imposing classifications prematurely, since new relations are daily discovered. If, then, we classify elements according to resemblances and analogies and in a natural way, we may have a good foundation for induction, which is the subject matter of the next two chapters. In this chapter, then, Herschel is, with very slight qualifications, true to his Baconian methodology; and it is here that he prepares one for the discussion of induction.

In his two chapters on induction (Herschel 1966, Chapters VI and VII, Part II) Herschel begins quite in accord with his Baconian views but, as we shall see, seeks to extend and modify these views to take better account of scientific practice. He begins by noting that the first task which a scientist faces is to find causes – *vera causae* – of specific events. Each event should be explained by reference to an immediate producing cause. If we do not know what such a cause may be, we may, of course seek to uncover it. But how? Here Herschel presents a quite Baconian point of view, that is, we begin by classifying events to note the agreements and differences or change in intensity. He here provides Mill with the basic material for his four methods, which are a slight reworking of Herschel's proposals for the discovery of proximate causes.

The chief difference between Herschel and Mill is that Mill makes the use of his four methods the whole of scientific method and their sole purpose the discovery and proof of laws. Herschel views them as often a mere beginning. He explains, for example, that in its then advanced state science is chiefly promoted by the discovery of so-called residual phenomena (Herschel 1966, p. 156). Residual phenomena are those phenomena which are discovered when the use of the four (or five) prime methods reveal events which are inadequately explained. Such residual events are often new facts, which would not have been discovered without the use of the methods. And it is this discovery of the limits of existing explanations which leads to new discoveries of *vera causae*.

There is a problem for Herschel's theory, however, which he does not explicitly point out but, on my reading of his text, he was quite aware of. This problem is that in his examples of the discovery of residual effects he presents cases of the discovery of mere counterexamples, that is, some theory or law is presumed and is found incapable of explaining this or that event. The four methods then become quite irrelevant and the connection between scientific method and Baconian theory is broken. I think that Herschel saw the difficulty because, after his more sophisticated examples of the discovery of residual causes from chemistry and physics, Herschel presents a simpler example in which he tries to connect explicitly the discovery of residual causes with the four methods. His example is that of Wells' theory of the discovery of the cause of dew. This example later became of central importance for Mill and, when reading Mill, one can easily come to the conclusion that he used this example merely because he knew so little science. But Mill can here appeal to the authority of Herschel who says,

> We have purposely selected this theory of dew, first developed by the late Dr. Wells, as one of the most beautiful specimens we can call to mind of inductive experimental enquiry lying within a moderate compass. It is not possible in so brief a space to do it justice; but we earnestly recommend his work . . . for perusal to the student of natural philosophy, as a model with which he will do well to become familiar. (Herschel 1966, p. 163)

Now the theory does have the advantage that one can pose various alternative explanations of dew as the mere product of the observations of agreements and differences. And one can then see whether such agreements and differences are adequate to explain all phenomena including some not previously considered. Is this a model for all such discoveries? As we shall see below Herschel does not believe that either. Herschel gives a further discussion of the verification of scientific theories or laws. He describes an empirical law, for example, as a mathematical formula which includes the results of various measurements. These, he says, cannot be trusted unless they are independently verified. And he stresses the importance of independent tests. According to Agassi he was the first to note the importance of such tests. We have now come quite far from the methods which are based on classification of events in accord with agreements and disagreements. But there is more to come.

The basic Baconian methods of science are based on the proposition that the facts can be observed and classified. Theories are then derived from and explain the observed relations. Herschel knew that this was a quite inadequate portrayal of scientific method, because science has depth. We cannot directly observe, he notes, the universe because it is too big, and we cannot observe many other processes, because they are too small. And when we explain, for example, electricity as a stream, we have not observed this stream directly. Rather we have observed some analogy between the phenomena produced by electricity and that produced by a stream. Science of this sort requires methods of a quite different kind than that accounted for by methods of classifications of observed phenomena, which are all that Baconian methods allow for. Herschel discusses, then, in his second chapter on induction the problem of "the higher degrees of inductive generalization, and of the formation and verification of theories." One of the most important examples of such a theory is the wave theory of light. As Agassi has pointed out, it was the success of Young and Fresnel in overturning Newton's theory of light in favor of a wave theory which was one of the most important events in science at the beginning of the 19th century and which posed a problem for the standard methodology of science. One of the shocks was, of course, that Newton had been proven to be wrong, though Herschel never mentions that. He is, however, concerned to explain the methods which are used in such cases and which do not fit the Baconian model.

In accord with the foregoing chapter Herschel puts forth as a foundation the view that all attempts to form theories must presume particular *vera causae* found by induction. But how do we proceed from there? There are, he says, three methods. We can proceed as before by induction; we can form at once a bold hypothesis; or we can use a third process which combines the benefits of the first two. He describes this process as follows. We can arrive at the laws which regulate the action of our primary agents (*vera causae*)

> by assuming indeed the laws we would discover, but so generally expressed, that they shall include an unlimited variety of particular laws; – following out the consequences of this assumption, by the application of such general principles as the case admits; – comparing them in succession with all the particular cases within our knowledge; and, lastly, on this comparison, so modifying and restricting the general enunciation of our laws as to make the results agree. (Herschel 1966, p. 199)

I take the difference between the second and third methods to be that in the first there is no induction to a higher law, whereas in the second there is an induction to a higher law which is then – as a matter of conjecture – extended to other instances and tested on them. Herschel has here apparently admitted that induction is not needed as in the second method. But he is quick to point out that it is the second method which is of most importance in science. As to the second he says,

> But the difficulty and labour, which, in the greater theories, always attends the pursuit of a fundamental law into its remote consequences, effectually precludes this method from being commonly resorted to as a means of discovery, unless we have some good reason, from analogy or otherwise, for believing that the attempt will prove successful, or have been first led by partial inductions to particular laws which naturally point it out for trial. (Herschel 1966, p. 200)

Herschel thus admits conjectures into science, but only grudgingly. They should only be made when we have inductive grounds for them and in most cases we may use a modified version of induction to discover theories or laws which are not obtainable by simple inductive methods. He points out further the importance of tests and of some methods of observations such as the use of averages when observations cannot be made precise. But the main points have already been made.

We come back, then, to the problem I posed above concerning the relation between the methodologies of Herschel and of Whewell and, we may add, to that of Mill. I noted there that it is not uncommon that the methodologies of Herschel and of Whewell are mentioned as if they were the same. There are, it is clear, differences which are so glaring that no one can deny. The only possible defense of the view that the methodologies of Whewell and of Herschel are the same is that the aspects of methodological theory on which they differ are unimportant. Let us look then at some of these differences.

The primary and quite general difference between Herschel and Whewell is that Herschel insists that we begin with observation and that we use the inductive methods later made standard by Mill. One of Whewell's major intentions in his work and in his dispute with Mill was to show that these methods had no place in science. It is true, of course, that Herschel treats these methods more restrictively than Mill does, who treats them as all of science and sought to reduce all other methods to them. Herschel does not do that, because he emphasizes the importance of independent tests, whereas Mill denied such tests were of any importance to science, because they were not reducible to his four methods. But Herschel restricts these further methods as much as he can and presents them all as based on inductions which precede them.

One can hardly say that a thinker, who asserts the primacy of observation or collection of facts, has the same methodology as one who denies that the collection of facts without theory plays any useful role in science at all. And one can hardly say that someone who says that bold conjectures are of limited value and should be only made when we have good inductive grounds for them had the same methodology as one who recommends that one should look to make bold conjectures and refute them. Nor can one say that a thinker who views the fact of the impregnation of descriptions with theory as a problem of observation, which should be overcome as best it can, has the same methodology as one who asserts the desirability of using theory-impregnated language in describing events – even if the theories are false. And, we may ask further, whether someone who believes that theories are justified by sufficient verification has the same methodology as someone who believes tests and verifications are merely methods we use to construct theories which may be seen to be true because they are clear and distinct.

There are, then, drastic methodological differences between Herschel and Whewell, which anyone who would want to maintain that they defended the same views would want to take into account. It appears, indeed, that Herschel is, as one might expect, quite close to Mill. They agree on the importance of the collection of facts as a basis for induction; they agree on the methods of induction; they agree on the need to limit conjecture and they agree on the verification theory of the justification of scientific theories. They disagree only in that Herschel adds qualifications and makes allowances for occasional use of other methods and emphasizes the importance of independent tests – he knew the impact of these in science, say, of the confirmation of the unexpected prediction of Fresnel's theory that there should be a light spot at the center of the shadow of objects under certain conditions. Mill denied all that in the name of consistency. We may conclude from this that Mill deemed Herschel's methodology inconsistent as did Whewell and that each sought to remove either the Baconian aspect (Whewell) or the newly found interest in conjectures and independent tests (Mill).

It would seem that a brief look at the methodologies of Herschel and of Whewell is sufficient to show how different these methodologies really are. It may be, of course, that we make the task too easy, so let us look at a couple of examples.

### 7.2.2. *Michael Ruse: Keeping Score of Successes Only*

Michael Ruse (1975) is one of those who maintain that there was a nineteenth century methodology which was shared by Herschel and Whewell. He states,

> Herschel and Whewell came to differ quite considerably over what I have called "metaphysical" aspects of science, Herschel inclining more to empiricism whereas

> Whewell was much influenced by Kant. However, they differed little, if at all, with respect to "methodological" questions, the kind of theory a scientist should aim to produce and the way he should try to confirm it. (Ruse 1975, p. 162)

Although Ruse's primary interest is in Darwin and Darwin's method, his analysis of Herschel and Whewell is of some importance for his discussion. This analysis is a presupposition of his discussion of the influence of philosophy – of the philosophies of Herschel and Whewell – on Darwin. So, let us see how Ruse argues for this point of view. We will no doubt have to begin by noting what Ruse considers methodology and whether his conception is adequate, since the quote just given seems rather narrow. Can Ruse make his case in this way?

Ruse begins by seeking to sort out Herschel's methodology from any "metaphysical speculations." He states,

> I shall now sketch those tenets of Herschel's philosophy which might have been of interest to a budding scientist; that is, I shall ignore Herschel's metaphysical speculations on the ultimate nature of science and concentrate exclusively on his methodology. I shall consider what, in Herschel's opinion, was the kind of theory a scientist ought to aim for and the kind of evidence a scientist ought to offer. (Ruse 1975, p. 160)

The evidence which Ruse gives for his analysis is, indeed, rather short and astounding. Herschel, he says, accepts a hypothetico-deductive model for scientific theories, that is, the view that scientific systems are distinguishable from other systems, because they contain laws and that a law supports counterfactuals. Herschel further, he says, distinguished upper level laws – "fundamental laws" – from lower level (derived) laws – "empirical laws." Although empirical laws are important, scientists aim at the production of fundamental laws and the best kind of fundamental law is quantitative. He finds that Herschel also hints at the distinction between "observable and unobservable concepts" of the modern logical empiricists, though what this has to do with Whewell or the common methodology is somewhat mysterious: Whewell's position here was certainly clear and quite different from Herschel's. Further, he says, Herschel emphasizes the need for *vera causae*. This need is interpreted in various ways by Ruse. He says that it means we should not use ad hoc explanations and that when we have tested a theory in various situations we may be sure we have found a *vera causa*. A *vera causa* thus appears to be any non-ad hoc explanation and the true explanation. Finally, Herschel emphasized the importance of the confirmation of theories by surprising and unexpected events in convincing scientists that their theories are true (Ruse 1975, pp. 161-162).

This catalogue may now be compared against Whewell's later researches and Ruse finds that the two thinkers are, for all intents and purposes, in complete agreement. "Hence," he says, "both with respect to theory-nature and with respect to theory-proof Herschel and Whewell spoke with almost one voice" (Ruse 1975, p. 164). Whewell adopted Herschel's view that quantitative laws are the best, he advocated a hypothetical-deductive ideal for science and he distinguished as Herschel had done before him "between two kinds of laws, speaking of 'formal' or 'phenomenal' laws and 'physical' or 'causal' laws, the models for this division being, once again, Kepler and Newton" (Ruse 1975, p. 163).

Ruse completes his case by pointing out that Whewell, just as Herschel, gave especial value to confirmation of theories by new, surprising cases as in his theory of the consiliences of inductions.

Ruse has argued that Herschel and Whewell had the same methodology, because they agreed on two central issues, that is, on the nature of theory and of its confirmation. One might concede that not only did they both hold that theories were deductive systems, but also that one might call theories on both views – to use modern terminology – hypothetico-deductive systems. But where differences arise, Ruse is quiet. He does not note, for example, any difference between Herschel's and Whewell's distinctions between empirical laws and, in Whewell's terminology, Fundamental Ideas or, in Herschel's terminology, the laws which regulate the action of our primary agents. (I find Ruse's terminology misleading here, since for Herschel laws of phenomena are also causal; they are, indeed, *vera causae*). Empirical laws are for Whewell stages on the way to the discovery of a fundamental idea. They become known with certainty as knowledge grows and are no longer (merely) empirical laws. For Herschel, on the contrary, empirical laws are the certain foundation on which we build further laws which encompass them.

It is quite the same in regard to verification. Both thinkers emphasize that theories should be independently testable and that confirmation by surprising independent tests are of especial value. But this is only part of their respective views of confirmation. Herschel emphasized the importance of methods which later came to be known as Mill's methods and Whewell denied them any role at all. For Whewell confirmations did not prove a theory true. Rather, it was proven only when it was intuitively seen to be necessary. For Herschel the confirmations alone were important.

In the two aspects which Ruse has selected to compare the methodologies of Herschel and Whewell, he does find some points on which they agree, and even on which they stand together and disagree with Mill. There are omissions which, if discussed, would reveal important differences. But far more glaring in Ruse's account is the failure to even consider the problem of how scientists proceed, when they make discoveries and/or how they should proceed. Is this not the core of methodology? Here Herschel sides nearly fully with Mill and Whewell breaks new ground by rejecting the inductive methods, the need for fact gathering before theory formation, and the need for a neutral language to describe events. In its place he recommends the steady use of bold conjectures and refutations until the facts are truly accounted for.

The differences between Herschel and Whewell and the similarities between Herschel and Mill which Ruse simply ignores are enormous. Even if one could reduce methodology to the study of the nature of theories and of verification, leaving fully aside all procedural questions as irrelevant for methodology, he has difficulties. But a proposal to leave these aside is surely absurd.

### 7.2.3. *David Hull: Confusions about Essentialism Blur Differences*

Due to obvious differences between them, the methodologies of Herschel, Whewell and Mill cannot be deemed the same. But one might still suggest that the differences between them are so slight from today's perspective, that one can hardly learn from the disputes which these thinkers had with one another. One can see that they have all been superseded, due to knowledge which renders all these views equally out of date. Their disagreements depended on common assumptions which are now rejected, and which, as a consequence, are of little interest for contemporary methodology. From an historical perspective, we can see that the decisions made then in favor of one or the other were dubious, whichever way one decided, since all alternatives were confused. If not his major thesis, this is certainly the tenor of comments offered by David Hull. He states, for example, "The doctrine of existence of static, immutable natural kinds underlay all of the previously discussed topics – inductive proof, occult qualities, and teleology – and evolutionary theory struck a strong blow against this doctrine," (Hull 1973, p. 68) and "In reading the scientific and philosophic works of the period, the modern reader often finds himself wondering how essentialism could have had the slightest appeal or could have seemed in the least plausible" (Hull 1973, p. 70) and

> Contemporary philosophers of science commonly concede that early philosophers of science were hopelessly confused on such key issues as the differences between discovery and proof, between deduction and induction, and between inference and concept formation. Herschel, Whewell and Mill raised important questions and made significant contributions to our understanding of science, but theirs was not the last word in the philosophy of science. (Hull 1973, p. 29)

and

> Herschel, Mill, and Whewell all concurred at least verbally in Bacon's confused distinction between deduction and induction in terms of increasing and decreasing generality . . . (Hull 1973, p. 23)

and

> That they (such philosophers as Herschel, Whewell, and Mill) rejected evolutionary theory, a theory which has outlasted many of the theories judged to be exemplars of scientific method, says something about the views of science held by these philosophers and scientists. (Hull 1973, p. 14)

and

> . . . theoretical innovations in science from Darwin to Einstein have necessitated a complete re-evaluation of the notion of natural kinds. The result of this re-evaluation has been the recognition that separate stories must be told for geometric, biological, and chemical species. Philosophers of science contemporary with Darwin tried to tell one story which would be adequate both for the "deductive sciences" like geometry and arithmetic and the "inductive sciences" like physics and biology. (Hull 1973, p. 71)

The tenor of Hull's remarks, then, is: first, that the philosophies of the nineteenth century, that is, those of Herschel, Whewell and Mill, failed to give adequate accounts of various aspects of science such as the aim of science, confirmation, or occult qualities; second, that these failures were due to a common error, that is, essentialism or the doctrine that there are entities such as species with specific and unchanging properties which may be – even must be – identified in science; third, that these methodologies were not up to the task of evaluating Darwin's theory; and, fourth, that Darwin's advance made all these theories passé by breaking the bounds set by essentialism.

All three methodologists rejected Darwin's *Origin of the Species*. All three failed to appreciate the strength of the evidence, which Darwin brought forth in favor of his theory. This failure indicates the inadequacy of their theories of scientific proof and/or confirmation, Hull claims. Whatever differences they may have between them, we may see that in this crucial respect they may all be (equally?) rejected. They failed to appreciate Darwin's theory, because Darwin's theory rejected the view that there were natural kinds: it presumed that species changed. But all three theorists held that there were entities with essences, and thus that species could not change.

Hull's proposals glitter with hope of decisive progress, due to the breakthrough which is apparently above all due to Darwin. According to Hull the overthrow of the view that species cannot change was at the same time a rejection of the view that science should seek to understand the essences of objects. If an adequate account of science is to be given, a new, non-essentialist aim of science must be postulated. The breakdown of the criticisms of Darwin offered by Herschel, Whewell and Mill is simultaneously a clear demonstration that the views of such nineteenth century thinkers as Herschel, Mill and Whewell are hardly relevant any longer.

We have seen above how important and interesting some of the innovations made by Herschel and Whewell were. And even Hull suggests they may have made some. The tenor of Hull's remarks may thus seem surprising, since on his view any advances they might have made could not enable them to appraise theories such as Darwin's adequately. On the crucial point they all broke down. Let us look, then, at Hull's analysis of the relations between Darwin and our three thinkers to see if these relations show the breakdown of 19th century methodologies. Is Hull's dismissal of these philosophies based on a failure to appreciate the criticisms which they offered of Darwin, the significant differences between their methodologies, and the relative strengths of their methodologies? Is his dismissal too quick and/or too easy? Or, has Hull identified the crucial weakness found in all 19th century methodology? Has he thus offered not only a decisive reason for rejecting them, but a challenging analysis for contemporary philosophers of science as well?

In order to answer these questions we may look at a few issues which we may now resolve, provided Hull is right in maintaining that these views all failed for the same reason. Three questions which Hull's discussion raises are: one, What is the relationship between Darwin's theory that species change and the essentialist view that there exist unchanging entities with definite properties? two, What is the relationship between essentialism and the quest for verification or some other form of proof? three, What is the relationship between essentialism and the quest for knowledge of deeper layers of reality? By discussing these questions we may see how far our three thinkers came and how far Hull has gone beyond them.

Our first question, then, is: What is the relationship between the view that species change and the (essentialist) view that there exist entities which have specific unchanging properties or essences? At least in some passages Hull suggests that essentialism must lead to a rejection of Darwin's theory and that this shows that the methodological theory is false. He states,

> Any evolution at all, regardless of the mechanism, as long as it is gradual, conflicted with the essentialist notion of natural kinds – and on the necessity of such natural kinds, empiricists and "idealists" were in agreement. The doctrine of the existence of static, immutable natural kinds underlay all the previously discussed topics – inductive proof, occult qualities, and teleology – and evolutionary theory struck a strong blow against this doctrine. (Hull 1973, p. 68)

Hull wishes to illustrate this direct connection between essentialism and the rejection of Darwin with Whewell as one example. On his view Whewell rejected Darwin's theory, because on Whewell's view there could be no knowledge, if evolutionary theory were true. Hull does not spell it out, but if I read him correctly this consequence is allegedly due to Whewell's essentialism (Hull 1973, p. 68). Is this true?

When Whewell rejected Darwin's theory he appealed neither to essentialism nor to the impossibility of knowledge given Darwinian theory. Hull claims that Whewell thought species had to be real and that without natural kinds there could be no knowledge. But in the passage he cites, in which Whewell discusses real entities, there is no reference to evolutionary theory, and in the passage on evolutionary theory there is no reference to epistemology. On the basis of the texts he cites at any rate Hull's claim that Whewell rejected Darwin's view because, if Darwin's theory were correct, there would be no knowledge is simply false. Whewell did not see this consequence and explained, indeed, how he thought they could be reconciled.

Whewell saw no need to respond to Darwin's theory at all, since Darwin's text added on his view nothing which would require him to modify his earlier critique of evolutionary theories (Todhunter 1876, vol. II, p. 434). In this earlier critique Whewell does not offer explicit metaphysical grounds for rejecting evolutionary theory, say, that it conflicts with some doctrine of natural kinds. Rather, he argues that any explanation of changes of species – which he would prima facie allow – would have to be complex. He also found other difficulties concerning the geological record, for example, and the origin of life. But his central difficulty was that, if one postulated the change of species, which he allowed, then he saw the need a deeper explanation of this change. He had thought about how such a deeper theory of how species change could be constructed, but he found them so implausible that he deemed it simpler and preferable to suppose that species were real (Whewell 1967a, vol. III, pp. 480*ff*). His rejection of Darwin, then, was a product of his analysis of the possibility of providing a deeper, realistic theory of how species change and not of either a methodological or metaphysical analysis alone. Perhaps, however, Hull deems this demand for a deeper explanation essentialist. We need to return, then, to this demand below. We will see that Hull rejects the need for such a theory in order to justify evolutionary theory.

In his discussion of Mill, Hull attempts to further make his case that 19th century methodologies failed due to their essentialism, but with a different argument.[1] Hull argues that Mill's essentialism or his theory of natural kinds was the foundation of a false theory of verification. This led him in turn to reject Darwin's theory as mere hypothesis. We come, then, to the second question: What is the relationship between essentialism and

---

[1] Really, he speaks of Herschel and Mill but leaves Herschel aside and discusses Mill alone (Hull 1973, pp. 68*ff*).

standards for verification? Hull avoids those difficulties for his argument which would arise if he took the various theories of Herschel, Whewell and Mill into account by concentrating on the most naive of the three, on Mill. Mill's mistake, Hull suggests, was due to his theory of natural kinds. He argues that the four (or five) arguments Mill allows in science presume their existence. They make this presumption, because the purpose of the use of Mill's methods is to identify the agreements and differences between various phenomena. This aim presumes in turn that certain entities always have the same (essential) properties. Thus, no theory such as Darwin's which rejected this presumption of essential properties could ever be justified by Mill's methods.

Mill did not argue that Darwin's theory could not be proven, but only that it was not. He himself did not concede that his methods could not be used to prove such a theory. Is Hull nevertheless right that it could not and Mill did not notice that? In a trivial sense of course he is: Mill's methods cannot be used to prove anything. As Whewell pointed out, they have hardly anything to do with the practice of science. Even the example of the discovery of the cause of dew, Whewell suggests, is not the discovery of some new cause, but the discovery of which among known causes is operative in a specific situation. But it is clear that Mill thought that changes in phenomena could be explained, if his methods could be applied. He saw, I presume, no problem in using his methods to discover a deeper theory which would explain changes in phenomena, say, in species. So, even his theory of natural kinds as well as his theory of confirmation would be reconcilable with Darwin's theory and in the same way that Whewell's was, that is, a deeper theory will explain the changes in species.

Although Hull explains how he wishes to make his case regarding the role of essentialism for Herschel, he nevertheless suggests that nineteenth century methodologists – in effect, Herschel, Whewell and Mill – sought to develop an adequate theory of confirmation or of proof and each failed to provide such as a consequence of the alleged fact that they never overcame their (common) confusion over induction and deduction. (Is this confusion due to essentialism?) For Bacon, Hull claims, induction was supreme. Herschel, he continues, already recognized the importance of deduction and was followed by Whewell and Mill, but the succeeding discussion did not serve to render things clear and, indeed, we seem still to be in the dark. So, there was no progress beyond Herschel.

This proposal will hardly do without an analysis of Whewell's methodology. Hull merely asserts that Whewell was confused about deduction and induction. Whewell allegedly did not grasp the quite general point that deduction and induction cannot be identified with ascent and descent (Hull 1973, p. 23). Does he think that Whewell failed to distinguish between a theory of deduction and a theory of science as deductive? Does Hull fail to appreciate the difference? It is clear that Hull's broadsides against the alleged common confusions, which he asserts without analysis of individual theories, cannot lead to anything like a plausible argument for his view.

Hull does note that Whewell's view of induction was quite different from Mill's (Hull 1973, pp. 69*ff*) but that a belief in "a finite number of discrete kinds of entities underlay both Mill's and Whewell's notions of verification" (Hull 1973, p. 70). But he never analyzes this relation for Whewell. He does suggest that Whewell's philosophy of science is "equivocal" in regard to Darwin's theory (Hull 1973, p. 36). I do not know what this means except, perhaps, that it did not lead to a decision about it. But then it must be consistent with it. But Hull denies this, when he says that, given Whewell's theory, if Darwin is right, there can be no knowledge. Secondly, he claims that the major fact which the essentialism of Mill should explain is that he rejected Darwin's theory, that is, that he deemed it a mere hypothesis (Hull 1973, pp. 7-8). Yet Hull says himself that Darwin's theory was faced with serious difficulties (Hull 1973, p. 451). What, then, should be explained by reference to essentialism, when the skeptical reaction to Darwin's theory could be justified by appeal to scientific arguments acceptable to Hull?

Hull's attempt to bring the theory of verification into close connection with essentialism requires that he explain the relationship between, on the one hand, the possibility of and need for explanations with depth or explanations which explain phenomena as the products of real processes going on behind them and, on the other hand, verification. We come, then, to the third question: What is the relation between essentialism – the view that there exists a final or fundamental layer to reality – and the quest for knowledge of deeper layers of reality? This question arises, since, contrary to Hull's view, essentialism is at least then reconcilable with change of species when change of species is explained as a product of some underlying process, which is in turn described in terms of some (deeper) unchanging entities. Should one seek to do that? Should one seek to explain the change of species, but without appeal to such entities?

We come here to a one of Hull's central themes: occult qualities. Now occult qualities have been castigated for centuries and become crucial in the debate over Newton's view: was force an occult quality and thus unacceptable? Hull sees here a close relationship between essentialism and the use of occult qualities. Essential attributes, he says, tend to become occult. Scientists, he says, hope to discover true, universal generalizations, but they seek to take exceptions into account, rather than to dismiss them as essentialists do (Hull 1973, pp. 40-41). Now Herschel and Whewell are supposed to be essentialists. But Herschel does not dismiss exceptions and

Whewell builds his philosophy around their discovery and use as impetus for improvements. Hull offers no explanations for these apparent discrepancies between his analysis and the views he analyzes. He avoids discussion of Whewell's view vis à vis occult qualities entirely.

On Hull's view Darwin did not offer any view of entities that are unacceptable or occult. In this respect he is even better than Newton, who might be deemed to have done so due to his use of the notion of force (Hull 1973, p. 49). Hull does not think, however, that Newton's introduction of the notion of force is illegitimate, that is, that it is not, as Hull says a scientific theory must be, "empirically meaningful." It is acceptable because it has a high explanatory power (Hull 1973, p. 48). Darwin, however, can achieve high explanatory power without appeal to anything which even slightly resembles an occult quality and is thus better.

What, then, is Hull's view of the quest for knowledge of layers of reality behind the phenomena? It would seem that he thinks it an essentialist mistake. The test of the acceptability of a scientific theory apparently is whether a theory is "empirically meaningful" (Hull 1973, p. 38). Is this Popper's test of refutability? Or does "empirically meaningful" mean something else? Hull does not tell us. The upshot, however, seems to be that the quest for knowledge of layers of reality behind the phenomena is questionable but, perhaps, not per se bad. It depends on the "empirical meaningfulness" of the theories which postulate them.

Is Hull in favor or opposed to the attempt to explain the change of species in any way which would be an addition to or substitute for current evolutionary theory? He can hardly be opposed in principle even if he thinks the pursuit misguided, since that would be an extension of our knowledge. He says that different stories should be told for geometrical, chemical and biological "species." Does this mean these fields cannot be in any way unified by deeper theories?

Hull apparently endorses what he takes to be a modern view of scientific method (Hull 1973, p. 72). He calls this view the currently received view and seems to say thereby that it is the best. But this is merely hinted at. One would be inclined to expect that on this view one could now see that Darwin's view is empirically justified. If Hull has no such view, why does he complain that 19th century philosophers deemed it a mere hypothesis? If he has one, why does he not spell it out?

On Hull's view there are apparently at least two reasons which explain why we might better appreciate Darwin today. In the first place one may now see that in order to justify a theory one need not answer questions about underlying processes, as Whewell claimed. In the case of Darwin's theory one need not explain the change of species or explain creation in order to show that evolution took place. In the second place, one can now see that the ability to explain a wide range of phenomena is sufficient, even though there are gaps in the evidence, such as the failure to exhibit a change of species or gaps in the geological record. Hull thinks we can do better, then, if we lower standards for justification. But he admits that his view is not yet adequately developed due to unsolved problems of induction.

Hull deems "the" modern view to be an advance due to the recognition that theories cannot be proven but can only be rendered probable (Hull 1973, p. 30). But does Hull believe that if one believes in the existence of natural kinds or if one endorses "essentialism" one must believe in the possibility of proof? I do not know, but this is quite obviously not the case, since it is certainly consistent to hold that entities of this sort exist, but cannot be identified by humans. And if proof is not required by essentialism, then the modern view may have other sources than this view alone. Still further, Whewell had already adopted a stronger hypothesis concerning empirical evidence: empirical evidence could not even show that a theory was probable. This view is also a modern view and, as Hull has at least implicitly conceded, is not refuted, since no method for the demonstration of the probability of scientific hypotheses has been demonstrated.

My discussion of a few aspects of Hull's proposals thus far may be summed up as follows. Hull's view of the relationship between essentialism and Darwin's theory, which says that essentialists cannot reconcile their views with change of species, is false: Whewell considered this possibility and rejected it as false, but not as inconsistent with scientific method. Hull seems to recognize this in other passages where he notes that Whewell thought evolutionary theory possible (Hull 1973, p. 64) and where he notes that one could respond not by rejecting the theory of natural kinds but by rejecting specific candidates (Hull 1973, p. 76). In his discussion of Whewell's rejection of Darwin's view, however, he ignores that.

Herschel or Mill also did not offer the kind of metaphysical objection to Darwin, which Hull finds so important. Allegedly their theories of verification embodied essentialist assumptions which lead to the rejection of Darwin. But Herschel should have had neither metaphysical nor methodological trouble with such a view. Hull never explains his case in regard to Herschel. Mill could also, perhaps, avoid conflict between his methodological view and evolutionary theory. He thought so at any rate. But his view is so confused and naive, one can here, at any rate, agree that the view is too primitive to bother with.

These results which we found in the discussions of the first two questions led us to the question of whether the rejection of essentialism requires us to abandon all quest for knowledge of layers of reality behind the phenomena. Hull does not explicitly tell us, but his answer seems clearly to be no, since he suggests that this process sometimes has been successful. It is apparently better to avoid doing that, however, since one wants theories which are "empirically meaningful." Darwin offers us a better example of this than Newton did. But how then are we to judge theories? We need a new theory of induction or a hypothetico-inductive model which we do not yet have. But, Hull says, we know it is the correct (received) view anyway. We find here, then, that Hull simply has very little to say about this crucial aspect of his analysis.

Before concluding my discussion of Hull, a few more words about his view of essentialism and its importance for methodology may be of interest. Early on in his studies of the questions discussed here Hull apparently felt encouraged in appraising the situation in the way he did due to the philosophy of Karl Popper. He quotes Popper with praise for having pointed out that the progress of methodology is the story of the overcoming of the doctrine of essentialism of Aristotle (Hull 1965, pp. 314*ff*). This is the beginning of an essay in which he argues that in discussions of Darwin's theory this doctrine is still not overcome: it is continually found in taxonomy or in attempts to define species. Progress in this area, then, would require that it be dispensed with here as well, as Mayr had noted before him. Hull, then, continues to attempt to develop Popper's suggestion in his later writings by tracing the defects of various methodologists to their essentialist doctrines and advocating that we learn from Darwin to reject this view and develop alternative views. Popper's proposal, then, is, indeed, the foundation for Hull's work.

Though Hull begins his discussion of methodology and essentialism on the foundation of Popper's proposal, he ignores Popper's own development. In the *Poverty of Historicism and The Open Society and Its Enemies* it is hard to tell whether Popper wishes to reject all attempts to discover real entities behind the phenomena or only those attempts which rely on definitions (Wettersten 1992, pp. 172*ff*, 212*ff*). Hull quotes one of these early statements where the attempt to discover deeper layers of reality is seemingly identified with essentialism. Popper later developed a somewhat different position. It was important for him then to offer a realistic view of science and still reject essentialism. The result is a modified essentialism which is found in his essay, "Three Views Concerning Human Knowledge" (Popper 1965): it is essentialist in that it asserts that we do seek knowledge of deeper levels of reality, but rejects any explanation as absolute.

Hull also finds a close relationship between essentialism and the attempt to gain knowledge of layers of reality behind the phenomena. I do not know what his view is, however, unless it is the view that "empirically meaningful" theories only should be allowed in science. But this is obviously too little as Popper – whose view in this regard Hull apparently follows ever so closely – clearly recognized. Hull seems to hint that a good theory of induction would solve the problem of developing a non-essentialist view of science. But how and why? He has no answer unless, perhaps, he wishes to reject depth. But, as we have seen above, he does not do that either. Whewell had already gone farther than Hull in seeing that induction would not do but that theory depth was still needed.

Popper's suggestion that the central difficulty of the philosophy of science has been to overcome essentialism is, no doubt, interesting. An analysis of the effect of this doctrine in a specific field such as evolutionary theory might also have been interesting. But Hull gets the historical analysis wrong by moving too quickly from the claim that the view that species change is incompatible with essentialism to the view that Herschel, Mill and Whewell were all essentialists, to the view that essentialism stood in their way, that it prevented them from recognizing the value of Darwin's theory. Nor does he explain its alleged methodological consequences. He also does not explain the relation between the quest for knowledge of layers of reality behind the phenomena and essentialism and does not even note that Whewell first and Popper after him have offered views of the quest for discovery of deeper layers of reality, which meet many objections against essentialist views. Both rejected the view that final layer or complete explanation is possible. He falsely suggests (or hints) that proof – rather then mere probability – is a consequence of all essentialist views and thereby fails to note Popper's combination of a modified essentialism with the view that no proof is possible.

The kind of mistakes which Hull makes in his historical analysis and methodological proposals might have been (or can be) avoided, if we take the views of Herschel and of Whewell and their differences somewhat more seriously than Hull has bothered to do. In order to understand and avoid the defects of essentialism, it is useful to understand attempts to do that even if they have failed. Whewell offers one attempt: he did not seek definitions and noted that entities could have an endless number of properties which had to be taken account of by theories, but no special set of these had priority. Hull himself says that we need different "stories" for geometrical, chemical and biological species. But his view is here so metaphorical and primitive that it cannot be deemed useful.

Hull represents the continued attempt to save the program of Mill and of Herschel even while seeming to reject them along with that of Whewell. The innovations of Whewell are first conflated with the views of Herschel and Mill. Then all views are rejected as out of date. Then the program of inductive proof of Herschel and Mill is reintroduced as new. The problem of depth of scientific theories is pushed aside or depth is deemed harmful or equated with old-fashioned inductive proof even though it is just this problem which the rejection of essentialism leads to. It is hard to say just what Hull thinks. Yet it is here, as Popper well knew already in the sixties (if not the thirties), that major difficulties and opportunities are to be found. As philosophy Hull's views and the discussion of them do not bring us far. They do serve as a bell-weather, however. One can here see how the rejection of Whewell is apparently rendered of no import and the following of Mill and of Herschel vindicated: we should seek better theories of induction rather than abandon this attempt in favor of the pursuit of deductivist alternatives as Whewell and Popper have done. The resulting confusions which are introduced or continued in today's discussion are, of course, the major damage that such discussions as Hull's do.

### 7.3. Joseph Agassi: Whewell as a Progressive Inductivist

Joseph Agassi was one of the very first to notice that even though Whewell is among most important philosophers of science in the whole history of the subject, he is still neglected (Agassi 1975b, p. 42). He deems his philosophy to be in various respects a forerunner of Popper's (Agassi 1975b, p. 337). And he has often commented on the originality, quality and perceptiveness of Whewell's philosophy. He has thereby done much to make good the wrong done to Whewell and to bring his contribution to our attention.

Yet Agassi also deems Whewell an inductivist philosopher of science and attributes to him many of the defects of this view (Agassi 1981, p. 28; 1963, pp. 28, 64). He gives us then two views of Whewell. He is a progressive, activist theorist of science, as well as a conservative, sometimes stodgy, inductivist. Can these two images of Whewell be reconciled? Or must we deem Whewell to have futily sought to combine two quite different and incompatible attitudes. We would then have a third view, which deems him confused as the standard one also does, but which sees his innovations as exceedingly valuable even though he unfortunately did try to pour this new wine into the old inductivist bottle.

At times Agassi describes Whewell as a Kantian inductivist (Agassi 1963, p. 28) or as a Kantian variant of inductivism (Agassi 1981, p. 174). In another place he attributes his attraction to his combination of activist and contemplative traits. But he deems the attraction of this combination so uncritical as to require explanation. His allure, then, Agassi says, is due to the success of Newton (Agassi 1975b, p. 391). In still another place he portrays Whewell's view of science as questions plus induction (Agassi 1975b, p. 254).

Such a characterization of Whewell as a deeply divided personality is prima facie puzzling. But Agassi seeks to explain his attribution of two attitudes to Whewell, while avoiding charging him with incoherency in the following way. He attributes to him quite progressive views of the process of scientific discovery. Whewell is here deemed an *avant garde* philosopher, because he emphasized the need for conjecture, for criticism, for tests and for corrections in science. But, since Whewell nevertheless deemed the final theory, that theory which in the end withstands all criticism, to be true, Agassi deems him an inductivist. In the end, on Agassi's view, Whewell falls back on empirical justification and probability. Whewell's theory of discovery, then, is progressive and powerful; his theory of justification is nevertheless old fashioned inductivism. Agassi's view of how to reconcile the Kantian and non-Kantian aspects of Whewell's philosophy thus also accords to a degree with the tradition, which sees his view of discovery rather favorably, but his views of justification an inadequate inductivism. The tradition deems him not inductivist enough, however, whereas Agassi deems him too inductivist.

If we do not press too hard, Agassi's interpretation of Whewell provides us with a basic, simple and correct view of Whewell's philosophy of science: on Whewell's theory we seek to eliminate theories until one is left and that is the true view (Agassi 1975b, p. 42; 1985, p. 36). So far so good. But when Agassi deems Whewell's theory inductivist, we might expect him to run into trouble: we have seen above that throughout his career he was opposed to inductivism. In order to check this expectation, it will be useful to look at Agassi's various discussions of Whewell's theory.

Agassi offers no developed view of Whewell, at least not in any single discussion. He offers, on the one hand, capsule presentations of Whewell's philosophy, which should serve to make his position vis à vis alternatives – especially Popper's – clear. On the other hand, Agassi has discussed separately various aspects of his philosophy, say, his history or his historiography of science. We may seek, then, to fill out the details of Agassi's capsule presentations of Whewell's philosophy with his numerous but scattered discussions of it. In his commentaries we find Agassi criticizing and praising Whewell. Whewell is an inductivist who defends old fashioned views, who relies on verification to prove scientific theories probable, who endorses Bacon's ladder of axioms, who writes a naive, black-and-white history of science and who reacts dogmatically and a bit cowardly in defending Newton against the new ideas of the adventurous Faraday. But since Agassi deems Whewell's view a mixture of Baconian and Kantian views, we also find him treating Whewell not as a member of the inductivist tradition, but of a critical one. He writes, to be sure, inductivist history of science, but when he did it, it was the heyday of such efforts and was written to exemplify his philosophy (Agassi 1963, pp. 10, 27-28). His methodology was progressive, because it saw the need for conjectures and criticism, for theory in order to make discovery and for intuition in scientific research. He saw that we had to learn from our mistakes. He had a sophisticated theory of observation as theory-dependent. And his progressive influence in science and its philosophy through such figures as Claude Bernard and Pierre Duhem or in geology was substantial.

When we seek to put these comments together puzzles arise. Whewell is both progressive and reactionary. He is both a thinker far ahead of his time and a typical expression of his age. Agassi seeks to lessen this tension by explaining how the progressive aspects of Whewell's view are infected by his conservative inductivism. This puzzling picture might, of course, be true. The strained unification of such traits is paradoxical, but individuals often fail to adequately reconcile differing aspects of their own views. Agassi's analysis might even be a corroboration of the standard view of him as confused. Before jumping to this conclusion, however, it would behoove us to look at the two sides of the picture to see if Whewell is so paradoxical a figure as Agassi renders him.

### 7.3.1. *Whewell, the Inductivist*

The characterization of Whewell's view as inductivist requires more than the mere observation that according to him the true theory is the last one remaining after attempts at refutation. This view does not include (1) the standard inductivist hypothesis that scientific theories emerge from the facts, or (2) the inductivist view that we infer the truth of theories from (independently established) facts or (3) the inductivist view that theories are rendered merely probable by inference from the facts. And we have seen above that Whewell rejects such views. The truth is discovered by conjectures and criticism, but identified as the truth by educated intuition.

Let us review, then, four aspects of Agassi's picture of Whewell as an inductivist which might justify his characterization. These are (1) Whewell's alleged endorsement of verificationism, (2) his endorsement of Bacon's ladder of axioms, (3) his alleged black and white historiography of science which fails to take account of mistakes and (4) his alleged view of the impossibility of going beyond Newton to a deeper theory.

### 7.3.1.1. *Whewell, the Verificationist*

Whewell is said by Agassi to be a verificationist. According to Agassi Whewell deemed theories which are well corroborated to be true or probably true (Agassi 1975b, p. 42). Agassi explains and/or elaborates this interpretation in various ways. One way is by attributing to Whewell the explanation of the verification of theories as due to the truth of the verified theories. For this reason he (Whewell) deems them according to Agassi probably true. Agassi attributes to Whewell – as well as to Bayes, Laplace and Helmholtz – the view that theories are probable, since certainty is not attainable (Agassi 1975b, p. 17). The correspondence of a theory with the facts could be explained, according to Whewell according to Agassi, as accidental or due to the fact that it is true and the latter explanation is more probable. The more improbable the evidence the less likely a theory will be overthrown. Agassi goes on to explain that this is the reason why the 19th century did not discuss the difference between high probability and certainty: they were quite satisfied with high probability.

This interpretation of Whewell is clearly wrong. Whewell dispensed with probability or attempts to achieve it by inductive inference as futile and insisted on necessity – or certainty – which was achieved in a different way. And Agassi also says that, when, for example, he contrasts Whewell's demand for finality with the probability demanded by contemporary inductivists (Agassi 1975b, p. 288). The cornerstone of Whewell's philosophy was his demand for certainty (necessity). He does not justify the assertion that scientific theories are justified, but rather deems them justified and asks how. His answer is not by verification but by intuition. I think this interpretation of Whewell has been demonstrated above and is reviewed once more below. So much for Agassi's central reason for calling him inductivist.

Agassi insists further that Whewell deemed a theory to be confirmed, and only then confirmed when it explains a new fact it was not designed to (Agassi 1975b, p. 261). Now Whewell did emphasize both the importance of tests and of independent corroborations. He deemed them as a matter of fact very important events in the history of science and he criticized Mill for failing to see this. So far we are in agreement. But this view of the importance of independent tests and of such confirmation does not need to be read as evidence of Whewell being an inductivist in either of the two central senses given to this word, i.e. that theories emerge from the facts or that we infer from facts or corroboration to theories. Although Whewell's methodology included the confirmation of theories by surprising events, these events served to educate the mind and not to infer the truth. Nor are they deemed as Agassi claims the only test or corroboration of a theory: a theory is justified only by its completeness and by the impossibility of any other solution. A confirmation by an independent test is neither necessary nor sufficient, even if it is often very important in refining a theory and eliminating alternatives. One must concede, as Blanché pointed out, that Whewell is sometimes misleading but also follow Blanché in his observation that his rejection of inductivism is nevertheless clear.

Agassi goes on to criticize the view he attributes to Whewell (Agassi 1975b, p. 45) on the grounds that intentions are irrelevant for the confirmation of a theory. This criticism is, I think, wrong on another count. Whewell did not view the confirmation of theories by independent tests important because of the intentions of the person suggesting a theory but rather because it extends the scope of a theory or, possibly, achieves a consilience of inductions. The unification of different types of events in one theory is quite objectively speaking an increase in explanatory power and a demonstration of it. The unified and precise explanation of the facts with a simple and clear theory was needed to identify the truth: confirmation by surprising events furthered this aim.

Agassi also interprets Whewell's problem as the problem of how experience validates theories (Agassi 1975b, p. 70) and claims that he sought canons of verification (p. 214) and even that he, along with Herschel, sought to show that a foolproof scientific method exists (Agassi 1981, p. 389). But since Whewell rejects inductive inference, these assertions cannot stand. The questions he put, as I see it, were how we learn from experience and how we justify theories. But experience does not validate theories. Intuition does. Ironically, Agassi also says that Whewell's problem is Popper's (Agassi 1975b, p. 71). Here I agree with him half way: the first of Whewell's two problems (How do we discover the truth?) is the same as Popper's.

The closest Whewell gets to offering canons of verification are his tables of the consiliences of induction. He does explain the conditions which must be achieved for us to see the truth and he does explain or portray the development of science through his table of consiliences. But these seem a long way from "canons of induction." This phrase seems to suggest he wanted something like Mill's methods of inductive inference. Agassi does suggest that Whewell and later Popper offered a formula for an acceptable theory according to which a theory should be subjected to and pass independent tests (Agassi 1975b, pp. 292, 300-301, 337). But even if this interpretation is true, and I think it is not, it offers no canon of induction. Agassi suggests that Whewell dodged the question of whether there is an algorithm for scientific discovery (Agassi 1981, p. 221). I think he said no, there is no algorithm. In his introduction to his discussions of "methods employed in the formation of science" he states,

> But, in truth, we must acknowledge, before we proceed with this subject, that, speaking with strictness, an Art of Discovery is not possible; – that we can give no Rules for the pursuit of truth which shall be universally and peremptorily applicable; – and that the helps which we can offer to the inquirer in such cases are limited and precarious. (Whewell 1967b, vol. II, p. 335)

### 7.3.1.2. *Whewell's Ladder of Axioms*

A second reason gives Agassi's for deeming Whewell an inductivist is that Whewell endorses a Baconian view according to which there is a ladder of axioms and one must ascend this ladder in an inductivist way (Agassi 1981, p. 286). One must proceed from the lowest to the highest and no step on this ladder can be skipped. Now Whewell does say that science only progresses if it works inductively. In his reply to Mill Whewell explicitly endorses Bacon's view concerning the ladder of axioms and claims to integrate it into his own view (Whewell 1971, pp. 276*ff*). Whewell claims that Mill ignores the fact that science progresses by succeeding levels of generalization.

This claim of Whewell's alone, however, will not make Whewell Baconian in the sense intended by Agassi: Popper also conceives the direction of scientific progress in this way, that is, from theories of lower to those of higher generalization. But Whewell includes a caution here which is, perhaps, more to the point. Science, on Whewell's view, should not try to go too far too quickly, since it may then fall into vague and useless speculation. It must explain the facts at hand. And to do that it must not try to go too quickly. Popper endorses Whewell's aim but thinks that testability can be best obtained by broad adventurous theories rather than too conservative ones. And this, Agassi will say, is due to the fact that Whewell seeks tests plus verification but Popper seeks tests plus refutation.

But Whewell offers no rules for new conceptions or for their scope. He stresses that we must find out with conjectures and tests which facts belong together. We must also note the context of this discussion. Whewell was arguing against the inductivist Mill because Mill wished to use a deductive method to further knowledge, that is, he wished to extend the application of existing ideas. Whewell's response is that we have much to learn: we need to try out new ideas in new contexts. His caution then is also an expression of openness to trying out new ideas and not a (Baconian) caution to refrain from doing so.

The caution recommended by Whewell does not reveal him defending any of the major doctrines used to identify inductivism, such as the view that ideas emerge from the facts or that they are justified by them. The difference between Bacon and Whewell is deep: for Bacon the ladder is an imposition to avoid error when theorizing; for Whewell, it is an imposition to formulate theories which can be tested and improved by comparison with the facts, in contrast to vague speculations such as those of Schelling, and to try new ideas in new contexts. It is an opportunity to find a foothold somewhere by enabling one to correct errors. And these errors are not condemned but deemed useful. Since Whewell speaks ambivalently, however, about Faraday's speculations, we may find old inductivist prejudices coming through. But even here we find Whewell working his way free of them and seeking to find a path which avoids empty speculation on the one hand and no theorizing on the other. Agassi finds inductivist whiffs in Whewell, but then we find those in Popper too.

7.3.1.3. *Whewell's Inductivist Historiography*

Agassi has also a view of Whewell's historiography as inductivist. On this reading Whewell's inductivist historiography is seen in his failure to recognize mistakes or gray zones in the history of science. Whewell fails, along with other inductivist historians of science, to see, says Agassi, that many scientific theories are neither quite right or wrong but represent progress even though they are to this or that degree vague, confused or wrong. Agassi suggests that Whewell follows Bacon in dismissing errors as insignificant (Agassi 1963, p. 64) and that he seeks to explain away conflicts such as that between Kepler's theory and Newton's which he "suddenly notices" and explains away with an Hegelian view (Agassi 1963, p. 172). He elsewhere goes so far as to say that Whewell had contempt for mistaken theories (Agassi 1975a, p. 216).

This view underestimates Whewell, the historian. He emphasized, indeed, the necessity of mistakes and of gray ideas on the way to truth. He spoke of them as such and even spoke with appreciation of the development of failed attempts such as the development of Cartesian alternatives to Newton. Agassi speaks in his *Historiography* of an all-or-nothing, black or white attitude as being exemplified by the refusal of discoverers to give the proper recognition to their sometimes confused or mistaken predecessors. The paradigm case in the history of science is the refusal by Lavoisier and some of his followers to give credit to Lavoisier's predecessors who defended phlogistonist theories. But this has nothing to do with Whewell, who recognized the importance of the gray predecessors of the white progeny. The progeny are to be sure white. And the history of science is built around such discoveries. And even though the gray is largely forgotten after the truth has been found the gray is neither neglected nor condemned. There are, according to Whewell, three stages in major scientific discoveries which are the process of error and correction leading to them, the discovery of the true idea, and its development. He surely did not leave out the process of development. He recognized Kepler's errors and had respect for them. It is true that

Brewster criticized Whewell for writing a history of science which was too nice, which neglected the sometimes difficult lives and fights for recognition which leading scientists have had to endure. But to classify Whewell in this fashion is to see the history of the historiography of science as black or white and to neglect the importance, originality and uniqueness of gray figures such as Whewell.

Whewell also wrote his history, as Agassi notes, to exemplify – really to justify – his philosophy. He sought to explain the growth of science. This makes his historiography far better than those other attempts criticized by Agassi as inductivists. These other attempts seek to merely relate the facts and to record the heroes of science. Whewell's history was theoretical, explanatory and critical just as his philosophy was.

#### 7.3.1.4. *Whewell, the Dogmatic Defender of Newton*

Agassi's fourth ground for deeming Whewell an inductivist philosopher and/or typical for his age is his alleged conviction that Newton could not be superseded and his alleged unwillingness to seriously consider Faraday's theoretical innovations. Agassi reinforces the standard view of Herschel, Whewell, and Mill when he states,

> Sir John Herschel, William Whewell, John Stuart Mill – these were the leading 19th century methodologists, and they repeatedly insisted that Newtonian mechanics, at least, is entirely perfect, is in no need of improvement or modification to the slightest, and cannot be superseded even in being explained by a still deeper or more general theory: Newtonian mechanics was the last word on all questions it answers – and in every sense in which it could be construed as the last word. (Agassi 1975b, p. 13)

I find no such insistence anywhere in Whewell. I do find the following:

> It is no doubt conceivable that future discoveries may both extend and further explain Newton's doctrines; – may make gravitation a case of some wider law and may disclose something of the mode in which it operates; questions with which Newton himself struggled. (Whewell 1967a, vol. II, p. 137)

Whewell discusses these difficulties elsewhere (Whewell 1967b, vol. I, pp. 257*ff*). He also indicates his openness to progress in general elsewhere as in his reply to Mill, where he states,

> I trust . . . that we have many new laws of nature still to discover; and that our race is destined to obtain a sight of wider truth than any we yet discern, including, as cases, the general laws we now know, and obtained from these known laws as the must be, by Induction. (Whewell 1971, pp. 284-285)

Agassi suggests elsewhere that Whewell thinks that Newton's theory could not be superseded because there could be no additional facts which could be the common explanans with Newton's theory which some new, deeper and broader theory could be (Agassi 1975b, p. 437). I do not know what passage Agassi has in mind so I cannot check this claim conclusively. I do not find it. I will return below to a more general claim of Agassi's about the impossibility of making discoveries by refutation which seems closely related to this claim.

A corollary of Whewell's view of Newton's mechanics is his view of Newton's optics. Here, Agassi explains, Whewell sought to explain the rejection of his optics as legitimate (Agassi 1981, pp. 413, 382). This explanation was needed in order to maintain the view that the acceptance of the mechanics was not mistaken. Whewell developed the subsequently accepted view, says Agassi, according to which Newton's *Optics* were ad hoc (Agassi 1975b, p. 189). This seems to me correct and interesting. It does not, however, justify the stronger comments Agassi makes about Whewell's attitude toward Newton; perhaps it was not intended to.

The subject of Whewell's attitude toward Faraday is much more complex. I do not have anything to add to it other than to note how divided Agassi's comments are. On the hand Agassi deems Whewell's reaction to Faraday a function of his clinging to Newton as the final truth. He says that Herschel and Whewell refused to entertain his views – even tentatively (Agassi 1981, p. 25). Even if this claim were true it might make him a dogmatist or justificationist, but not necessarily an inductivist. But here Agassi's picture is also not clear. In other places he points out how advanced relative to his age Whewell was. Although Agassi notes that Whewell clearly stated, for example, that Faraday had refuted results of Coulomb (Agassi 1971a, pp. 146, 271) he is still not satisfied, because it took him too long to do so. Agassi also points out that Faraday appreciated Whewell (Agassi 1971a, p. 34) and that Whewell was no dogmatist and an admirer of Faraday (Agassi 1971a, p. 146) and that both Whewell and Faraday valued bold hypotheses even though Whewell, but not Faraday, thought that certainty was possible in the end (Agassi 1971a, p. 31). Perhaps Agassi has reason for complaint, but even he can nevertheless not put Whewell in the same class with his contemporaries. He does so at times but then talks quite differently elsewhere.

### 7.3.2. *Whewell, the Progressive Thinker*

Agassi has a further quite non-inductivist view of Whewell. This view gives Whewell due credit for being, along with thinkers such as Galileo, Boyle and Kant, a member of a critical tradition in opposition to the inductivist tradition of Bacon, Newton and Mill (Agassi 1975b, p. 50). Agassi praises his intellectualist methodology (Agassi 1975b, p. 152). He notices his recognition of the need for conjectures and refutations (Agassi 1975b, p. 51) and that the distinction between criticism and controversy was a cornerstone of his philosophy (Agassi 1971a, p. 148). His perception of the need for intuition in science (Agassi 1971a, p. 176) is deemed important and his sophisticated theory of observation (Agassi 1971a, p. 129) is given due credit.

One major concern of all Agassi's comments on Whewell, however, is to form a good bird's eye view of the philosophy of science. In order to do that he wishes to draw sharp and succinct lines of demarcation between the view of the best inductivist philosopher of science – Whewell – and the view of Popper which he deems in many respects a further development of ideas of Whewell (Agassi 1975b, pp. 68-70, 74). In carrying out this exercise Agassi criticizes Whewell's progressive elements as inadequate. This inadequacy is largely due to his inductivist or, at any rate, to his justificationist views. Let us look at Agassi's praise and criticism to see how and whether his plan works.

### 7.3.2.1. *Whewell on Imagination*

Agassi praises Whewell for seeing the central role of intuition in science. He further explains that intuition can be either critical or not and says that Whewell's was, indeed, critical. Agassi notes that Whewell was rejected on account of his intuitionist view (Agassi 1981, p. 311). But Agassi has Whewell's view of discovery in mind and not Whewell's view of justification. Agassi does not note that Whewell had a second theory as well. On this second view intuition was the product of study attained by an educated mind and did offer justification, as explained above. It was Whewell's alternative to inductivism. Since Agassi portrays him as an inductivist, it is perhaps no wonder that he does not explain Whewell's alternative to it.

According to Agassi Whewell was improperly rejected by Mill and others because they deemed him to assert that intuition was authority or rejected his progressive view. This defense of Whewell's theory of the path to truth is nice. But, this criticism is not quite right. It fails to take into account Whewell's two theories of intuition. Agassi might deem the rejection of Whewell for his theory of justification quite reasonable; he has apparently little respect for the combination of Whewell's activist methodology with his contemplative view of the aim of science and the justification of scientific theory. Perhaps he would even deem Whewell's view stronger should one deem him inductivist. But this is not clear, since at one place he describes a view which he attributes to Galileo and Einstein which is identical to that view which I attribute to Whewell. He states,

> Some holders of modified or critical intuitionism, notably Galileo and Einstein, have claimed that in science finality can be achieved, but their theory of finality is not based on any criterion of the validity or authority of intuition; rather the claim that the finality of some product of our intuition can be ascertained by reason alone, by the thorough critical examination of our theories, as against facts or otherwise. (Agassi 1981, p. 316)

Whewell emphasized again and again that we saw the truth only after critical study. If Einstein's view and Whewell's are in this respect really identical it is surely remarkable.

Two more, somewhat different aspects of Agassi's discussion of Whewell's theory might be mentioned here. The first is Agassi's discussion of the need for a metaphysical theory of causality. Before Popper and Hempel, Whewell deemed causal explanation to be deductive (Agassi 1975b, p. 260). But although Whewell also recognizes that metaphysics is valuable for research (Agassi 1975b, p. 399), he sought, as Popper and Hempel did later, to do without a metaphysical interpretation of causality. Agassi thinks we need such a theory. The second aspect is Agassi's discussion of the problem of the choice of hypotheses. Whewell deemed imagination and guesses important factors in the development of science. But, according to Agassi, he ignored the problem of the choice of hypotheses (Agassi 1981, p. 249). Even though Whewell noted that Kepler was impressed by the *Timaeus* (Agassi 1981, p. 259), he saw no possibility here of explaining the choice of hypotheses. In spite of his activist view of science, he could not solve the problem of the choice of hypotheses (Agassi 1981, p. 240). Agassi apparently sees Whewell approaching his own theory of the role of metaphysics in the selection of hypotheses, even if he does not come very far toward it. Each of these criticisms is interesting. I will not discuss them in detail, however, because it does not touch on our varying readings of Whewell and because it is quite modern: Whewell and Popper are hit equally.

### 7.3.2.2. *Whewell on Discovery*

Agassi finds Whewell's theory of scientific discovery and especially of factual discovery extremely interesting because it is intellectual: it emphasizes the need for ideas before factual or experimental studies can be made. On this view we need ideas to guide experiments or even to notice events which may be harbingers of new discoveries. Agassi conjectures that Whewell wrote an anonymous review of Faraday which recognized the importance of Faraday's theory for his discoveries (Agassi 1971a, p. 201). But Whewell, Agassi says, has a problem in seeking to explain discovery: some discovery is accidental. Since, however, there is no way to discover new facts other than by predicting them (Agassi 1975b, pp. 72-73), no discovery should be accidental. Or, since there is no way of finding facts to go with theory other than by predicting them (Agassi 1975b, p. 202), we cannot explain such discovery. Whewell's solution to this problem, according to Agassi, is that we invent new ideas. After we have invented a new idea we may predict – and discover – new events. These predictions are surprising and in some respect accidental.

But they are also to be expected from the point of view of the new idea and surprising from the point of view of the old (Agassi 1963, pp. 63-64).

Agassi nevertheless finds this theory too narrow (Agassi 1963, pp. 64*ff*). The reason he does so is that on his reading Whewell deems all discovery as due to correct prediction. This theory, Agassi says, is too narrow because we often find new phenomena when our expectations are not realized or when our theories are refuted. According to Agassi, then, Whewell correctly emphasizes the importance of ideas, but incorrectly deems all discovery confirmation.

Now Agassi is no doubt right in pointing out that some discovery does not take place by surprising true predictions but rather by surprising refutations. But does Whewell deny that? I think not. His theory of the path to truth includes, as Agassi emphasizes, the view that we need to make conjectures and refute them. So, why should we not make discoveries by noticing unexpected results? Whewell discusses his view of the necessity of theory for discovery in his discussion of accidental discoveries (Whewell 1967b, vol. II, pp. 23*ff*). Let us look at this discussion to check Agassi's interpretation.

At first blush it seems, indeed, that Agassi's interpretation is the correct one. Whewell does claim that no discovery is accidental and that correct and precise ideas are needed in order to make discovery. I still find his interpretation inadequate, however. I do so because Whewell distinguishes between the apparent accident which leads to discovery and the discovery. The trained observer will notice something unexpected, something not predicted. In his example Malus noticed accidentally that light reflected through the windows in the Louvre became dim. The person lacking the proper ideas would explain such a surprising phenomenon away as a mere accident. The person with proper ideas may explain them. Malus did and discovered the polarization of light on Whewell's account. A discovery occurs, then, only when surprising or accidental phenomena are explained. The explanation is a prediction but the phenomena themselves may be noticed because they run counter to expectations.

Whewell thus explains (accidental) discovery as due to unexpected phenomena plus the explanation of them. The difference between Agassi and Whewell is not nearly so great as Agassi portrays it, if one translates Whewell's view into Agassi's language. For Agassi as well for Whewell some theory and not merely an unexpected observation is needed for discovery. For both this unexpected observation need not be predicted. It would seem as if Whewell requires that the theory be present first even if it is not used to deduce the (surprising) phenomena. But this is also not quite his view: he makes clear in other places that theories are developed to explain facts.

### 7.3.2.3. *Whewell on Observation*

Closely connected with Agassi's interpretation of Whewell's view of discovery is his interpretation of his view of observation. Just as Agassi attributes to Whewell a very strong view according to which all discoveries are predicted, he attributes to him a strong view according to which facts are only observed through the perspective of some theory and thus no facts not conforming to a theory can even be observed! (Agassi 1975b, p. 437). Whewell defended the view that all observations are theory-dependent. This is a quite modern view, which is defended by Popper and Agassi as well. Yet Agassi finds here too a defect in Whewell's view which once again should be due to his defense of a justificationist position. On Whewell's theory, so Agassi, all evidence is interpreted by a theory. So, only evidence which is interpreted by a theory can count as a refutation of it. And if it is reinterpreted, then it must count for it. So, although Whewell not only allows for refutations in theory of science, his theory of the interpretation of evidence prevents there from being any.

I find this argument no better than the standard ones for the incommensurability of scientific theories or other suggestions – by Butts for example – that there is no refutation on Whewell's view of science. It is true that at the final stage, when all has been settled, then there can be no more refutation and we reinterpret the facts to conform to the theory. But even then we may, by the consilience of inductions, increase the scope and depth of some (final) theory. And in the process of the development of any theory things are by no means so clear. In order to account for evidence we may not have to change only our view of the evidence but also our view of the theory.

So, I find Whewell's view to be hardly different from Agassi's. Both thinkers allow for reconsideration of evidence as well as theory. Both see that old observations may have to be translated into new language. Both hold that things may be messy for a while until facts and ideas get sorted out. The theory that all evidence is interpreted evidence, that it requires ideas to be portrayed, no more blocks refutations on Whewell's theory than the modern theories of paradigms or frameworks do. The breakdown of an attempt to explain a fact, the discovery that a fact cannot be described by a theory, or the finding of direct conflict between fact and theory are all possible on Whewell's view of the use of theory to interpret facts. They are possible because we may discover conflicts between the description of them which we find available or at hand and some idea or explanation of them. These conflicts involve, of course, theory. But that does not block conflict nor the need to test a new idea according to its ability to explain or reinterpret events explained by other or previous theories.

### 7.3.3. *Whewell's Influence*

Agassi notices in his positive view of Whewell the debt of Duhem (Agassi 1981, pp. 283; 1957), of Merton (Agassi 1981, p. 175), of Chamberlain (Agassi 1981, p. 168) and of others to him. This view of Agassi's has more recently been developed by him and myself together (Wettersten and Agassi 1991). On this view of Whewell's influence Duhem developed his own non-inductivist and conventionalist view of science and its history as an alternative to Whewell. But it was from Whewell that he learned to dispense with induction. Following Whewell's lead conventionalism dispenses with inference from facts and substitutes the adjustment of theory to facts. Conventionalism does this by modification with interest in true predictions alone. Whewell does this by changes, large and small, which lead us to the true description of the world. Duhem, indeed, takes Whewell's third stage in the development of any idea, that is, the adjustment of ideas to render them more precise, to be the whole of science (Wettersten 1998).

Agassi sees Whewell's strongest and most interesting influence in the encouragement of active, non-inductivist views of science. I have here no quarrel with his view. It is powerful, fruitful and interesting and opens new perspectives on the history of the philosophy of science. This one-sided influence and neglect of Whewell can be better understood, however, if we see that he does not belong in the inductivist school and was in fact their most powerful and effective critic before the 20th century.

### 7.3.4. *Agassi's Method of Analysis*

Agassi seems to me to have three views of Whewell which he has not reconciled. On one view Whewell is no inductivist, but an eclectic mixture of the inductivism and Kantianism. This is the standard view of Whewell, which has led to the corollary that he is confused. But Agassi does not say he is confused nor does he explain why Whewell's non-Baconianism or Kantian elements do not fit with his inductivism.

On Agassi's second view Whewell is an inductivist like all others who insist on verification and a ladder of axioms, who deem scientific theories probable and who cling to the necessity of Newton's theory. This is the weakest aspect of Agassi's analysis. Whewell's theory of verification is not inductivist but intuitionist, his theory of ladder of axioms was used to defend the introduction of new ideas rather than to block them, he did not deem scientific theories probable but necessary, and he did not cling to Newton the way Agassi says. Agassi's central reason for calling him an inductivist which really corresponds to Whewell's view – he (Whewell) deems the last theory in the process of conjectures of refutation to be the true one – hardly makes him an inductivist. It only shows that he is a justificationist, that he thinks science finds the truth and justifies its claims to have found it. Such a characterization as this is never given by Agassi in his general characterizations of inductivists as sufficient to deem a thinker inductivist. Agassi needs something more, which is emergence from the facts or justification by them, which he claims to have but does not.

Agassi's third view of Whewell is that he is a powerful thinker with progressive and correct ideas in the methodology of science. On this view Whewell correctly emphasized the activity of the scientist and the importance of theories and mistakes even in making observations. He was also an excellent historian who not only represents the heyday of (inductivist) history but has had considerable influence on succeeding *avant garde* thinkers such as Duhem. This third view is accompanied by criticism as well. Sometimes the criticism goes awry, because it does not take into account Whewell's rejection of induction, his unique theory of justification and the difference between his views of the path to truth and the identification or nature of true theories. But sometimes it is the criticism of an *avant garde* thinker which is especially designed for him, such as the criticism that Whewell does not explain the choice of hypotheses or the unity of science.

These three different views of Whewell arise, I think, because Agassi seeks to offer a view of Whewell's place in the history of the philosophy of science and/or of the history of science quite independent of this historical role as a representative of one of three views of science. His two analyses should be complementary but they clash. The logical, non-historical classification which results from his attempt to put him in the same overview of the philosophy of science with Popper and Duhem and modern inductivists leads Agassi to put Whewell in a pigeonhole where he does not belong: he seeks to render him an inductivist and explains away or does not notice Whewell's protests to the contrary. But Agassi is often an appreciative and insightful reader of Whewell: he discovered his importance very early and has pointed to consistently over the years. Since Agassi appreciatively notices his *avant garde* views, however, we sometimes get a picture of Whewell as progressive and sometimes as confused.

I find Agassi's attempt to classify Whewell as a defender of one of three typical views in any case superfluous. If we have a history of the development of our problems, we do not need, I think, anything more. An historical overview is, in my opinion, better than the kind of logical scheme of alternatives which Agassi (and Popper) have sought to provide. We can, I think, better explain the innovation (or lack of innovation) of Popper's view if we see how the problems developed. We may thereby achieve an overview without seeking to give a (complete) list of possible alternatives which, independent of intentions, Agassi's three way characterization seems to be.

# PART FOUR

# WHEWELL REAPPRAISED TODAY

## CHAPTER 8

## AN OVERVIEW: WHEWELL'S PHILOSOPHY IN RETROSPECT

The foregoing critique of Whewell's critics allows us to build a new interpretation of Whewell's philosophy. This interpretation must reject that view of Whewell's philosophy which is presented in the *Dictionary of National Biography* (D.N.B.) and which remains dominant today. According to this view traditional view Whewell was extremely erudite and made perceptive methodological observations, but was also a confused and eclectic thinker. Historians and philosophers who concern themselves with Whewell's philosophy still seek to sort out the allegedly confusing mixture of elements – Kantian, Platonic, empiricist – in Whewell's philosophy. They often seek to save one aspect or the other – more recently this meant, for example, the theory of the consilience of inductions – while rejecting the systematic aspects of Whewell's approach, which gives sense to the details. On the basis of the foregoing discussion, however, we can see that the Kantian, Platonic and empiricist elements are not eclectically and confusedly thrown together. They are used to build a new, coherent view of science which solved deep and pressing problems.

We need to recognize as well that Whewell's view is false and that some of those criticisms of it offered in the 19th century may be seen today to be correct. But to see the strengths and weaknesses of Whewell's theory adequately, we must see even these criticisms in context: they showed weaknesses in his view, which he either shared with his contemporaries or which resulted from his attempts to solve problems they ignored. These very weaknesses often show which problems were then pressing. In order to appreciate Whewell's philosophy, then, we need to look at these problems as well as the defects which his attempts to solve them led to.

In order to present a new interpretation of Whewell's philosophy which grows out of the above discussion, there are three tasks which it will be useful to carry out. The first task is to sort out the wheat from the chaff among the objections to Whewell's view. In carrying out this task those problems may be explained, which Whewell's view sought to solve, but which were ignored by his critics, often under the erroneous assumption they knew the answers to them.

A second task is to show the coherence and the power of Whewell's theory and to explain these as due to its capacity to solve those – often new – problems posed by Whewell. This demonstration of its coherence and power may be accomplished by reviewing the answers which he gave – or which may be given – to his critics. Such a defense of Whewell's system must, of course, also take into account those places where it is inadequate. But even though it failed, we may show that when it failed, it failed because it uncovered and/or tried to overcome real difficulties which the opposing views had not even noticed. With hindsight it can now be said that in the 19th century it was the most impressive view, because it went the farthest toward posing and solving new problems.

Unfortunately the progress it made was largely ignored and as a result the deepest problems it raised were not generally put on the agenda for discussion. Progress Whewell made, then, has had to wait well over one hundred years – Whewell was active until he died in 1877 – before it could be once again noticed. A third task, then, is to explain some of these problems.

The interpretation of Whewell's philosophy offered here raises a further question as to what good the study of Whewell's philosophy today might still be. I will turn to this question in the final chapter.

## 8.1. Criticisms of Whewell's Philosophy from Herschel till Today: Hits and Misses

The reviews and criticisms of Whewell's philosophy from Herschel's review until today have developed a few themes which have been proposed, developed, modified or reinforced, but hardly ever questioned. And when they have been questioned the questioners have not been able to change the tenor of the general discussion. After following the development of these discussions in some detail, it is possible to see the broad picture of their development and to restructure our conception of Whewell on the basis of this overview. Virtually all of the criticisms of Whewell have their origins in the first reactions to his work. Later commentators have tinkered here and there, but their prime tasks have been to elaborate on the comments of the early reviews.

The standard picture of Whewell which has evolved out of the elaboration of the various early critiques portray him as a philosopher of science, who (1) neglected the basis of science in facts and who therefore had no adequate theory of proof (Herschel § 1.1, Mill §1.2.3.1, Butts § 5.5, Buchdahl § 7.1), who (2) used Kant somewhat poorly and vaguely and who moved over too close to Hegelianism (Herschel § 1.1, Mill § 1.2.3.1, Mansel § 1.5) and who even proposed that truth changed (Mill § 1.2.3.1), who (3) proposed an unrealistic view of justification by intuition, on the one hand, and of necessity as the inconceivability of any alternative, on the other hand, which were merely psychological (Herschel § 1.1, Mill § 1.2.3.1, Mansel § 1.3, de Morgan § 1.5, Butts § 5.5), who (4) confusedly combined Kantian with empiricist views (Herschel § 1.1, Mill § 1.2.3.2, Butts § 5.5, Laudan § 6.1) or who proposed mysterious views about the development of empirical into necessary truths and thus a confusing eclectic view of knowledge and its growth (Herschel § 2.1.2, Mill §§ 1.2.3.1–1.2.3.2, Stoll § 4.3, Butts § 5.5) who (5) confusedly combined the matter he took over from Kant with Platonic elements he added later in order to save his view (Herschel § 2.1.2, Mansel § 1.5, Clark § 2.1.1, D.N.B., Stoll § 4.3, Blanché §§ 4.4.1–4.4.3), who (6) was forced to do theology to give the needed justification for his view because it was incomplete (Stoll § 4.3, Blanché § 4.4.4, Butts § 5.5) and who (7) wrote an overambitious history which has some fine insights but which suffered from the haste and ambition of the author as well as his attempt to bring the entire history of science under one scheme (Herschel § 1.1, Brewster § 2.1.3, Clark § 2.1.1).

This is an impressive list of complaints which has been developed in the last one hundred and fifty or so years, and it has been done so with wide agreement. It is nevertheless often false and always misleading. As the above analysis shows, there are objections which have been sustained over the years which are simply false: he did not neglect empiricism but offered a substitute for defective inductivist views (see the discussion of Herschel § 1.1, Mill § 1.2.3.2, Butts § 5.5, and of Blanché's defense of Whewell § 4.4.4); he had a theory of justification (see the discussion of Mill §§ 1.2.3.1–1.2.3.2); his view of necessity was not merely psychological (see the discussion of de Morgan § 1.3); he did not need theology to justify his view, since history played that role (see the discussion of Stoll § 4.3, Blanché §§ 4.4.4–4.4.5 and of Butts § 5.5); he did not confusedly mix Kantian and Platonic elements but developed a coherent view which has in some respects similarities with Kantian and in some respects with Platonic views but cannot be classified as either without further ado (see the discussion of Mansel § 4.5, Stoll § 4.3, Blanché §§ 4.4.2, 4.4.3, 4.4.5 and of Butts § 5.5); he did not confuse empirical and necessary truths but described our increase of knowledge as in increase in the knowledge of necessary laws (see the discussion of Butts § 5.5); his view was quite openly incomplete, but only because he had explained why no view could be complete: reductionism explained knowledge away (see the discussion of Blanché § 4.4.5 and of Butts § 5.5; see also Whewell 1971, pp. 300*ff*).

Some of the criticisms leveled at Whewell over the years do hit his view and show it false and/or inadequate in this way or that. But the difficulties which he had and which led to these criticisms are for the most part due to his attempt to solve problems which were quite general, but which only he tried to solve. He did become Hegelian to a degree against his wish, but only because he tried to explain the fact, which others refused to notice, that an established theory at one time, such as Kepler's, may contradict an established theory at a later time such as Newton's (see the discussion of Mill § 1.2.3.1). He did offer (another) inadequate theory of justification, but only because he correctly insisted that neither induction from facts nor a priori forms alone could offer an adequate view of scientific knowledge: only intuition educated by the critical appraisal of theories was left (see the discussion of Mill § 1.2.3.2); he did face difficulties in justifying his view by his appeal to history, but his idea was new and powerful and the only new alternative available when all else had failed; his view of necessity and of intuition suffered from difficulties, but he was forced into these difficulties by his attempt to solve new problems growing out the inadequacy of inductivist attempts to reduce necessity to generalization or Kantian attempts to identify one necessary and a priori framework. In general, then, he was rejected not because his solutions were poorer than alternatives, but because his methods were new and different.

It might nevertheless appear here that an attempt to positively reappraise Whewell has gone awry, since his view suffered in any case from so many difficulties that the general rejection of his view was quite right, even if some of the details may have been mistaken. We may give him credit for being an adventurous thinker, who tried to solve problems that were too much for him, and at the same time corroborate the view of him as confused and overambitious. This might, indeed, have been the case if the views he developed had not been not very good responses to the problems he tackled. But this is not the case. They were very good indeed. He invented a new way of doing the philosophy of science, which overcame inadequacies found in all alternatives. He showed how within his framework new methods of justification and new approaches could be developed. His general thrust was imaginative and powerful, even if it broke down at important points as well as in some details. At the time of its publication it offered the most insight into the problems of understanding science and its history. This store of insight and enlightenment has not yet been exhausted. Let us first look once more at the various objections to Whewell's view to improve our perspective.

### 8.1.1. *Whewell Violates Empiricism and Neglects Proof*

Herschel objected to Whewell's philosophy immediately; shortly afterward Mill followed in his footsteps. Their central criticism has survived until today. Scientific theories, they said, must be justified, must be shown to be true, by induction. Whewell's arguments that induction could not play any role in science have never been adequately answered. A mere program for the reduction of all ideas to sensations or facts or observations was offered, with no pretense that it had been carried out, as has been explained above in the discussions of the criticisms of Herschel (§ 1.1) and Mill (§ 1.2.3.2). Whewell himself held that the use of inductive methods presumed the knowledge of an idea and were thus superfluous in the discovery of an idea and inadequate for the proof of an idea. In regard to the former he states,

> Upon these (Mill's) Methods, the obvious thing to remark is, that they take for granted the very thing which is most difficult to discover, the reduction of the phenomena to formulae such as are here presented to us. When we have any set of complex facts offered us; for instance, those which were offered in the cases of discovery which I have mentioned, – the facts of the planetary paths, of falling bodies, of refracted rays, of cosmical motions, of chemical analysis; and when, in any of these cases, we would discover the nature which governs them, or, if any one chooses so to term it, the feature in which all the cases agree, where are we to look for our A, B, C and a, b, c? Nature does not present to us the cases in this form; and how are we to reduce them to this form? You say when we find the combination of ABC with abc and ABD with abd, then we may draw our inference. Granted: but when and where are we to find such combinations? (Whewell 1971, p. 263)

(Whewell continues to show the worthlessness of Mill's examples for the purposes he intended.)

In regard to induction Whewell says,

> Thus in each inference made by Induction, there is introduced some General Conception, which is given, not by the phenomena, but by the mind. The conclusion is not contained in the premises, but includes them by the introduction of a New Generality. In order to obtain our inference, we travel beyond the cases which we have before us: we consider them as mere exemplifications of some Ideal Case in which the relations are complete and intelligible. We take a Standard, and measure the facts by it; and this Standard is constructed by us, not offered by Nature. We assert, for example that body left to itself will move on with unaltered velocity, not because our senses ever disclosed to us a body doing this, but because (taking this as our Ideal Case) we find that all actual cases are intelligible and explicable by means of the Conception of Forces, causing change and motion, and exerted by surrounding bodies . . . Hence in every inference by Induction, there is some Conception superinduced upon the facts: and we may henceforth conceive this to be the peculiar import of the term Induction. (Whewell 1967b, pp. 49-50)

When Whewell moves from the context of discovery to the context of proof (Whewell 1967b, vol. II, p. 74) it may seem as if he defends a version of inductivism in which one infers from the verification of a theory to its truth. He says, indeed, that the truth of a theory is established by verification and this is even like book-keeping. I think, however, that Blanché was right when he said that even though some passages in Whewell may sound that way and his language may be at times misleading, it is clear that that is not what he meant. And following such passages as these Whewell continues to explain himself and make clear that that is not what he means. He says,

> At first sight it might be supposed that the formula of Inductive Logic need only be of this kind: "These particulars, and all known particulars of the same kind, are exactly included in the following general proposition." But a moment's reflection on what has just been said will show that this is not sufficient: for the particulars are not merely included in the general proposition. It is not enough that they appertain to it by enumeration . . . This is, as we have seen, the mode of stating the evidence when the proposition is once suggested; but the Inductive step consists in the suggestion of a conception not before apparent . . . Thus in our formula, besides stating that the particulars are included in the general proposition, we must also imply that the generality is constituted by a new conception, – new at least in its application.
> Hence our Inductive Formula might be something like the following: "These particulars, and all known particulars of the same kind, are exactly expressed by adopting the conceptions and Statement of the following Proposition." It is of course requisite that Conceptions should precisely embrace the facts, according to the explanation we have already given of those conditions. (Whewell 1967b, vol. II, pp. 87-88)

Whewell explains further the significance of this difference. He explains that induction only then occurs when the inductive reasoner reaches the conclusion that only one proposition can explain the facts, when he sees it as necessary. But this is an intuitive process. He states,

> But this conviction, that the inductive inferences is not only consistent with the facts, but necessary, finds its place in the mind gradually, as the contemplation of the consequences of the proposition and the various relations of the facts, becomes steady and familiar. It is scarcely possible for the student at once to satisfy himself that the inference is thus inevitable . . . We may, therefore, without impropriety, renounce the undertaking of conveying in our formula this final conviction of the necessary truth of our inference. We may leave it to be thought, without insisting upon saying it, that in such cases what can be true, is true. (Whewell 1967b, vol. II, p. 90)

Whewell nevertheless did have a theory of the role of empirical research in proof. Empirical research is necessary, he says, to render an idea so clear and distinct that we may see its truth. He states,

> Experience is requisite to the clearness and distinctness of our ideas, not because they are derived from experience, but because they can only be exercised upon experience. And this consideration sufficiently explains how it is that experiment and observation have been the means, and the only means, by which men have been led to a knowledge of the laws of nature. (Whewell 1967b, vol. I, p. 663)

Why, Whewell asks, should one believe that the reductionist program suggested by Mill could be carried out? We apparently move in circles, since the only answer we can find is that science is based on facts alone. Whewell objected that induction cannot be used to generate or prove scientific theories. Descriptions of facts, he said, depend on theories. By the time we have descriptions which we might want to use for inductions, then, we already have the theories and proofs of them at hand (see the discussion of Mill § 1.2.3.2 and Whewell's criticism of Mill, Whewell 1971, pp. 238*ff*). The only half-way serious attempt to answer this argument was Mill's confused view of a supposed distinction between description and explanation (see the discussion of Mill, § 1.2.3.2). This view was largely abandoned in the 19th century and only in the 20th century do we find Buchdahl's futile attempt to refurbish it (see § 7.1). Nevertheless, Whewell's allegedly inadequate theory of, or neglect of, induction has become standard in the literature.

### 8.1.2. *Whewell Is a Hegelian*

It was also Herschel who introduced the idea that Whewell had moved much too close to the Germans. Such thinkers as Schelling or Hegel were obviously intended even if not mentioned by name. Herschel merely suggested this (see the discussion of Herschel, § 1.1). Mansel had something like this in mind as well when he warned about extending the range of a priori categories (see the discussion of Mansel, § 1.5). Mill did not even talk about Hegel, but found it possible to give a telling criticism of Whewell which revealed a Hegelian streak in his view: Whewell endorsed the view that contradictory theories could be true. Mill might have said, though he did not, that Whewell even added a definite Hegelian flavor when he (Whewell) said that theories at a higher level could be true when they contradicted those at a lower level. Whewell had here, of course, such mutually contradictory theories as those of Kepler and Newton in mind (see the discussion of Mill, § 1.2.3.1).

It is rather astonishing that this strongest argument against Whewell's view has not been put in the forefront of the criticism of him. It has, indeed, since Mill first raised it, been scarcely repeated at all. The problem which Whewell was trying thereby to solve had to be rediscovered in the 20th century (Wettersten and Agassi 1991, pp. 357*ff*). This is unfortunate because the problem which led Whewell to his concession to Hegelianism, that is, the problem, how we can explain the fact that the best theories in science are all true and nevertheless often contradict each other, was thereby ignored if not suppressed. Herschel and Mill and his competitors agreed that all scientific theories or all good scientific theories were true – certainly Kepler's and Newton's had to be true. They contradicted each other, since on Newton's theory the orbits of the planets do not turn out to be perfect ellipses. Nor is this a mere matter of details: Kepler's whole method depended on finding perfect geometrical figures. This needs at least to be explained away, if not explained. Whewell tried to explain the transformation of hypotheses in the history of science and got himself in trouble. His central example, unfortunately, was the transformation of Cartesian physics in the light of Newton's theory, which gives the impression it only applies to theories commonly deemed defeated. Yet as his discussion of Hegel's view of the relation between Kepler and Newton makes clear, on Whewell's view a new theory explains and corrects an old theory. He there criticizes Hegel for, among other mistakes, clinging to an Aristotelian theory in which the orbits should be shown to necessarily follow the path of a perfect figure (Whewell 1971, p. 512).

Whewell got himself in trouble, however, but only because he followed out the consequences of the assumptions he shared with his competitors. They refused to do so and then rejected his view because he went too far. There may be, we might note, one exception. This exception is Peirce. He saw that theories changed and moved toward Hegelianism himself. He wished, of course, to reconcile that with an empiricism more strict than Whewell's. But what can be said is that he took the problems more seriously and thus, though he too failed, his response is more interesting.

### 8.1.3. *Whewell's Justification by Intuition Is Illusory; His View of Necessity Is Merely Psychological*

The attack on Whewell's views of justification by intuition and of the necessity of scientific laws (fundamental ideas were seen to be the only possible explanation of the facts and thus both true and necessary) was developed in various ways by Herschel, Mill, Mansel, de Morgan and others. The former two insisted on applying the reductionist standards and sought to argue that any scientific method of justification had to be inductive and any adequate conception of necessity could only be derived from experience. Mansel joined forces by limiting necessary ideas more severely than Kant and by leaving induction to do the job for the rest. The extension of necessary ideas and the methods of their justification were unacceptable for him. De Morgan found Whewell's approach inadequate, but at least flirted with the idea that new methods for the justification of scientific theories and new views of the necessity of scientific laws were needed: Bacon's view of method would not do and necessary truth did grow, as de Morgan emphasized.

The demand that any view of necessity be reduced to experience or observation was merely the insistence on an associationist- inductivist program. It could not be carried out and Whewell could not, therefore, be too impressed by the demand that it be carried out. The limitation of necessary ideas as Mansel sought to do, could not account for science. In each case he had good reason to move beyond either view. The attack on his view of justification as either non-existent (Mill, §§ 1.2.3.1–1.2.3.2) or as merely psychological (Mansel, § 1.5, and de Morgan, § 1.3) and as quite inadequate by all, since intuition could be and often was misguided, was more serious. His answer was stronger than he was given credit for. He argued that everyone agreed that scientific theories were in fact justified. They could not be justified either by inductive inference from facts or by some transcendental, Kantian method: the former could justify nothing – as explained above in my discussion of his view of proof – and the latter was not that method used to justify scientific ideas, i.e. those ideas whose justification must be explained to account for science, for chemistry, for example, as well as space and time.

Whewell argued, then, transcendentally from the facts of the history of science and from the agreed-upon premise that some method must be available. He argued that he was quite aware that intuition could be misguided, but that the educated intuition of scientists was, as everyone agreed as a matter of simple fact, not misguided. Since no other method of proof was present, it must be intuition. This may a mystery or a gift from God, but this observation does not hit his argument for its truth. Even de Morgan's argument, that he could perfectly well conceive that Newton's laws should be false, could not impress Whewell unless he could make it plausible in the context of the knowledge of the facts which were to be explained. De Morgan made no such attempt. Whewell's problem situation was once again ignored. His argument that this was the only explanation left after one had critically reviewed all alternatives and that this view could account for the history of science was ignored.

The same can be said for his view of the necessity of scientific laws. His view is quite simple: a law is necessary when the world could not be other than as this law describes it. This is not a psychological view of the matter and is only deemed so because Whewell claims that the only way we can know that a law is necessary is that we cannot conceive the world to be other than as this law describes it. The achievement of the educated mind which sees this inconceivability is, however, the method by which we gain knowledge of the necessity of scientific laws and not a description of the nature of the necessity. Whewell's critics demanded more, but they had no standards for what, in addition to that which Whewell offered, was necessary (see the discussion of de Morgan, § 1.3).

#### 8.1.4. *Whewell's Empiricism Is Confused; Whewell Confusedly Combines Kantian and Empiricist Views*

Whewell's view of justification, it was said or intimated immediately after the publication of Whewell's *History* and *Philosophy*, is not adequate because Whewell offers no theory of inductive inference. This view has, however, for the most part rested on the interpretation of Herschel and Mill according to which Whewell's view of empirical method must provide for (inductive) proof. Since Whewell's view of proof is not inductive and his view of empirical method does not include proof, it is no wonder that critics have trouble finding this view. It is also no proper criticism. As Blanché has clearly pointed out, Whewell had dispensed with inductive inference. His view of empirical method was deductive: one tested ideas until they could explain all the facts and no other could. But the end result was not a valid or convincing inductive inference from facts to theory, but clear and distinct ideas from which facts could be deduced.

This idea of Whewell's, the idea that the purpose of empirical research was not the justification of our ideas by induction but their proper development has not been appreciated by any of his critics. But it has been appreciated by those who sought to use Whewell's ideas. Bernard in France, Lange in Germany and Peirce in America all sought to be build theories of science using this idea. The question of the adequacy of this idea to provide for a realistic theory of science arose out of this work. It was Stoll, then, who came to Whewell against the background of the new-Kantians in Germany and the work of Lalande in France – she missed the close connection with Peirce – who asked whether Whewell's theory could provide a way of avoiding idealism, that is, a way of explaining the development of ideas which was not idealist. She argued that Whewell could not, but she insisted here and in regard to the explanation of knowledge that his system be complete, and this is just what he rejected. She did not ask whether Whewell's theory of empirical research was adequate for his purpose, and thus she did not explain this purpose. The alleged inadequate mixture of empirical and Kantian elements was then picked up by Butts, who also did not explain Whewell's theory of empirical research. He did not distinguish it from a theory of inductive inference.

In addition to the traditional attacks on Whewell's theory of justification, his view of necessity and his view of empirical method, it has more recently been contended that his view of empirical method cannot be coherently combined with his view of necessary truth (Butts, § 5.5). The incoherence of Whewell's view has, indeed, been the upshot of the discussions of Whewell since the D.N.B. summarized the reaction to him. Early critics as well as recent ones have criticized Whewell for his view that necessary truth could grow. This view has been picked up in the modern literature by Butts, who was worried about how empirical truths could become necessary. De Morgan had already answered that: mathematicians had known that any map could be drawn with four colors but could not show that it was necessary. So, a truth known as a true generalization can be later shown to be necessary and known as a necessary truth. Only, on Whewell's view, when it is shown to be necessary is it properly justified. Empirical truths may be known, but are never scientifically justified. Thus Whewell's view of empirical method and of empirical propositions can be nicely integrated with his view of necessary truths or fundamental ideas.

### 8.1.5. *Whewell Confusedly Combines Kantian with Platonic Views*

Whewell was criticized by Mansel for his illegitimate extension of Kantian a priori frameworks of space and time to include other scientific theories in his theory of fundamental ideas. This criticism was quickly picked up by Clark and Herschel in their obituaries of Whewell, when they called him Platonic. It then found its way into the judgment of the D.N.B. The most important attempt to develop this view was made by Blanché, who sought to show in detail that in each aspect of his work Whewell's indecision between Kantianism and Platonism led to contradictions or confusions. This objection has been important for others, before and after Blanché as well. The critique of his view of justification and of necessity as merely psychological rests on that interpretation of him according to which he slides between that which is (merely) imposed by the mind and that which is real.

There is, however, no contradiction between, on the one hand, the assertions that ideas grow in the mind, are imposed on the world, and reach fruition as the fundamental ideas of science and, on the other hand, the assertion that these ideas truly describe the world. That the process by which these ideas grow is one of actively imposing ideas on the world and of improving them poses no problem of coherency. Nor does the view that this process leads to the truth. Whewell can thus combine quite well the Kantian idea that ideas are imposed on the world with the Platonic idea that the true ideas are discovered and not merely invented or proposed (see the discussions of Mansel, § 1.5, and Blanché, § 4.4). We may ask, of course, why all that should be. But Whewell relies once more on his transcendental argument: it is a fact that scientific knowledge is gained in this way. This, he says, is the only possible explanation. One may argue about that. But if this claim is correct, then his theory is well defended, indeed. In order to meet this argument, instead of criticizing it for its incompleteness, one should argue about its explanatory power.

Whewell's explanation is, to be sure, incomplete, since we have no explanation of why it is the case that knowledge may be obtained in this way. But this is not fatal. He argued that such incompleteness was unavoidable, if we wanted to avoid futile metaphysics. We know, he might also have argued, that Newton's theory is the only theory capable of truly describing the mechanics of moving bodies even though we cannot yet explain the force of gravity, that is, why that is the case. Completeness, as everyone agrees, is not necessary for proof in science.

But, one might say, this view will still not do, since Whewell is in fact Hegelian: he speaks as if the imperfect scientific ideas we apply to the world are really those ideas which he elsewhere characterizes as Platonic. He is in fact forced to do that and we come back to his Hegelianism. But his Hegelianism, as was explained above, was rendered necessary by facts of the growth of science, which no one else sought to explain and was not due to his use of Kantian and Platonic elements.

Whewell is often charged with having offered a confused Kantianism. But it is hard to know what standards are being applied here. It was Kant's intention to give a detailed deduction of the categories of science. He here set himself a task which Whewell did not seek to carry out. But there was no reason why he should. He did not try to deduce the categories of thought from the conditions for knowledge but rather tried to explain how science discovered correct ideas. The task of deducing the a priori categories as conditions of experience does not arise on Whewell's view. The task of discovering these categories, including those of space and time, is on Whewell's view not a philosophical one but mathematical and scientific ones, even when space and time should apply to all experience and other ideas should not. The philosophical task, or better the task for the science of science, is to explain how these ideas are developed and discovered by the use of a critical method. The categories of space and time when described correctly are not conditions of experience as in Kant, but the correct ideas of space and time. These are developed and discovered as are all ideas, even if the methods in the case of space have been above all mathematical.

### 8.1.6. *Whewell's View Is Not Justified*

One of the most surprising aspects of the reaction to Whewell's philosophy is that his own proposal for justifying any philosophy of science, that is, that one test it as a scientific theory of science, was nearly totally ignored. In the 19th century the question was hardly raised. No one asked – with the exception of a few remarks by Brewster – whether Whewell's theory was the best explanation of the history of science. They merely went on as before and criticized his view for not meeting the standards of inductivists and/or Kantians. Later it was proposed that Whewell had a system but that this system must be, but was not, complete. This view was specifically proposed by Stoll and taken up by Blanché and Butts.

I have explained above (Stoll, § 4.3, Blanché, § 4.4.5, Butts, § 5.5) that, although Whewell agreed his view was an incomplete portrayal of how knowledge is possible, he did not view this to be a defect in his theory. Rather, he deemed any attempt to give a complete explanation to fall into the many and poor systems which were exemplified by German philosophy of the sort everyone in England, at least, wished to avoid. Still further, he deemed it quite possible with his own method to justify his view of science without any metaphysical additions. This could be done, he said, because it could explain the history of science, as he explained for example in the preface to the second edition of his *Philosophy* of the

*Inductive Sciences*, and no other theory could, as he explained in his criticisms of his competitors, especially Mill.

Nobody, from Herschel to Stoll to Blanché (who saw a need for theology and found an inconsistency in his theological views) to Strong (who defended his historical perspective) to more recent commentators such as Laudan have given this view its due. It was imaginative and new. It provided both a new method for the critical appraisal of theories of science as well as one for their justification. Brewster employed Whewell's standard against him and gave a good criticism. (Sometimes the best ideas are proposed, passed over and later rediscovered, as the history of the wave theory of light shows). But no one took this standard seriously enough to judge Whewell's theory well or poorly by it and to then offer alternatives. Perhaps the new-Kantians were the first to do that, but they, just as Duhem and others afterward, did not debate cases enough to tell. Such debates have had to wait until the 20th century and the aftermath of Popper's innovations.

There is a further and intimately related problem which has hardly been discussed but which is nevertheless quite important for an understanding of Whewell's philosophy, the following appraisal of it, and for a conception of the philosophy of science. This problem is: how can the philosophy of science progress? Whewell saw the problem and tried to develop his own view of it. Since he deemed his own philosophy of science a science of science, it should develop by incorporating the truths of previous philosophies of science, while correcting their mistakes and adding to them just as a good scientific theory does vis à vis the foregoing theories in its field. This seems a good and powerful idea (Wettersten 1994). But nevertheless his theory was poor: he had no interpretation of progress in the philosophy of science which enabled him to compare his view vis à vis its predecessors in an adequate way. Although theories should be compared vis à vis their capacity to explain the history of science, his predecessors had not even attempted to do so. And his attempt to present predecessors as approximations to his own view was exceedingly strained. A good example is his discussion of Newton's comments on method which he sought to reinterpret in the light of his own views (Whewell 1971, pp. 181*ff*). He had further Hegelian problems concerning the truth of theories of scientific method, since he had to reinterpret and move to a "higher" level with hardly any rules for translation.

### 8.1.7. *Whewell's History Is Inadequate*

Whewell's history of science was given high marks by early commentators such as Herschel or de Morgan for his detailed insights into the process of scientific discovery or the psychology of science. But the method of writing history after the fashion of Whewell was quickly, at least officially, rejected. Whewell's history was too full of mistakes it was said. And this was due to the fact that Whewell's work was hastily done and due to the fact that Whewell was overambitious: no single scheme could be constructed which would encompass the entire history of science. This was just what he hoped to contribute. He states,

> As will easily be supposed, I have borrowed largely from other writers, both of the histories of special sciences and of philosophy of general. I have done this without scruple, since the novelty of my work was intended to consist, not in its superiority as a collection of facts, but in the point of view in which the facts were placed. (Whewell 1967a, vol. I, p. x)

His new interpretation of the history of science was the first to have portrayed the growth of science without induction, was the first to have emphasized the role of error in the growth of scientific knowledge, and the first to have pointed out the importance of the refinement of ideas once they have been found, ideas which are found in his view of the three stages of the development of any fundamental idea.

The sharp criticism of Whewell's *History* has concealed just how important Whewell's *History* was for the history of the historiography of science. In spite of his rejection of inductivism, he was, perhaps, responsible for the inductivist style of writing the history of science so sharply criticized by Agassi (Agassi 1963). His justificationist view led him to single out, for example, the final successful scientist. But Whewell took this justificationist style further than most. He did not just report on successes and failures and give medals to the good guys and condemnation to the bad guys. Rather, he noted the struggle, the importance of mistakes which precede a discovery, as well as the importance of the small modifications and refinements which follow a discovery. The former was largely dismissed in the wake of the criticism of Whewell. But the latter became the approach followed by Duhem in his attempt to offer a comprehensive alternative to Whewell (Wettersten 1998). Whewell thus showed not only that the quest for the true idea was important and that the modifications of scientific theories after they have been discovered is important, but also that comprehensive histories of science written from a particular point of view are valuable.

We can see here how inductivist views prevented Whewell's *History* from playing the role it might have played. Instead of complaining that Whewell's scheme was false or inadequate and offering a better one, most commentators rejected all attempts to explain the history of science with theory as overambitious. His followers did not bother to note their debt to him. Whewell's theory was not at first allowed to compete with others for the honor of being deemed the best explanation of the history of science, because all such attempts were rejected out of hand. And this was done on the basis of just those inductivist prejudices which Whewell was trying to combat. Later his view was not taken seriously. Whewell's view was not rejected because it was an inadequate contender, but rather because no contenders were to be allowed at all. After new contenders appeared on the scene his theory was deemed out of date and never taken into consideration at all.

## 8.2. Whewell's Innovative Problems

The criticisms which were offered of Whewell's theory were by and large of two sorts. On the one hand there were those criticisms which were merely external criticisms and which were delivered from the point of view of inductivist or Kantian philosophers. Whewell was criticized for not reducing all ideas to sense impressions even though one of his main points was that no such reduction was possible. He was criticized for extending the range of Kantian, a priori ideas unjustifiably even while he explicitly rejected the relevant Kantian positions and explained the need for such extension. He was criticized for offering no inductive proof, even though he had shown the irrelevance of inductive arguments for science. These criticisms hardly added anything of value. That Whewell's theory contradicted these established views was emphasized by Whewell himself. They did not succeed in showing that the Whewell's view was wrong, since they offered programmatic remarks instead of refutations which either ignored the problems Whewell had raised for them or failed to respond to them adequately.

On the other hand, there were criticisms which did hit Whewell's theory. But these criticisms were offered as clear-cut refutations, while the problems which led Whewell into his difficulties were ignored. He was criticized for deeming contradictory theories true, but the reason for his doing so – various scientific theories which everybody agreed should be deemed true contradict each other – was ignored. He was criticized for offering an incomplete theory, but his argument that any attempt at completeness would lead to futile metaphysical systems was ignored. He was criticized for offering no justification of his view, but his own method of justifying his theory by explaining the history of science was ignored. He was criticized for offering no good theory of the justification of scientific theory, but the argument for the necessity of this view – since induction was irrelevant and Kantianism too narrow and no other alternative was possible, only he could explain the growth of scientific knowledge – was ignored. He was criticized for being too ambitious and hasty in developing an explanatory scheme for science, but his argument that a new scheme had to be found, since existing ones were inadequate, was ignored or rejected out of hand, because it was not inductivist.

In order to see Whewell's view in a new light, then, we need to review the various problems which he wanted to solve and how he attempted to mold his solutions to them into one explanatory system. The foregoing analysis of Whewell's critics provides the foundation for such a review and a reappraisal, because it has revealed those places where they merely found inconsistencies between their views and Whewell's, where they missed or ignored Whewell's problems, and where they found real deficiencies in his view. We have discovered not only that the picture of Whewell's work built up over the years is misleading, but where and how it is. Let us try to make some corrections and see how the picture changes.

### 8.2.1. *Ideas Are Not Reducible to Observations; Perception Is Dependent on Perspective*

The foundation of the most dominant views of science before Whewell's philosophy was the view that all knowledge has its source, when speaking psychologically, in observation or, when speaking epistemologically, in facts. The rejection of this crucial thesis, which was defended so vigorously by Herschel and Mill, was the cornerstone of his philosophy. There are at least three aspects to his rejection of this view. One concerns the theory of perception or the foundation of science in observation. The second concerns the content of scientific ideas or the possibility of reducing scientific theories to statements about observations. I discuss these here. The third concerns the method of induction or of inductive proof. (The discussion of this aspect is found in § 8.2.3 and § 8.2.4.)

Whewell endorsed the Kantian view that ideas could not be derived from, or reduced to, observations. He did so from both epistemological and psychological points of view. From epistemological investigations he not only knew that all theories alleging that scientific theories were proven by inference from facts to theories were false, but, just as importantly for Whewell, scientific ideas were claimed to be necessary but no necessity could be derived from observations. This epistemological point of view had psychological consequences: ideas must be generated in some other way. And Whewell argued that the content of ideas always added to the material observed and was never reducible to it. He argued, then, that perception is never pure, that is, it is never a simple picture of that which is presented to the senses. Rather perception always involves interpretation of that which is presented to the senses. Whewell deemed this process to involve unconscious inference and backed up his view with the physiological researches of Charles Bell who found rapid eye movements as an essential aspect of visual perception. The mind constantly interpreted what it saw and tested these interpretations in order to correct them.

This theory had the further consequence that all descriptions of facts were theory-laden. The frameworks or theories which we used to describe the world determined to some unknown extent just how the facts were described or, perhaps we should say, conceived. There could be no method for sorting out the two aspects – the ideational and the observational – directly. One could not then simply compare theories with given facts. Our improvement of theory and our improvement of descriptions of the facts have to go in hand. We have to use each to improve the other. Whewell built on the work of Kant by translating his view of the necessity of certain a priori ideas into a psychological and epistemological theory according to which ideas were imposed on the world and actively tested and modified until the correct one was found. He sought in his theory, then, to explain how knowledge was possible given these facts of the generation of ideas, of the perception of the world, and of the active process of correction and change of ideas.

### 8.2.2. *Ideas Change as Knowledge Grows*

Whewell apparently saw from the beginning that a successful theory of science had to explain how scientific theories were discovered and developed. This problem arose quite naturally out of his rejection of the traditional inductivist views of how this process proceeded. He took as the foundation for his own alternative, then, the history of science and, thereby, the facts of scientific change. He found in this history that ideas are not merely accumulated, but change in the history of science. This view parallels quite nicely his view of the psychological process of discovery. A common explanation of both is possible: the mind is active in producing ideas and testing them; the history of science shows the co-operative appraisal and sorting out of the suggestions of individual scientists.

But change in the history of science raises new and interesting problems as well. Whewell not only wished to explain how this process proceeds; he also wished to explain why the various stages of this process should be deemed knowledge. He here had to account not merely for up-to-date science, but also for out-dated theories, which had since been replaced by better ones. The difficulty lay in the fact that these previous theories had firm places in the history of science as great discoveries. The theories of Galileo and of Kepler were here prime examples. But even phlogiston theory had to be accounted for as well. How could the succeeding corrections and replacements of these theories be reconciled with the view that science found the truth?

Whewell replied that truth changes. This was not a frivolous move, but a serious attempt to account for that change which occurred in the history of science. It was developed so as to explain how the empirical content of the theories of Galileo and of Kepler could be deemed true even though these theories contradicted the theory of Newton and even though the description of the facts is theory-laden: no sharp and easy separation of false theory from true fact was possible. Its prime virtue lay in the fact that it took these deep problems seriously. Its advantage over alternatives lay in the fact that they only ignored Whewell's difficulties by ignoring the problems which Whewell had uncovered.

### 8.2.3. *Scientific Theories Are Justified by Clarity and Distinctness*

Whewell had rejected all attempts to justify theories by induction or by Kantian means. Induction offers no proof and is irrelevant for scientific practice, since it presumes knowledge of the facts before this knowledge was at hand, that is, we can only know what the facts are after we have the correct theory in hand. We could not, then, gather (certain) facts in the hopes of finding the correct theory. But we do not need facts for inductive inference in any case, because such inferences from facts could only be used after we already knew the idea and then they are, of course, superfluous. And if we know the idea and seek to prove it with a deduction from the facts, we will find no inference possible. We can also not provide any a priori or transcendental deduction of the necessary framework for science. Such a deduction depends on science. And science is constantly growing. We have no adequate foundation for such a deduction. If, on the other hand, we have some foundation in science, then we need nothing further, since the "fundamental ideas" have already been found. Such a deduction would also be superfluous.

A new method for, or new explanation of, the justification of scientific theories was thus needed, which relied neither on induction nor on established scientific theory. Whewell arrived at his own theory by eliminating alternatives and asking what was left. After he eliminated inductive and then existing Kantian alternatives, what he found was the fact that scientists come to clear and distinct ideas concerning the truth of scientific theories. If these ideas are not justified by induction or some other method, then they must themselves be capable of identifying the truth. He could show that as a matter of fact those educated in science, those who had conducted research and mastered the facts, attained clear and distinct ideas of the truth. And this, he said, was their justification. This theory had serious disadvantages, which were quickly pointed out. But it had the advantage that it accorded with the facts and relied on assumptions shared by all his competitors such as the assumption that scientific theories were necessary and known to be so. And it did not appeal to defunct theories. Whewell's argument that only his view survives after all views have been tested against the history of science was ignored.

### 8.2.4. *Empirical Method Is Method of Discovery*

Whewell and his contemporaries were agreed that the desired and/or proper method for the discovery and/or the justification of scientific theories should be empirical. But all theories of empirical research then offered relied on induction and/or the achievement of inductive inference as the goal of empirical research. Yet neither one goal nor the other could be reasonably pursued. How, then, should one conduct empirical research? And what should be the goal of such research?

Whewell solved these problems by making empirical research the path to truth instead of its justification. There was no other path. One had to study the facts and one had to see that only one theory could explain them, if one wished to attain the truth. The empirical method could then be one which used a deductivist approach. One had to impose ideas on the facts, test them, replace them and refine them, if one wanted to find the truth. This theory of empirical method reconciled very nicely Whewell's Kantian view of ideas, his theory of the psychological process of the generation of ideas, his theory of the growth of scientific knowledge, and his theory of the justification of scientific theories with the acknowledged need to explain the primary role of empirical research in science.

The criticisms of it ignore its role in Whewell's theory. They deem it a theory of inductive justification, which it is not. Or, they say, if it is not, then Whewell has no theory of justification. These criticisms not only ignore its role, however. They also ignore the pressing problems which Whewell was trying to solve: how could science be empirical when inductive proof played no role in science?

### 8.2.5. *The Possibility of Discovery and of Truth Explained*

The theory of scientific discovery and the theory of scientific truth, which were developed by Whewell in answer to the questions discussed above, raised new problems concerning the possibility of the growth of knowledge. Whewell claimed that if (1) science proceeded by an active processing of imposing ideas on the world, of testing and improving them, and (2) if justification occurred through the discovery of clear and distinct ideas, and (3) if science had no foundation in sure facts other than that provided by correct theories, and (4) if the empirical method was merely the path to truth, how was it possible that those theories which appeared to humans to be clear and distinct should also be true?

Whewell had apparently blocked the way to an adequate solution to this problem when he argued that no complete system of knowledge and no complete analysis of the relations between ideas and facts, which Whewell called the Fundamental Antithesis of Philosophy, was possible. We could know, then, that science proceeded in this way, and that science attained the truth. But an explanation of why this is possible or of why we can approach the truth in this and in no other way was beyond the scope of human reason. Whewell argued, then, that no complete explanation should be sought. If one did so, one ran the high risk that one would construct empty philosophical systems such as those produced in Germany in the aftermath of Kant. But if no complete explanation is possible, how should we decide what is good enough? And what is good enough? Whewell could here only offer his arguments, that if one sought a complete explanation one would explain knowledge away. This might be dissatisfying, but if it should be deemed inadequate – as it was – it would behoove its critics to explain how completeness could be achieved without the undesired consequences – which they did not.

Whewell was, indeed, dissatisfied himself. He offered an explanation, which was more theological than philosophical. God had made the world in such a good way. This was deemed by his critics as a concession that theology was needed to defend his philosophy of science. But this was not so. Rather, it was an addition open to those who believed. One was free to reject it as one chose. But this did not mean that an alternative was needed to "complete" or to justify his philosophy. His philosophy was justified, he thought, because it explained the history of science and not because it could offer a complete system.

### 8.2.6. *Justification in the Philosophy of Science Rendered Possible*

If the theory of the justification of scientific theories needed to be changed, then the theory of the justification of philosophies of science had to be changed as well. This problem of how to justify a theory of science had hardly been noticed. The inductivist versions of theories of scientific method had presumed that, after Newton at least, no justification for inductivist theory was needed: one needed merely to fill in the details or carry out the program. For Kant the problem was more difficult and he invented a new method which was his transcendental deductions. In Germany others such Jakob Friedrich Fries followed with their own alternatives. In England the problem was not taken so seriously.

Whewell saw, however, that the inductivists had no solution, that Kant's solution was inadequate, and that succeeding attempts by the Germans left science in favor of mere metaphysical systems. He needed then a new method which would be close to science. He found such a method by making his theory of science a science of science: a philosophy of science is then justified when it can account for the history of science as the growth of knowledge and no other theory can.

This ingenious solution was not taken seriously by Whewell's critics. For the most part they merely applied inductivist and/or Kantian standards without noticing that alternatives were needed. The attempt to justify a theory by its ability to account for the history of science was then taken as inadequate. Presumably it left out problems such as proof. This was not the case however, since scientific proof was among those facts of the history of science which his theory should take account of. Whewell's view was underestimated, because the requirements it set were not appreciated. It was erroneously deemed unneeded, because the difficulties of competing views were ignored. It does raise some interesting and challenging problems of rationality due to the fact that it is self-reflexive. Does this open the doors for "justification" too wide? Is any self-reflexive theory of justification justified or only some? Such problems were not even posed.

### 8.2.7. *A New System; A New Historical Explanation*

Whewell wrote his history and his philosophy of science together in the hope that one would justify the other. His history was criticized for being too ambitious in its attempt to bring the whole history of science together into one scheme. Yet Whewell's contention, that in order to develop an adequate and complete theory of science one needed to do just that, was ignored. His philosophy of science was criticized for being incomplete, while his arguments for the limits of reason which demanded such incompleteness were ignored; but his history of science was criticized as too ambitious because it attempted to be complete, while his arguments for his method of doing the philosophy of science and the resulting need for a complete account of the history of science were ignored.

In order to develop integrated solutions to his array of new problems Whewell found it necessary to devise new methods for the philosophy of science. How could historical, epistemological, psychological, methodological and metaphysical explanations of science or of the growth of knowledge be integrated? His solution was that a philosophy of science should take into account the various aspects of science which we find in its practice and its history. In order to do that, and in accord with his theory, the facts of the history of science had to be completely, precisely and systematically portrayed. These facts had then to be accounted for. But the description of the facts and theory could not be entirely separated: all descriptions of facts are theory-laden. Nevertheless, he held that some correct system could be found in this manner, that his, indeed, was that system.

Whewell's challenge to appraise competing theories of science according to his proposed standard was rejected and his proposal ignored. Instead it was evaluated by the (inadequate) standards of the day. This meant the serious problems in both detail and in his system were ignored while poor criticism was used to reject it. The result was that much of the progress Whewell had made in the development of his theory was passed over and is still in the process of being recovered.

## 8.3. The Aftermath: Progress in Every Field

Whewell had constructed a new and far-reaching systematic philosophy of science. It offered integrated solutions to a variety of problems. And just where there were weaknesses it opened new and interesting problems. Its systematic unity and the new problems it opened were ignored in favor of its rejection on the basis of criticism which was sometimes correct, but always more or less shallow. Whewell had integrated (1) his psychological theory of perception and the origin of ideas according to which the mind creates ideas and uses them to perceive and interpret the world, with (2) his theory of the growth of knowledge according to which new ideas are responses to the weaknesses of the old and which reinterpret them, with (3) his theory of justification and necessity of scientific ideas, according to which the necessity of true ideas is clearly and distinctly perceived after the effort has been made to impose ideas on the world and correct them until no more correction is needed, with (4) his theory of the empirical method of science as an active process of the discovery of new ideas until the truth is found, even though it alone is no process of justification, with (5) his theory of truth according to which true ideas are not given innately in the mind but the mind is so constructed that it is capable of generating the true ideas and of seeing them when it does so, with (6) his theory of the justification of philosophy of science, according to which the philosophy of science is the science of science and a philosophy of science is justified when it can account for whole history of science and no other can, with (7) his overview of the history of science, which served to justify his philosophy of science or, better, his science of science.

Although, contrary to the tradition, this view was highly coherent and well thought out, it did lead to difficult problems and there were good reasons for rejecting it. But these reasons for rejecting it could only be fairly used to reject it in favor of alternatives, if the problems which led to them were taken account. One needed to ask whether other theories could do any better and whether one could learn from Whewell's difficulties. This was hardly done. But let us see what it might have offered the inquisitive mind.

### 8.3.1. *Psychology: Ideas and Perception*

According to de Morgan Whewell was the leading theorist of the psychology of scientific research. De Morgan was no doubt right. And his psychology was revolutionary because it not only emphasized the activity of the mind and the learning which this activity rendered possible but fully dispensed with associationism. But it was not developed. Whewell's studies in the practice of science had opened up new possibilities for psychology. These possibilities were not explored. Rather, there were new and concerted attempts to defend and develop associationist views and these attempts became known as the beginning of "scientific" psychology. It was not until Oswald Külpe saw the inadequacy of his teacher Wilhelm Wundt's attempt to explain how ideas were the product of sensations alone that the problems of a non-reductionist psychology which took account of the phenomena described by Whewell – instead of explaining them away – were put on the agenda. And they have had a very difficult time to remain on this agenda ever since. I have described this aspect of the reaction to Whewell elsewhere in my *The Roots of Critical Rationalism* (Wettersten 1992).

### 8.3.2. *History and Theoretical Change*

Whewell had discovered that in the history of science ideas change. There was not only addition to knowledge in the history of science, but later theories replaced and contradicted earlier ones. He tried to take this into account and could do no better than to offer an Hegelian view according to which truth changes when a new and better theory replaces the old theory. He was criticized for this view and correctly so. It strained his realism extremely and opened the door for much that would not be wanted. But everyone – with the possible exception of Peirce who was unfortunately too reluctant to discuss his debt to Whewell in detail – ignored this problem. It was Karl Popper who rediscovered and presented it as a new discovery in the 1960s, whereby I do not mean to suggest that Popper's suggestion that his result was new was implausible: it had been so well suppressed that it had to be said clearly again.

### 8.3.3. *Justification in Science*

Whewell was often criticized for his theory of the justification of scientific theories by the clear and distinct intuition of their necessity. It was pointed out that many did not seem to share such intuition and that intuition was often wrong. Whewell explained that he meant educated intuition and that agreement really was at hand in the history of science. This response may now be deemed inadequate, though Agassi suggests

Einstein endorsed something like it. Whewell's contemporaries seemed to have shared it in one way or other. They accepted at any rate the premises of Whewell's argument, that is, that scientific ideas are necessary and known to be so by the scientific elite. Whewell's other reason for deeming intuition the source of justification was ignored: all other possibilities had been eliminated and this was left. Now, one could say that other possibilities were available, but no one could show how. One could also say that all possibilities were no good and we must look further – de Morgan can be read, perhaps, that way – but then we must grant how far Whewell has come in reformulating our problem. And this was not done. The new shape of the problem was ignored.

### 8.3.4. *Empirical Method*

The identification of the empirical method with proof is very old. When Whewell proposed that they be separated, at least in as much that justification was not inference from the facts, he was accused either of having no method of proof or a confused view of empirical method. The possibility of a theory of science which deemed empirical method the only path to truth, but not in itself a justification, was not even given a hearing. Yet it offers a new approach to old problems which in conjunction with the problem of justification offered new perspectives for tackling the problem of improving Whewell's theory of science. Instead, the so-called context of discovery was separated from the so-called context of justification in such a way that empirical justification was still demanded. Whewell had separated discovery from truth to account for empirical method, for discovery and for justification in a unified way. Instead of taking up this challenge the theory of discovery was separated from the theory of proof, so that all else could remain as before.

### 8.3.5. *Discovery and Truth*

Whewell offered a genetic theory of the discovery of ideas: the mind developed ideas until it hit on the right one and it was so constructed that, if it tested ideas vis à vis their ability to explain the facts, it would hit upon the correct one. He was criticized for mixing a Kantian theory according to which ideas are imposed on the world with a Platonic theory according to which ideas have an independent and real existence and may thus describe the world the way it truly is. His metaphysical theory of the generation of ideas and of their relation to the world might be improved – as empirical psychology for example. But in order to do that, one has to see its strengths and this was that it could explain the psychological, the historical and the epistemological process in a coherent way. Even though Peirce took up this idea, merely explaining the correspondence between ideas and the world in a different way, this was never acknowledged and, instead of leading to the improvements in the philosophy of science it could have supplied, it was hastily dismissed by nearly everyone.

8.3.6. *Justification of Philosophies of Science*

Whewell not only saw that his new approach to the philosophy of science demanded a new theory of the justification of a philosophy of science, but also that he had to avoid the metaphysical excesses, which the Germans had been led to in the aftermath of Kant, as they wrestled with similar problems. He found an ingenious solution, which was true to his Kantian attitude even while sticking close to the facts. His method of justifying a philosophy of science by its ability to account for the history of science rendered, indeed, the philosophy of science more empirical than it had ever been before. The method was rejected anyway. And no better one was offered. Only in the 20th century has the idea been once again examined.

8.3.7. *Rationality*

The coherence of Whewell's system was based on the self-reflexivity of his theory of science: his theory of science was developed and justified just as any scientific theory should be if this theory itself is true. This offered a powerful solution to Whewell's problem of how to introduce new standards in the theory of science. Other attempts to do this had placed philosophy over science as an authority merely. Others had sought to conform with science but had merely conformed to standard views about science. Whewell found an alternative.

But this alternative was not without serious problems of its own. How could one explain this change of standards? He here needed a theory of progress, which applied to the progress in the theory of science. This theory had to explain the change of standards for scientific research and their improvement. He hardly developed such a theory. His lack of a solution is evident in his discussions of his predecessors in the philosophy of science. Even though he has appreciation for Condillac, Newton and others, he has trouble explaining the nature of progress in the philosophy of science or, how and why his predecessors should be deemed approximations to his own view. Such problems as whether self-reflexivity is enough or whether science should contain standards which compete with one another were not formulated. Although these problems which grow directly out of a Whewellian theory, they have been first put on the table in the 20th century. The reason for this has been that Whewell's system and its coherence were not taken seriously enough to discover its limits and the new problems it raised.

## CHAPTER 9

## THE GOOD THAT WHEWELL CAN STILL DO

Whewell's path-breaking studies could have done even more good than they did, if his contemporaries had only been willing to take his philosophy seriously. It was not enough to merely accept his impressive results in the history of science, the methodology of science, and, to some extent, in the psychology of learning in order to explain them away. (Without his work the theory of unconscious inference would likely not have been seen to be necessary and would thus not have become a problem or a problem so soon.) He had not only shown that there were deep problems in the philosophy of science, which his contemporaries could not solve, but had also devised a bold new method for solving them. He offered challenging new solutions. Even when his views raised difficult problems, these problems were far deeper and important than those which became standard fare in the philosophy of science.

Since the advent of Popper's theory of scientific method, problems which Whewell first raised have been rediscovered and put on the agenda once more. And even many of Whewell's proposals for how to deal with them have been revitalized. Much of this work has been done without direct knowledge of Whewell's work. But Whewell's impact has no doubt made such rediscovery of old problems easier. It would, of course, have been still easier had Whewell's research been generally known and appreciated. He had, after all, laid much of the groundwork for the development of the most modern and difficult problems in the philosophy of science: How one can explain scientific growth in such a way as to explain the history of science as well? and, How one can explain scientific growth without a theory of inductive inference? and, How can one explain the complex relation between theory and fact? and more.

Since we have now rediscovered these problems, we may find it interesting to see and understand Whewell's path-breaking work but its relevance today might appear minimal. Both his theory of scientific laws as necessary and his theory of the intuitive recognition of the necessity of scientific laws would, for example, have no appeal for the *avant garde* of today. But, even though Whewell's problems led him to results which we today find clearly unacceptable, these results were not the product of mere sloppy thinking or incoherent philosophy. They were the product of intelligent and daring moves designed to solve quite intractable problems which are still with us today. Should we now, with a greater awareness of his problems, be tempted to repeat them, we may see where they lead quite easily: we have an overview laid out for us.

## 9.1. Science as a Struggle to Gain Knowledge

Whewell put forth for the first time the very modern view that science was a struggle. Science developed, he said, by making daring moves, by taking risks, which inevitably were full of errors and falsehoods. But it could advance, if one accepted this fact and sought ways to remove the error so as to get closer to the truth until, on his view, the truth was finally attained. Surprisingly enough this view was accepted by his contemporaries as largely true. They accepted much from his historical accounts; they thought, indeed, that they had little choice. But at the same time they said this was not really so new. They had already known that science used hypotheses and that science was contemplative. Whewell had not, on their view, introduced radically new views into the philosophy of science. Rather, he had given a lively and more comprehensive presentation of views which were more or less established.

This acceptance of Whewell's descriptions of science did not lead them to take Whewell's philosophical or methodological problems seriously. They had a quite different agenda. Whewell's historical studies had encouraged him to pose new and important methodological and philosophical problems. These problems took his historical results as facts which one needed to explain and not to explain away. Whewell's contemporaries took them as facts – or appearances – which one needed to explain away in terms of the traditional views. Today, philosophers still try to implement this program. It leads, however, to lame compromises and confused results. We need here a return to Whewell's program. Many, of course, due to Popper's views, have already done so.

## 9.2. Methodology as a Struggle to Improve Science

Philosophers are often ambivalent about the history of their subject. It is a grand and fascinating history and no one can quite do without it. But it is not necessarily or obviously a history of progress. It is certainly not so in the sense in which science is, whatever that sense may be. The history of the philosophy of science is as problematical as philosophy itself. There really should be no history of what is now known as the philosophy of science proper, the theory of the methods and structure of science. This is so because scientific method has been known for so long – since Copernicus at least. It seems rather absurd that it should take us over three centuries to say correctly what scientists of the early seventeenth century already knew. But we still have not succeeded clearly enough to allow us to agree on this point and go on to something else.

Today the obvious thing to say is that the history of the philosophy of science has brought us theories of science which are ever better or closer to the truth. This is, indeed, exactly what Whewell did say. But this view was not only rejected by all of his contemporaries – I do not know of a single thinker who supported Whewell – it is still rejected by the vast majority of thinkers. The rejection today is not so clear cut as then. And we might ask if this is an improvement. First, however, let us look a little at what Whewell discovered.

Whewell saw that the theory of science needed improvement. The standard Baconian view and even Herschel's or Brewster's more sophisticated theories were not good enough. In such a situation one could, as most today do, simply argue that one theory of science is poor and that one can offer a better one. This is, indeed, unobjectionable to a point. But it leads us quite unavoidably to two problems. One problem is to explain our progress of the knowledge of science. The second is to explain the relationship of this progress to the development of science. Whewell solved the first problem by deeming the philosophy of science a science of science. The philosophy or theory of science should explain science as its particular domain in the same way that mechanics should explain laws of bodies in motion. And, he says, this theory of science can be progressive. We have in fact, he says, made progress by making conjectures as to the nature of science. We have further found refutations of these views by appealing to the facts of scientific research. On the basis of this new knowledge we have formed better theories of science, the culmination of which, of course, is that of Whewell.

This view of the theory of science leads one inevitably to the view that science can be improved. If we have better theories we may streamline research, we may introduce new and more powerful methods and we may eliminate mistaken approaches. Whewell thought that all that was possible. He had, as it were, already a theory of the interaction between science and the theory of science. Each can learn from the other to not only get closer to the truth but to attain it. This view had influence through the work of Duhem and Peirce. But for the next part it was ignored. The history of the philosophy of science was deemed of no special interest for the study of science itself. The study of science was deemed to be either confined to the domain of (inductive) logic which the province of philosophy or to the domain of the facts of science which the province of history and these two were deemed quite distinct. The erroneous conjectures of the past could play little role in this inductive view of the study of science. This is apparently still the most widely held view.

The overview of problems in the philosophy, history, and methodology of science which Whewell provides us with is thus not merely an historical curiosity or an experiment in the philosophy of science, which we may turn to see the logic of particular types of moves. It is an important part of our heritage and in two quite distinct though closely related ways. On the one hand, we can hardly understand how the discussion has arrived at its present state without understanding the role which Whewell has played in its unraveling. This role has been largely underground, that is, his influence and the power of his system has never been acknowledged, even though his ideas have been used in many different contexts, and even though much work of subsequent thinkers was designed to meet the difficulties he raised for standard views. He changed permanently our conceptions of the history of science and this change went deep. He posed the background for the philosophical work of philosophers such as Duhem, Peirce or even Mach. He prepared the way for methodological perspectives such as that offered by Bernard. And he had great influence in psychology by pointing out the problems which an associationist theory of learning faced: how, for example, could it be developed so as to explain the inferential and probing quality of even simple observations?

In order to understand these problems today we need to understand their history. This history has been misleading because the problems which have been on the agenda have been treated for the most part as if they have a background merely in the dominant views. The fact, that defenders of these dominant views have been seeking to meet desiderata which have been developed by competing views, and that the major competitor is often the view of William Whewell, has been suppressed. But without this knowledge one can hardly evaluate the proposals which have been given intelligently: in order to evaluate a solution to a problem one must know what the problem is. We need now to appreciate the strength of Whewell's view to understand the history of theories of science as well of the historiography and methodology of science. And this knowledge is needed to explain how and why contemporary problems happen to be posed the way they are.

Viewing the philosophy of science as a historical and progressive discipline which interacts with science gives it, after the fashion of Whewell, a more important function than that granted to it by standard philosophies of science: it serves to preserve and improve science itself. Philosophers of science have, from Herschel to the present, given themselves this task. But, having accepted it, they have become uncritical due to the perceived need to conform to the doctrines defended by scientists. Whewell has shown the way to a critical approach, which still accepts the need to take account for both the practice of scientists and the views scientists hold about their work. It may thus serve to be critical of science and to seek to improve it without falling into the obvious traps of seeking to dictate how science should be conducted: science and the philosophy of science become partners in which the philosophy of science should explain the successes and failures or inadequacies of science, should criticize the practices of science, where these seem dubious, and offer proposals for the improvement of science, where these seem possible. All of these need to be, of course, tested against and with the experience of scientists and rejected where they cannot survive such tests.

### 9.3. Science and Methodology as a Struggle to Improve Rationality

This first kind of knowledge which we can gain from studying Whewell, that is, the knowledge that scientific practice is a struggle which needs to be accounted for in any philosophy of science, is, as I have just explained, complemented by a second kind of knowledge, that is, the knowledge that scientific methods are progressive and must be studied historically. Those are complemented by a third: the knowledge that the study of practice and development of scientific methods requires study of the growth of rationality. The study of the modes of criticism found in the history of the reaction to Whewell offers us a study in the practice of rationality. This practice is sometimes quite poor and thus not very enlightening. But the inadequate reaction to Whewell which we can now, with hindsight, better understand offers us another opportunity to ask what went wrong and if the trouble has some cause in the very standards which were applied, and not merely in the ambitions and vanity of men.

Criticisms of Whewell's theory were often convincing to his contemporaries, even though we might not find these very criticisms convincing today. They often met the accepted standards of the day. Whewell had already tentatively moved beyond these standards. But he himself was often incapable of using his most powerful responses, because they violated these standards. In either case we must ask, if this is true, how the application of traditional standards of rationality may serve to limit rationality.

Before dealing with this question let me repeat why it seems clear that this did, indeed, occur. It seemed then clear to anyone that a theory of science had to be in some sense a theory of proof. But Whewell's theory was seen by nearly all critics to offer an inadequate theory of proof. There was no need to look further into new and abstruse methods of proof, such as that offered by Whewell of his own theory. Indeed, even he seems to have been reluctant to defend his view by pointing to the coherence of the various parts, but instead, unwisely in my opinion, he sought to defend each part of it independently and by meeting then current standards which he had already surpassed. Each part of his view thus appeared to his contemporaries to be not merely new but strange: it did not fit these standards well. So, in regard to various points, he had to lose by the then established standards.

That this result was a block to the growth of knowledge is now evident. For, whatever we think of Whewell's views, we now see that he saw and developed original problems. And he tried, as best he could, to deal with them. Though here again we find him losing his nerve somewhat. He knew, for example, that Kepler's theory contradicted Newton's and that this had to be accounted for by a theory of science. But he never challenged his opponents to do just that. And the reason, I conjecture, why he did not do that is that just on this point, though he tried to deal with it and even made it a central part of his theory, he was unsure. He wished to show in his discussion of this point that Hegel's view of the importance of Kepler and insignificance of Newton due to the derivability of Newton's theory from Kepler is misguided, because Newton's theory is not only not derivable from Kepler's but inconsistent with it. But even while criticizing Hegel he moved to an Hegelian position in spite of himself. He was much more candid about the problem than his contemporaries, who refused to mention it, but not candid enough: he did not bring it into the debate with Herschel and Mill.

The application of the then-reigning standards, according to which theories should be true and proven to be true and scientific theories were examples of such, led in practice to a suppression of problems. It has taken over one hundred years to put these problems back on the agenda. But even here the establishment often simply refuses to acknowledge their existence as witnessed by the quixotic attempts discussed above to discuss Whewell, while continuing to suppress his most interesting problems. If this should be doubted let me offer an example. The official organ of the American Philosophy of Science Association publishes virtually no discussions in their journal of the problems which Popper has sought to solve. (I leave unargued the claim that these are often problems which Whewell first put on the agenda.) One essay was explicitly rejected because it did not attempt to solve the problem of induction. The author – Joseph Agassi – complained that this was improper but the governing board endorsed this act of discrimination. Thus standards of rationality demanding proof – and if not proof then the quest for proof – led then and lead now to the irrational commitment to views that are protected from criticism by suppressing the discussion of problems they seem not to be able to solve.

We need, it would seem, better standards of criticism which come not merely to identify failures to meet this or that high standard. Rather, we need lower standards which will allow us to learn from mistakes so as to better formulate our problems (Wettersten 1995b). Our failure to have learned more quickly from Whewell should certainly be caution enough. This was to some degree understandable, since Whewell himself probably did not grasp the depth of the change he sought to bring about and not merely in the theory of scientific method but in the theory of rationality as well. Whewell himself still sought to meet the highest standards and it was just here, indeed, where he needed lower standards to account for change and for his uncertainty about whether, for example, Faraday could supersede Newton. So, he was led to implausible claims about the certainty of scientific intuitions and some vacillation.

We can now see, however, in the case of Whewell what the costs of the implementation of such high standards are and we now have alternatives. These alternatives need to be discussed. But to do so we need to discuss the standards by which to discuss them. And here we come to Whewell's proposal to apply their standards to themselves.[1] And this leads to new problems which we need to discuss. If the search for inductive proof is anything other than an irrational commitment to a fading dream of the 17th, 18th and 19th centuries, the defenders of this search need to discuss their rationale for continuing this search. And to do so they will surely need something other than induction. A serious treatment of Whewell is thus still a deep problem for anyone who seeks a theory of induction and a source of insight for anyone who seeks to take his problems seriously.

[1] Bartley rediscovered the power of this move and applied it to Popper's view, see Bartley (1962).

# REFERENCES


Agassi, J. (1957). Duhem vs. Galileo. *British Journal for the Philosophy of Science* **8**, 237-238.

Agassi, J. (1959). On the Discovery of General Facts, *Impulse* **10**, 1-3. Reprinted in: (1975), pp. 73-80.

Agassi, J. (1963). Towards an Historiography of Science. *History and Theory*, Beiheft 2. Mouton & Co.

Agassi, J. (1971a). *Faraday as a Natural Philosopher*. Chicago: University of Chicago Press.

Agassi, J. (1971b). John Herschel's Philosophy of Success. In: R. McCormach (ed.), *Historical Studies in the Physical Sciences*, pp. 1-36. Philadelphia: University of Pennsylvania Press. Reprinted in: Agassi (1981), pp. 388-420.

Agassi, J. (1975a). The Nature of Scientific Problems and Their Roots in Metaphysics. In: (1975), pp. 208-239.

Agassi, J. (1975b). *Science in Flux*. Dordrecht and Boston: D. Reidel.

Agassi, J. (1981). *Science and Society*. Dordrecht, Boston and London: D. Reidel.

Agassi, J. (1985). *Technology. Its Philosophical and Social Aspects*. Dordrecht: D. Reidel.

Agassi, J. (1988). *After Lakatos. On the Gentle Art of Philosophical Polemics*. LaSalle: Open Court.

Anonymous (1840). Review of the *Philosophy of the Inductive Sciences. Athenaeum* **672**, 707-709.

Apelt, E.F. (1854). *Theorie der Induction*. Leipzig: Wilhelm Engelmann.

Bartley III, W.W. (1976). On Imre Lakatos. In: Cohen, Feyerabend and Wartofsky (eds.), *Essays in Memory of Imre Lakatos,* pp. 37-38. Dordrecht: D. Reidel.

Bartley III, W.W. (1962). *The Retreat from Commitment*. New York: Alfred A. Knopf.

Bernard, C. (1957). *An Introduction to the Study of Experimental Medicine*. New York: Dover Publications.

Bell, Ch. (1824). On the Motion of the Eyes. In: *An Exposition of the Natural System of the Nerves with a Republication of the Papers Delivered to the Royal Society on the Subject of Nerves*. London.

Blanché, R. (1935). *Le Rationalisme de Whewell*. Paris: Libraire Félix Alcan.

Blanché, R. (1955). *L'axiomatique*. Paris: Presses Universitaires de France.

Blanché, R. (1966). *Structures Intellectuelles*. Paris: Libraire Philosophique J. Vrin.

Blanché, R. (1969). *La Méthode Expérimentale et la Philosophie de la Physique*. Paris: Libraire Armand Colin.

Brewster, D. (1831). *The Life of Sir Isaac Newton*. London: John Murray.

Brewster, D. (1838). Whewell's History of the Inductive Sciences. *The Edinburgh Review* Vol. **66** (133), 110-150.

Brewster, D. (1842). Whewell's Philosophy of the Inductive Sciences. *The Edinburgh Review* Vol. **74** (150), 265-306.

Brewster, D. (1866-67). Obituary of William Whewell. *Proceedings of the Royal Society of Edinburgh*, 29-33.

Brown, J.R. and J. Mittelstrass, eds. (1989). *An Intimate Relation*. Dordrecht: Kluwer.

Buchdahl, G. (1971). Inductivist vs. Deductivist Approaches in the Philosophy of Science as Illustrated by some Controversies between Whewell and Mill. *The Monist* **55**, 343-367. Reprinted in: Fisch and Schaffer (1991), pp. 311-344.

Buchler, J., ed. (1940). *The Philosophy of Peirce*. London: Routledge & Kegan Paul Ltd.

Burtt, E.A. (1954). *The Metaphysical Foundations of Modern Science*. Garden City, NY: Doubleday.

Butts, R. (1965a). On Walsh's Reading of Whewell's View of Necessity. *Philosophy of Science* **32**, 175-181.

Butts, R. (1965b). Necessary Truth in Whewell's Theory of Science. *American Philosophical Quarterly* **2**, 161-181.

Butts, R. (1968). Introduction. In: R. Butts (ed.), *William Whewell's Theory of Scientific Method,* pp. 3-29. Pittsburgh: University of Pittsburgh Press.

Butts, R. (1970). Whewell on Newton's Rules of Theorizing. In: R.E. Butts and J.W. Davis (eds.), *The Methodological Heritage of Newton,* pp. 132-149. Toronto: University of Toronto Press.

Butts, R. (1973). Whewell's Logic of Induction. In: R. Giere and R. Westfall (eds.), *Foundations of Scientific Method: The Nineteenth Century,* pp. 53-85. Bloomington: Indiana University Press.

Butts, R. (1977). Consilience of inductions and the Problem of Conceptual Change in Science. In: R.G. Colodny (ed.), *University of Pittsburgh Series in Philosophy of Science; Logic, Laws and Life,* vol. **6**, pp. 71-88. Pittsburgh: University of Pittsburgh Press.

Cassirer, E. (1974). *Das Erkenntnisproblem in der Philosophie und Wissenschaft der neuen Zeit, Dritter Band, Die nachkantischen Systeme.* First edition, 1920. Darmstadt: Wissenschaftliche Buchgesellschaft.

Charpa, U. (1987). Herschel's Methodologie der Erfahrungswissenschaft. *Philosophia Naturalis* **24**, 121-148.

Chevalier, J. (1954). La Philosophie de Claude Bernard. In: J. Chevalier (ed.), *Claude Bernard, Philosophie: Manuscrit Inédit*, pp. 45-62. Paris: Éditions Hatier-Boivin.

Clark, W.G.C. (1866). William Whewell in Memoriam. *Macmillan's Magazine* **13**, 545-552.

Cohen, L.J. (1968). An Argument that Confirmation Functors for Consilience are Empirical Hypotheses. In: Lakatos (1968), pp. 247-250.

de Morgan, A. (1859). Review of *Novum Organum Renovatum. The Athenaeum* **1628**, 42-44.

de Morgan, A. (1860). Review of *On the Philosophy of Discovery. The Athenaeum* **1964**, 501-503.

de Morgan, A. (1915). *A Budget of Paradoxes*. Second edition. Edited by D.E. Smith. Chicago and London: Open Court.

Ducasse, C.J. (1951). William Whewell's Philosophy of Discovery: I and II. *The Philosophical Review* **60**, I: 56-59, II: 213-214.

Duhem, P. (1978). *Ziel und Struktur der physikalischen Theorien*. Hamburg: Felix Meiner Verlag.

Ellis, R.L. (1838). Review of *Mechanical Euclid. The Edinburgh Review* **67**, 81-102.

Erdmann, J.E. (1857). Deutsche Philosopheme und Britische Philosophen. *Zeitschrift für Philosophie und philosophische Kritik* **30**, 90-100.

Fisch, M. (1985). Whewell's Consilience of Inductions: An Evaluation. *Philosophy of Science* **52**-3, 239-255.

Fisch, M. (1991). *William Whewell: Philosopher of Science*. Oxford: Clarendon Press.

Fisch, M. and S. Schaffer, eds. (1991). *William Whewell: A Composite Portrait*, Oxford, Clarendon Press.

Forster, M.R. (1988). Unification, Explanation, and the Composition of Causes in Newtonian Mechanics. *Studies in the History and Philosophy of Science* **19**, 55-101.

Fries, J.F. (1831). *Neue oder anthropologische Kritik der Vernunft.* Second edition. Heidelberg: Christian Friedrich Winter.

Grote, J. (1900). *Exploratio Philosophica.* Cambridge: The University Press.

Harper, W. (1989). Consilience and Natural Kind Reasoning. In: Brown and Mittelstrass (1989), pp. 115-152.

Herschel, J. (1841). Whewell on Inductive Sciences. *Quarterly Review* **68**, 177-238. Reprinted in: *Essays from the Edinburgh and Quarterly Reviews,* pp. 142-256. London: Longman, Brown, Green, Longmans and Roberts, 1857.

Herschel, J. (1866). Obituary of William Whewell. *Proceedings of the Royal Society* **16**, li-lxi.

Herschel, J. (1966). *A Preliminary Discourse on the Study of Natural Philosophy.* New York and London: Johnson Reprint Co. (This is a facsimile of the 1830 edition.)

Hesse, M. (1968a). Consilience of Inductions. In: Lakatos (1968), pp. 232-246.

Hesse, M. (1968b). Replies. In: Lakatos (1968), pp. 254-257.

Hesse, M. (1971). Whewell's Consilience of Inductions and Predictions. *The Monist* **55**, 520-524.

Hull, D. (1965). The Effect of Essentialism on Taxonomy – Two Thousand Years of Stasis(I). *The British Journal for the Philosophy of Science* **15**, 314-326.

Hull, D. (1973). *Darwin and His Critics.* Cambridge, MA: Harvard University Press.

Köhnke, K.Ch. (1986). *Entstehung und Aufstieg des Neukantianismus.* Frankfurt: Suhrkamp.

Külpe, O. (1912). *Die Realisierung.* Vol. I. Leipzig.

Külpe, O. (1920). *Die Realisierung.* Vol. II. Leipzig.

Külpe, O. (1923). *Die Realisierung.* Vol. III. Leipzig.

Lakatos, I., ed. (1968). *The Problem of Inductive Logic.* Amsterdam: North Holland.

Lalande, A. (1929). *Les Théories de l'Induction e de l' Expérimentation.* Paris: Boiven & Co. Cie.

Lange, F.A. (1873). *Geschichte des Materialismus,* Zweite Auflage. Iserlohn: J. Baedeker.

Laudan, L. (1970). Review of Robert Butts' *Whewell's Theory of Scientific Method. British Journal for the Philosophy of Science* **21**, 311-312.

Laudan, L. (1971a). William Whewell and the Consilience of Inductions. *The Monist* **55**, 368-391.

Laudan, L. (1971b). Reply to Mary Hesse. *The Monist* **55**, 525.

Leuckfeld, P. (1898). Zur logischen Lehre von der Induktion. *Archiv für die Geschichte der Philosophie* **11**, 374-390.

Lewes, G.H. (1857). *The Biographical History of Philosophy.* London: John W. Parker and Son.

Liard, L. (1878). *Les Logiciens Anglais Contemporaires.* Paris: Libraire Germer Baillere et Cie.

Lugg, A. (1989). History, Discovery and Induction: Whewell on Kepler on the Orbit of Mars. In: Brown and Mittelstrass (1989), *An Intimate Relation,* pp. 283-298. Dordrecht: Kluwer.

Mach, E. (1917). *Erkenntnis und Irrtum,* Dritte Auflage. Leipzig: Johann Ambrosius Barth.

Mackie, J.L. (1968). A Simple Model of Consilience. In: Lakatos (1968), pp. 250-253.

Mansel, H.L. (1851). *Prolegomena Logica.* Second edition London: Rivingtons.

Mansel, H.L. (1853). *The Limits of Demonstrative Science.* Oxford: William Graham.

Mansel, H.L. (1856). *A Lecture on the Philosophy of Kant.* Oxford: John Henry and James Parker.

Mill, J.St. (1973). A System of Logic, Ratiocinative and Inductive. In: J.M. Robson (ed.), *Collected Works of John Stuart Mill,* Volumes 7 and 8. Toronto: University of Toronto Press.

Peirce, C.S. (1960). *Collected Papers of Charles Sanders Peirce.* Vols. I-VI. Second edition. Edited by C. Hartshorne and P. Weiss. Cambridge, MA: The Belknap Press of Harvard University Press.

Peirce, C.S. (1958). *Collected Papers of Charles Sanders Peirce.* Volumes 7 and 8. Edited by A.W. Burks. Cambridge, MA: Harvard University Press.

Popper, K. (1957). *The Poverty of Historicism.* London: Routledge & Kegan Paul.

Popper, K. (1945). *The Open Society and Its Enemies.* London: Routledge & Kegan Paul.

Popper, K. (1982). *Logik der Forschung.* Siebente Auflage. Tübingen: J.C.B. Mohr (Paul Siebeck).

Popper, K. (1965). Three Views Concerning Human Knowledge. In: *Conjectures and Refutations* (second edition), pp. 97-119. London: Routledge and Kegan Paul.

Richards, J.L. (1987). Augustus de Morgan, the History of Mathematics, and the Foundations of Algebra. *Isis* **78**, 7-30.

Rescher, N. (1978). *Peirce's Philosophy of Science.* Notre Dame–London: University of Notre Dame Press.

Ruse, M. (1975). Darwin's Debt to Philosophy: An Examination of the Influence of the Philosophical Ideas of John F.W. Herschel and William Whewell on the Development of Charles Darwin's Theory of Evolution. *Studies in the History and Philosophy of Science* **6**, 159-181.

Schagrin, M.L. (1973). Whewell's Theory of Language. *Studies in the History and Philosophy of Science* **4**, 231-240.

Seward, G.C. (1938). *Die theoretische Philosophie William Whewell's und der Kantsche Einfluss.* Tübingen: Buchdruckerei Christian Gulde.

Smith, S., ed. (1899). *Dictionary of National Biography.* Vol. LX. London: Smith Elder & Co.

Spencer, H. (1853). The Universal Postulate. *Westminster Review* **60**, 513-550.

Stair-Douglas, J. (1881). *The Life and Selections from the Correspondence of William Whewell.* D.D. London.

Stein, G. (1946). How Writing is Written. In: W.R. Benét, N.H. Pearson (eds.), *The Oxford Anthology of American Literature*, Vol. II, pp. 1446-1451. Oxford: Oxford University Press.

Stephen, L. (1900). *The English Utilitarians.* New York.

Stoll, M.R. (1929). *Whewell's Philosophy of Induction.* Lancaster, PA.

Strong, E.W. (1955). William Whewell and John Stuart Mill: Their Controversy about Scientific Knowledge. *Journal of the History of Ideas* **16**, 209-231.

Todhunter, I. (1876). *William Whewell, D.D.: An Account of His Writings.* London: Macmillan and Co.

Walsh, H.T. (1962a). Whewell on Necessity. *Philosophy of Science* **29**, 139-145.

Walsh, H.T. (1962b). Whewell and Mill on Induction. *Philosophy of Science* **29**, 279-284.

Wettersten, J. and J. Agassi (1991). Whewell's Problematical Heritage. In: Fisch and Schaffer (1991), pp. 345-369.

Wettersten, J. (1992). *The Roots of Critical Rationalism.* Amsterdam and Atlanta: Rodopi.

Wettersten, J. (1993). Rethinking Whewell. *Philosophy of the Social Sciences* **23**, 481-515.

Wettersten, J. (1994). William Whewell: Problems of Induction vs. Problems of Rationality. *British Journal for the Philosophy of Science* **45**, 716-742.

Wettersten, J. (1995a). Preliminary Report on Efforts by Psychologism to Gain Influences in Proper Epistemological, Methodological and Psychological Societies. In: I.C. Jarvie and N. Laor (eds.), *Critical Rationalism, the Social Sciences and the Humanities*, pp. 129-152. Dordrecht: Kluwer.

Wettersten, J. (1995b). Styles of Rationality. *Philosophy of the Social Sciences* **25**, 69-98.

Wettersten, J. (1998). Welche Probleme stellen ad hoc Hypothesen heute? *Zeitschrift für philosophische Forschung* **52**, 589-609.

Wheatstone, Ch. (1838). Contributions to the Physiology of Vision: Part I, On Some Remarkable and Hitherto Unobserved Phenomena of Binocular Vision. *Philosophical Transactions of the Royal Society* **128**, 371-394.

Whewell, W. (1830). Review of John Herschel's Preliminary Discourse on the Study of Natural Philosophy. *Quarterly Review* **90**, 374-407.

Whewell, W. (1836). *Newton and Flamsteed, Remarks on an Article in Number CIX of the Quarterly Review*. London: John W. Parker.

Whewell, W. (1837). *Mechanical Euclid*. Cambridge.

Whewell, W. (1839). *Astronomy and General Physics Considered with Reference to Natural Theology*. London: William Pickering.

Whewell, W. (1845). *Elements of Morality including Polity*. London: J.W. Parker.

Whewell, W. (1967a). *History of the Inductive Sciences*. Third edition. London: Frank Cass & Co. This is a reprint of the 1857 edition.

Whewell, W. (1967b). *The Philosophy of the Inductive Sciences, Founded upon Their History*. Second edition. London, Frank Cass & Co. This is a reprint of the 1847 edition. The first edition appeared in 1840.

Whewell, W. (1852). Letter to the Author of the Prolegomena Logica, by the Author of the History and Philosophy of the Inductive Sciences. Printed by William Whewell. Republished in: Whewell (1971), Chapter XXVIII.

Whewell, W. (1853). *Of the Plurality of Worlds: An Essay*. London.

Whewell, W. (1858). *Novum Organon Renovatum*. London: John W. Parker and Son.

Whewell, W. (1971). *On the Philosophy of Discovery*. New York: Burt Franklin. (This is a reprint of the 1856 edition.)

Williams, P. (1991). Passing on the Torch: Whewell's Philosophy and the Principles of English University Education. In: Fisch and Schaffer (1991), pp. 117-148.

Wundt, W. (1862). *Beiträge zur Theorie der Sinneswahrnehmung*. Leipzig and Heidelberg: C.F. Winter'sche Verlagshandlung.

# COMMENTARIES

**Michael Segre**

# WHEWELL'S LEGACY AND THE ART OF ARGUMENTATION

ABSTRACT. In his masterly study, Wettersten has presented in detail the views of Whewell's opponents, those who prevented him from doing good. To do him full justice and also for the benefit of lay readers, it would now be helpful, to let Whewell speak, or at least outline his contribution to history and philosophy of science.

John Wettersten has presented us with a masterly and meticulous study of the vicissitudes of Whewell's philosophy of science. He has chosen a particular style of dealing with Whewell, letting his critics speak, rather than discussing his legacy directly. The result is an essay similar to a dialogue, in the glorious tradition of Plato, Giordano Bruno and Galileo's writings on science and philosophy. It is thanks to this unconventional presentation, involving prominent philosophers and men of science from Whewell's day to the present, that one learns of many philosophical, scientific, metaphysical and methodological nuances related to the development of modern science. The book is an adventure into the meanderings of philosophical argument.

Wettersten's main purpose, as I understand it, is to draw attention to the importance of Whewell's work for science, its philosophy and history, and for scholarship in general. He rightly laments that this importance has been neglected, whether deliberately or not. In other words the book proposes an intellectual rehabilitation of Whewell and one of its many merits is to present a series of writers that have considered Whewell's ideas – a helpful starting point for any philosopher or historian of science endeavoring to find out more. It is thanks to Wettersten and a number of other authors that Whewell is being given increasing attention. However, the message of the book, in its very style, takes on a broader scope than the mere study of Whewell's seclusion. Wettersten's is a more general and ambitious study in philosophical dialectics and development. It is primarily as a lay reader and secondly as a historian of science that I would like to make a few remarks on this presentation with the object of

In: John Wettersten, *Whewell's Critics: Have They Prevented Him from Doing Good?*, edited by James A. Bell (*Poznań Studies in the Philosophy of the Sciences and the Humanities*, vol. 85), pp. 307-314. Amsterdam/New York, NY: Rodopi, 2004.

proposing and encouraging a line for future research: let Whewell speak for himself and let him, as far as possible, answer his critics.

Let me add to Wettersten's treatment a few words (and raise a few queries) related to Whewell's work and its importance, particularly concerning his contribution to the philosophy of science. I do this because I feel this has still not been outlined, either by Wettersten or by many recent works concerning Whewell. Indeed, while Wettersten's treatment tries to do justice to Whewell, to a certain extent it unjustly pays him too little attention, at least when compared to the attention given to his critics. Let me end this short outline by making a few remarks on Whewell's significance for the history of science – a significance which is particularly neglected despite its being an integral part of his contribution to the philosophy of science.

### *The Importance of Whewell's Legacy*

Who was William Whewell? Why is he important? What was his contribution to the philosophy and history of science that "could have done good," and to what extent is it still relevant today? This last question, more directly related to the present book, seems trivial but is not. Even the recent literature following Whewell's "revival," amazingly, does not answer clearly and concisely.[1]

Whewell was a most versatile intellectual and philosophy was only one of his many interests.[2] He was a cleric, composed sermons and wrote on theology. He was a leading educator (the Master of Trinity College, Cambridge, and later Vice-Chancellor of that university) and was active in education and its theory. He contributed to various branches of science (he introduced, *inter alia*, the words 'scientist' and 'physicist' into the English vocabulary): mathematics, physics, crystallography, meteorology, geology and architecture.[3] He invented a meteorological instrument, the anemometer. He composed verses, translated Plato's *Dialogues* and German literary works into English. He enriched the scientific and technical vocabulary of the English language (the terms 'ion', 'anode', 'cathode', 'Eocene', 'Miocene', 'Pliocene'). He wrote on moral philosophy. The extent of his interest is well exhibited in the detailed intellectual biography, published in 1876 by Isaac Todhunter.

---

[1] One exception is John Herivel's (1967) short and lucid Introduction to William Whewell's *The Philosophy of the Inductive Sciences*.

[2] I rely on the first source I have available, Butts (1976).

[3] His pioneering use of the term 'scientist' in analogy to 'artist' is in an essay review (1834), p. 59 (I mention this article as a supplement to Wettersten's detailed bibliography). The article mentions no author and I have found this reference to Whewell in the excellent *Dictionary of the History of Science*, edited by Bynum, et al. (1981), p. 381.

What, then, is the importance of his philosophy of science in relation to his other works and in general? A very good reference work such as the *Dictionary of Scientific Biography* (with its well-written entry by Robert E. Butts) leaves the question open. Like all articles in this work, the article begins by generally classifying Whewell's contribution to "history and philosophy of science, physical astronomy, science education" (Gillispie 1976, p. 292). It ends, however, with the following appraisal:

> In physical astronomy Whewell's work on the tides ranks second only to that of Newton. Also of great importance was his lifelong effort to modernize and improve science and education at Cambridge. The achievement in history and philosophy of science probably is less significant, although recent revival of interest in Whewell has centered mainly upon his insights in philosophy of science and methodology. Interest is growing in the interrelations of history and philosophy of science; and so long as this interest continues to be fruitful, it will be well worthwhile considering what Whewell had to say on the nature of scientific discovery, inductive methodology, and the characteristics of scientific progress. (Gillispie 1976, p. 294)[4]

The importance of Whewell's contribution to history and philosophy of science and its relation to his other first-rank contributions to science and knowledge in general still remains to be clearly explained for the benefit of lay readers like myself. In this context it would also be interesting to know to what extent he himself considered his history and philosophy of science of central importance. The best and only way to answer this question is still to read his monumental work, sometimes described as "omniscience" (see Harivel 1967, p. xiv). As I understand it, Whewell is so important because, among other achievements, he was one of the first to break with that radicalism that characterizes the Enlightenment, so typical of philosophers such as Francis Bacon and René Descartes.

Both Bacon and Descartes dismissed earlier philosophy – mainly Aristotelian philosophy – and, in general, advocated eliminating errors from the root, without discussion.[5] Their views of the development of knowledge, Bacon's in particular, have dominated our culture for many generations, and to a certain extent the latter's views remain current today. Bacon influenced Whewell and, to an even greater extent, his contemporary critic John Stuart Mill, who attempted to formalize in logic the principle of induction. The Baconian school, in addition to breaking with previous philosophical views, advocated an empirical-inductive study of nature. This influence is directly seen, *inter alia*, in the titles of Whewell's major works in the history and philosophy of science, *History of the Inductive Sciences* and *The Philosophy of the Inductive Sciences*.

---

[4] The author has written a series of works on Whewell (see Wettersten's bibliography), yet his attempt on the same page above to describe Whewell's contribution to history and philosophy of science remains slightly confusing, justifying his call, at the end of the biography, for further studies.

[5] For a discussion of this radicalist view, see Agassi (1990), pp. 157-164.

Yet, just like one of his prominent contemporary critics, John Herschel, Whewell timidly suggested that, perhaps science is not only induction, or that induction is something different from the Baconian inference from particulars to generalizations. Whewell emphasized the importance of hypotheses in science, and I would be interested to know more on his concept of induction. This contradicts Bacon's radicalism, according to which there is no place for hypothesis in science.

Whewell, unlike Bacon, considered hypotheses formulated in ancient times. A break with radicalism, even if partial and vague, means not only looking at the past but, more generally, opening a dialogue with any counterpart, including opponents of one's views. In this context one might mention that Whewell lived shortly after Hegel had inaugurated the era of modern dialectics as a necessary tool for progress in human thought and history. Dialectics, the Greek term for the art of conversation or argument, has come to have a variety of meanings in the history of philosophy. Whewell barely mentioned Hegel, and then only relation to specific scientific problems (not in relation to Hegel's theory of history). Yet, independently from Hegel, in the very fact that he considered history and philosophy of science to be two sides of the same coin he pleaded a re-evaluation of his forerunners, which is a clearly anti-radicalist view: "Whewell emphasizes the vital role played by controversy, often of a metaphysical kind" says Herivel (1967, p. xxvi) in his analysis of Whewell's views of the advance of knowledge. In this sense Whewell can be said to have started paving the way that led, in our century, to Karl Popper's philosophy of science, encouraging criticism as the most appropriate tool of dialogue and advancement of knowledge. This is how Popper defines the method of philosophy:

> It is a variant of the (at present unfashionable) historical method. It consists, simply, in trying to find out what other people have thought and said about the problem in hand: why they had to face it: how they formulated it: how they tried to solve it. This seems to me important because it is part of the general method of rational discussion. If we ignore what other people are thinking, or have thought in the past, then rational discussion must come to an end. (Popper 1980, pp. 16-17; Preface to the First English Edition, 1959)

This, by the way, is the approach adopted and encouraged by one of the pioneers of modern science, Galileo Galilei. Whewell may have not liked criticism, but this did not mean that he discouraged it at a methodological level (Galileo, too, disliked criticism).

### *Dialectics and Science*

When Galileo wished to speak in favor of Copernicanism he did it in the form of a Dialogue. His famous *Dialogue Concerning the Two Chief World Systems* (1632) takes the form of a dialogue between three interlocutors: Salviati, who speaks for Copernicus and Galileo; Simplicio, who speaks for Ptolemy and Aristotle; and Sagredo, a "learned layman," who acts as an arbitrator of sorts, though he obviously has Copernican leanings. Galileo skillfully combines science, philosophy and rhetoric into a powerful propagandist composition, and, in a unique example of popular scientific and philosophical argument, diplomatically criticizes the Aristotelian cosmology, arguing that the sun is located at the center of the planetary orbits and that the earth revolves round it. Even after his condemnation Galileo continued using the same style of writing and his *Two New Sciences* (1638), which is more technical, and takes the form of a dialogue between the same interlocutors. It is thanks to Salviati's words that seventeenth century intellectuals could learn more about the complicated Copernican theory and thanks to Simplicio that the same readers could appreciate it.[6]

It is true that Galileo's opponent, Simplico, speaks less than Salviati or Sagredo, and that the latter is actually on Galileo's side. Yet he is not totally silenced. Galileo no doubt makes an effort, albeit inconsistent, to find out and present what other people have thought and said about the problem he treats, why they had to face it, how they formulated it and how they tried to solve it.

Whewell was at least as inconsistent as Galileo and it is hard to say whether he considered science before the seventeenth century scientific revolution as real science. Yet the fact that he took ancient hypotheses into account indicates his readiness to open a dialogue, even with views that were considered prejudiced and superseded. To what extent he was silenced and his influence weakened is explained at length by Wettersten.

Wettersten's work is certainly a fulfillment of Butts' wish quoted above, in that one considers what Whewell had to say on the nature of scientific discovery, inductive methodology, and the characteristics of scientific progress; however, Wettersten does not explicitly answer the question of the importance of Whewell's achievement in philosophy and history of science. Wettersten relies on the assumption that Whewell was a great historian and philosopher of science and successfully argues that his work on history of science has been partly misunderstood, partly misrepresented and certainly unjustly neglected.

Wettersten, unlike Galileo (when presenting Copernicanism) and Whewell (when speaking of modern science), grants space nearly exclusively to Whewell's opponents; in his book, Whewell's voice or spokesman is missing. It may not have been Wettersten's primary aim to present Whewell's thought, nor, of course, to write a dialogue as such. Yet the style he has chosen calls for Whewell's presence, and this is particularly emphasized by Wettersten's

[6] For English translations, see Galilei (1953/1967; 1974).

lament that modern philosophy had been unfair to Whewell. Even a highly subjective presentation of Whewell would have been a guide to understanding Whewell's philosophy of science, a task that no description of the unfortunate fate of his views can replace.

In the long bibliography presented and discussed by Wettersten, there certainly is one or more works that present either his history or his philosophy of science. Many of these, as Wettersten rightly argues, have serious disadvantages. Isaac Todhunter, for instance (§ 2.2), "worked conscientiously through all of Whewell's works . . . But there is no development of a point of view and everything is handled quite uniformly." Yet Wettersten's criticism of Todhunter sends one back to my question: What, then, was so important in Whewell's contribution? At the end of his Foreword, Wettersten says he has presented his view of Whewell toward the end of the book, together with criticism and a new interpretation of his philosophy (however I find the postscript more helpful in this sense). Nevertheless, I remain eager to know more: by concentrating on Whewell's critics – a most important contribution to scholarship – Wettersten has partly omitted to tell his readers what Whewell views were, and thus to a certain extent prevented himself from doing good. This may be the next immediate and necessary challenge to Whewell's scholars.

*A Case Study: The Historiography of Science*

Although Whewell's history of science is part and parcel of his philosophy of science, it has been totally neglected, even by Wettersten. Whewell's contribution to history of science is without precedent and becomes particularly relevant to modern history of science, if one considers his contribution to the historiography of science. Bacon's philosophy of science had a considerable influence not only on science and learning in general, but also on the writing of science and its history, and was the main source of the "Whig" historiography of science, whose disadvantages are widely known, though still partly neglected[7] Whewell attempts to break this tradition. Perhaps Whewell's work can be better understood by considering his followers, rather than his critics.

One of Whewell's leading followers, at least chronologically, was the historian and philosopher of science Pierre Duhem. Like Ptolemy, Osiander and Galileo's opponent Cardinal Bellarmine, Duhem believed that science can merely provide mathematical instruments, not describe reality. Science, for the instrumentalist Duhem, is not empirical, but a series of conjectured hypotheses which can be temporarily useful but are

---

[7] This has been argued in detail in Agassi (1963/1967). Agassi criticizes Bacon's and Duhem's historiographies of science.

by no means the truth. Therefore, at any time, science is no more than an instrument used as a convention till a better instrument can be found. One can agree or disagree with this view (most philosophers of science today would disagree with Duhem), but it remains of central importance for the development of philosophy of science, with important implications for the history and historiography of science.

It was Duhem's work that dealt a blow to Bacon's philosophy of science and, even more, to the historiography of science written in the wake of his approach (Although similarly to the "Whig" historiography of science, one does not know any more what is really important or relevant in scientific development). Duhem, unlike Whewell, was not put aside and his philosophy was allowed to do good. On one occasion Wettersten explained to me beautifully how Whewell could be considered to be a bridge between Baconian inductivism and Duhem's conventionalism. Such an explanation would have fitted well into his book.

More generally, all this suggests that Whewell recognized, perhaps better than any of his predecessors and contemporaries among historians and philosophers of science, the interdependence of history and philosophy of science, an interdependence that is still neglected today by some leading historians of science – a century and a half after Whewell's death. Whewell's message may therefore remains relevant today and this is why it is important to let him, at last, speak.

*University of Munich*
Museuminsel 1
80538 München
Germany
Segre@lrz.uni-muenchen.de

## REFERENCES

Agassi, J. (1963/1967). *Towards an Historiography of Science*. The Hague: Mouton; Middletown, CO: Wesleyan University Press.

Agassi, J. (1990). *An Introduction to Philosophy: The Siblinghood of Humanity*. Delmar, NY: Caravan Books.

Butts, R.E. (1976). Whewell, William. In: Gillispie (1976), vol. 14, pp. 292-295.

Bynum W.F., E.J. Browne and R. Porter, eds. (1981). *Dictionary of the History of Science*. London: Macmillan.

Galilei, G. (1953/1967). *Dialogue Concerning the Two Chief World Systems – Ptolemaic & Copernican*. Trans. Stillman Drake. Berkeley: University of California Press.

Galilei, G. (1974). *Two New Sciences*. Trans. Stillman Drake. Madison: Univ. of Wisconsin Press.

Gillispie, C.C., ed. (1976). *Dictionary of Scientific Biography*, vol. 14. New York: Charles Scribner's Sons.

Herivel, J. (1967). Introduction. In: Whewell (1967), pp. ix-xxxviii.
Popper, K. R. (1980). *The Logic of Scientific Discovery*. London: Routledge & Kegan Paul.
Todhunter, I. (1876). *William Whewell, D.D. An Account of His Writings with Selections from His Literary and Scientific Correspondence*. 2 vols. London: Macmillan.
Whewell, W. (1834). On the Connection of the Physical Sciences. By Mrs. Sommerville. *The Quarterly Review* **51**, 34-68.
Whewell, W. (1967). *The Philosophy of the Inductive Sciences*. 2 vols. New York: Johnson Reprint.

**Joseph Agassi**

# THE CASE STUDY AND ITS IMPORT: WETTERSTEN ON WHEWELL

ABSTRACT. John Wettersten's study of the history of the struggle against William Whewell's innovations belongs to the fallibilist revised history of philosophy. Whether Popper's demarcation of science should be replaced with his theory of rationality or not, his study of pseudo-science should be replaced with that of pretense within the commonwealth of learning. This should facilitate reform and enrich the understanding of history through the firm distinction between views of historians and those of their subjects.

John Wettersten's study of the case of William Whewell is admirable. It is a part of a very ambitious project, of a strictly fallibilist rewriting of the history of rationalist philosophy. He deviates from Karl Popper's version of critical rationalism, being more of a realist, keeping closer to extant records in explorations of the scientific establishment and their insensitive, closed-minded attitude to philosophic innovation. It is an intriguing aspect of the Popper school that every suggested modification of its central ideas opens the road to new and engaging reflections. The fallibilist criticisms of Popper's contributions are of course within the fallibilist circle of ideas, and should possibly be taken care of in the execution of this project. Other criticisms of fallibilism have to be taken account of differently; here they will be ignored.

Each of us create our past, said Jorge Luis Borges; few have discussed the question, how should this best be done? Popper said, it is best to try to view defunct ideas as the best available solution to problems that at the time were pressing. This is true only for problem-oriented thinking, consciously so or not. Wettersten criticizes this prescription: Popper's prescription holds for those who sincerely care about intellectual problems; the problems that challenge intellectual leaders are often political, not intellectual. The Baconian idea that the rationality of science

In: John Wettersten, *Whewell's Critics: Have They Prevented Him from Doing Good?*, edited by James A. Bell (*Poznań Studies in the Philosophy of the Sciences and the Humanities*, vol. 85), pp. 315-329. Amsterdam/New York, NY: Rodopi, 2004.

is its resting on the broad basis of carefully compiled empirical evidence is hard to study in accord with Popper's prescription. The advocates of Bacon's view who ignored the problems that it raises, clung to it, and successfully blocked progress, for them Popper's prescription is hardly relevant. Their problem concerned the prestige of science. These were irresponsible leaders and should be judged as such.

Wettersten's case study is detailed, meticulous and balanced; it shows a sad picture: Whewell refuted traditional views of science and offered a new alternative. What is today called the received view of science is an inferior version of a variant of Whewell's idea as presented by Claude Bernard and Pierre Duhem (without acknowledgment) well over a century ago. This calls for a reassessment.

Even were all of Whewell's suggestions objectionable, there still is his criticism. No rationalist recommends the oversight of criticism. The party advocating criticized ideas should at least acknowledge the criticism and ask for time out. This, after all, is what many of them did when considering Hume's criticism of inductivism and Berkeley's criticism of science. What did they do about Whewell's criticism? As Wettersten argues, they did try to respond to it; yet they did not acknowledge the criticism, pretending that it does not exist. In a sense this is no novelty: Berkeley's criticism led many to respond to it, yet he was scarcely mentioned, and when he was it was in derision. Whewell was treated less honorably, in times when the standard of recognition of just criticism was higher.

## 1. *The Logic of Scientific Pretense*

The problem that Popper initially studied concerns the demarcation of science from pseudo-science. In his first book he falsely identified metaphysics as pseudo-science, despite some exceptions that he noted. In his second book he studied the problem of rationality, as an extension of his earlier work: his first book was limited to empirical rationality, to the rationality of science. William Bartley suggested then to view the extension as a problem-shift: the problem of demarcation of science is now replaced by the problem of the rationality of science, and then by the problem of rationality as such. Jagdish Hattiangadi went further, replacing the study of the demarcation of science by the study of it as a set of rational activities aimed at the advancement of learning.

Both shifts leave a lacuna: the theory of rationality applies to all conduct, including that of adherents of any pseudo-science; hence, pseudo-scientific activity is rational, and so pseudo-science is not irrationality. Popper said a little about irrationality in the name of responsibility; he spoke against irrationalism; he never viewed as irrationalist pseudo-science as such: at times it is rationalist. The recommendation of Bartley and Hattiangadi, then, leaves behind the question, what characterizes pseudo-science? Whereas they have their own substitutes to the problem of the demarcation of science, they ignored pseudo-science.

This is not to endorse Popper's views. He took for granted the traditional view of science as a set of (synthetic) theories (a set of sets of synthetic statements); pseudo-science is then characterized the same way. To some extent Popper noted this when he discussed theories immunized to criticism. He labeled them "reinforced dogmas." These theories are often immunized by the inclusion of a construal of the source of all possible criticism of it as evil. Theories thus immunized are pseudo-scientific in principle, said Popper, regardless of intent. Frank Cioffi understood him, perhaps rightly, to have given up his early identification of non-science with pseudo-science, in favor of its identification of pseudo-science with the partial set of non-science, whose members are immune to criticism.

Popper was vague on this. He characterized science and pseudo-science as sets of theories, presenting the problem as what differentiates scientific theories from pseudo-scientific ones? At times, however, he admitted that science requires the scientific attitude: a theory, he observed, is rendered unscientific by the intent to save it from possible refutations. The same goes for metaphysics: in his second book he admitted that openness to criticism should do at times. In his third book he examined the openness of some metaphysical theories to criticism. He suggested that the question, what problem a metaphysical theory came to solve, opens the question, how successful is its solution? This is intriguing, since some metaphysical theories are valuable as they offer programs for scientific research. In this case, confusing the program with its execution is typical pseudo-science, one often called vulgar, and it is common within the scientific community. A gross example is vulgar Darwinism. Pseudo-science is thus a matter of attitude, of intent. It may be the wish to save a theory from refutation; it may also be the wish to elevate it to the position it does not have, possibly due to some confusion, or worse, some pretense.

The lacuna left by the problem-shift recommended by Bartley or by Hattiangadi is thus filled to some extent by Wettersten, as he suggests to replace the discussion of pseudo-science by that of pretense to rationality, as he studied the particularly pernicious pretense of the guardians of science. Obviously, pretense is often practiced by irrationalists, but it is the pretense of rationalists that really signifies, particularly the scientific leadership. This is not unusual, untypical or uncharacteristic. Wettersten illustrates this in a detailed, significant case history; the role of pretense in the history of the rationalism of the scientific community he finds painful, but too significant and interesting to ignore. Its study is possibly useful, as it may help reduce some prevalent pretense and the harm it causes.

The challenge of scientific pretense may be tricky, but the source of the harm is evident: it is both leading scientists who resist daring innovations with obviously lame excuses, and philosophers and historians of science who indiscriminately validate whatever leading scientists say. The very obviousness of their attitude discloses their pretense: combining intelligence and irresponsibility they reveal their motive: pretending to defend science they obviously disregard truthfulness: they are power-worshippers who take knowledge to be a power-base.

## 2. *The Logic of the Study of Intent*

Discussing pretense in the scientific establishment becomes critically rational, being simultaneously appreciative and critical of the commonwealth of learning.

What rules of historiography should be employed in discussing pretense? And how should I go about discussing this question? I do not know; only one simple, partial criterion has guided me: the historiography of science should be comparable to that of politics – especially when involving its social aspects. The limitation of this comparison is obvious, as scientific research is much more rational than political activity. This is no invitation to the history of science to be less rational than that of politics. The historiography of political conduct has achieved a high degree of sophistication that few historians of science notice, apparently because they see no need for it, possibly on the assumption that human weakness is the prerogative of political practice not of scientific research. Historians of science are aware of weaknesses of scientific researchers, of course; they even note some of them, though quite marginally; but the imperfect conduct of researchers is ascribed to them as individuals, not as researchers. Yet no part of human conduct is perfect. The best asset of critical rationalism is its sense of proportion: it views the degree of rationality of research as significantly higher than that of political practice but not incomparably so.

Wettersten's sense of proportion is conspicuous as he studies the pretense that is a part of the politics of science. The traditional historians of philosophy and of science may say the politics of intellectual activity is of no interest, and is better ignored. Wettersten shows that this policy is very shortsighted: political activity may take off of the agenda important items, so that those who care for progress should pay attention to impediments to it.

Some philosophers and historians of science advocate externalism. They insist that science cannot be free of rhetoric and of private interest and all that. Even they recoil from the idea that there is such a thing as the politics of science, which influences the course of science for the worse. Their wish is not to demote science but to promote their own output as not worse. This separates the intent of the externalists from that of Wettersten, who aims at the reform of the customs and institutions of the commonwealth of learning.

The rewriting of history is always exciting, especially when it is a large-scale revision that expresses a revolutionary change. Such a change has occurred; it is the rise of critical rationalism, named after Popper. It inherited the philosophical interests of two other revolutions, one in physics, named after Einstein, one in logic, named after Russell. Its central idea rests on the impossibility of proof or of certitude except in logic, and on of the empirical refutation of some of the very best scientific theories ever. The central idea of critical rationalism is that criticism is instructive, and the better the theory criticized, the more instructive its valid criticism. This central idea is sufficiently important to invite the rewriting of the history of both science and its philosophy, even of philosophy in general. This is the story of the translation from the framework of uncritical rationalism to that of critical rationalism; the one views criticism as disgrace and it dismisses refuted ideas; the other views criticism as a form of respect and it is appreciative of ideas that invite criticism. Thus, whereas the uncritical stress the agreement of the outcome of a crucial experiment with one theory, the critical stress the clash of that outcome with the other.

The study of scientific output (of the past or of our own times) initially presented uncritically offers an enormous challenge. It is like trying to explain how come people speak one language while thinking it is another. Critical rationalism says, following inductivism is barren; how come research and its outcomes were presented for centuries in accord with it?

This is where the critical rationalist school parts company with current analytic school. That school's central idea – due to Ludwig Wittgenstein – is meant to beat classical skepticism with: skeptics cannot reasonably deny the commonness of common language; skeptics cannot declare all members of a given group mistaken about their own customs; and so on. This sounds eminently commonsense. It is a kind of commonsense, as it is very widely practiced. Every social setting rests on presuppositions and in the name of commonsense any criticism of it is often rightly dismissed. The legal system, for example, forbids questioning laws in law-courts (except in extreme cases when their applications to concrete cases are obviously absurd). Fortunately in legislatures laws are criticized and altered, and in research even the most basic tenets are, even those declared unquestionable in Wittgenstein-style analysis. Freud questioned many basic psychological tenets; Whewell and Bernard and Duhem questioned notions about science received in the scientific community; Malinowski questioned most ideas about local institutions; and the overthrow was witnessed of common notions about what language is or its grammar or its role. These have all been repeatedly overturned.

As the analytic school offers replies to the skeptic's critique of classical rationalism, it should be taken as part of that camp. Yet this is hard to maintain: classical rationalism was always clear about the truth by convention, showing little interest in it and wishing to see it replaced with the truth by nature. The advantage of the critical rationalist school over the analytic one is that it renders superfluous exercises of the kind Wettersten is engaged in. Thus the bonus of this kind of exercise is that it exposes the limitation of the style of argument adopted in that school: it is commonsense to avoid raising basic questions, but for the sake of reform this avoidance should not be indefinite, certainly not a principle.

How can historians apply reforms in retrospect? Once, when I lectured in the philosophy seminar of the London School of Economics, with Imre Lakatos in the chair (it was soon after Popper's retirement there), I raised this question, relative to the prescriptions that he, Lakatos, was the first to conceive: how could these be followed by the eminent thinkers of the past whose ideas he presented as his examples? His response was surprising: he shrugged his shoulder impatiently: it is the well-known effect of the Cunning of Reason as described by Hegel, he said as a matter of course. I was taken aback not because this answer is odious hot air, but for a much deeper reason: were Hegel's idea of the Cunning of Reason a satisfactory description of intellectual growth, then there would be no need for rationalist historical studies of ideas, such as those of Lakatos. For these which rest on the assumption that to some extent thinking is reasonable, not merely the expression of passions, so that to some extent actions are guided by thought, hardly by mysterious goals nebulously attributed to Reason, whoever She might be.

This is intriguing: what makes Hegel's answer outside the rationalist camp and Wittgenstein's inside it? Hegel says, History thinks for us; Wittgenstein says tradition does. Hegel permits change but Wittgenstein permits thinking reasonably (at least within traditions that are reasonable). How then is reasonable change possible? This is answered by critical rationalism: it takes effort to criticize ideas extant to them and more so to seek alternatives. This way the elimination of error is better than justification, from common usage or from anything else. Justification is part and parcel of tradition. How is it to be explained? As commonsense in some specific cases and as errors when taken as a general method, one that takes justification to be rationality. What then should be said of dogmatism, of the error of clinging to errors? And what should be said of the dogmatic adherence to justification? And what should be said of the obviously fake justification of justifications? How should the intent to deceive in the name of reason be presented?

The ascription of pretense is morally uncomfortable; intellectually it is an escape from a serious problem: however unpleasant it is to declare some benefactors of humanity deceitful, it is easier to say that they knew what they were doing, so that they did deceive, rather than to say that they were honest so that they did not know what they were doing. This is a valid and correct corollary to Wittgenstein's position: it is difficult to see how people allegedly active in the one framework (of semi-classical rationalism, where positive evidence is what matters) can be described as working in the other (of critical rationalism where refutation always matters more). Newton and Pasteur, to mention two towering rationalist figures, deemed criticism an expression of intense contempt. How should critical rationalists approach them?

Since the question is historiographic, one may compare the many cases of uncritical history with the few cases of critical history. What rules do these follow?

### 3. *Critical Rationalist Remodeling of the Past*

The first example of a critical rationalist rewriting of history was the study of the case of Galileo by Alexandre Koyré, a contemporary of Popper who wrote before he knew of him and who met him and became his fan only in old age. (He cited as his inspiration Gaston Bachelard, who, in turn, was indebted to Einstein.) The second was John O. Wisdom, another contemporary of Popper who wrote before he knew him, and who later became his associate and disciple. Koyré showed the tremendous importance of the role of some errors and of criticism in the development of Galileo's enterprise as an endeavor to fit physics within Plato's metaphysics. Wisdom showed the immense impact of George Berkeley's criticism of the infinitesimal: for two centuries it was the chief challenge for mathematicians. Lakatos then expanded on the historiography of mathematics and I have done the same in the historiography of the physical sciences. His aim was to develop a fallibilist philosophy of mathematics. My aim was less ambitious: to present sets of competing criteria for historical studies of science and their workings. The historiography of the social sciences offers different challenges for fallibilists, and a few important scholars have taken it up; I will make do with the mere mention of John Watkins, Ernest Gellner, I. C. Jarvie, John Wettersten and Jim Bell. Moving to the historiography of philosophy, let me observe that it offers the most difficult challenge for fallibilists. John Wettersten has undertaken a pioneering effort in an attempt to respond to it. Adam Avshalom has also contributed to this, but with regret I will skip his work here: the questions he discusses are important, but not pressing here simply because of my starting here distinctly within critical rationalism, in order to examine Wettersten's position. This is not meant to be a dogmatic stance; on the contrary, taking a view seriously enough to apply it consistently and thoroughly amounts to inspecting it for any weakness that it may have and that may thus emerge.

The first step is the writing up of the easily available information on the significant role which criticism played in intellectual history. When pushed to the wall scholars admit that criticism is valuable, but, they add, it is a mere preliminary: it is the clearing of the ground for the builders, to use John Locke's celebrated metaphor: critics do perform valuable work, provided that builders add then new ideas that are above criticism. And tradition goes with them.

Let me skip the details of the criteria for the historiography of science. I have tried elsewhere to illustrate how *desiderata* for a good scientific theory posited by critical rationalists may guide historians of science profitably in their studies. Consider Newton's opposition to the idea that a scientific theory may be false (but near to the truth). Had he agreed, he would be satisfied with his ingenious proofs that Kepler's and Galileo's theories are approximations to his own. He was not satisfied with these proofs and he implied that these two theories, and even that of Copernicus, are absolutely true. The critical reading of Newton's tortured comments on his predecessors is more satisfactory than the traditional reading, simply because tradition shares with Newton the difficulty which the critical view resolves.

It was William Whewell, incidentally, who discovered that Kepler's theory is but an approximation to Newton's, not a corollary of it. He did so as he explained his dismissal of Hegel's assertion that Newton stole his theory of gravity from Kepler. Having explained that Newton's theory prescribes to a planet its Keplerian ellipse only in a two-body system, on the grounds of which he (rightly) dismissed Hegel, he was left with the terrible corollary that Kepler's theory is inconsistent with Newton's for a system with more than one planet. He dismissed this summarily as unimportant with a lame (Hegelian) excuse. The presence of this discovery in Pierre Duhem's writings led me to surmise that he was familiar with Whewell. It is hard to assume that such a great (and pious) scholar could report this important discovery without reporting that he found it in the publications of someone else, and it is hard to assume that this great scholar was unaware of Whewell or not interested in him. One thing is clear: he had contempt for the realist and for the English; Whewell was both.

A scheme for explaining the divergence of opinions may be useful, especially about the actions of historical figures – their own views and those of their historians. There are limited sets of possible actions that historians can apply to case studies: for better or worse, the scripts of historical writing are limited to a small (though hopefully growing) number of blueprints. For example, today historians do not ascribe to their subjects supernatural powers. They also refuse to view them as saints, even though saintliness is not quite supernatural. Nor are historical figures supposed to be able to step too far out of their historical settings. Moreover, histories of politics are limited to descriptions and discussions of political circumstances and actions, in accord with what historians consider political, though, of course, the political is not considered utterly divorced from other aspects of human life. Historians of any field describe conditions and activities in that field, in accord with their own views of it. From time to time they glide to side issues, from biography or from other fields. Side issues often are embellishments, and to begin with they may be overlooked. As the study progresses, Wettersten shows, this becomes ever harder. As embellishments may obscure the structure of a narrative in the history of science, it is advisable to ignore them if possible. What however is embellishment?

This depends on the theories that guide historians. For example, as long as historians of science deemed criticism unimportant, they ignored it. Thus, though a huge part of Newton's greatest book is devoted to the critique of Descartes, Bernard Cohen has noted, till recently most of them ignored it. History is full of problems forgotten as soon as they are solved. Not only history but autobiography too often fails to mention a problem solved. Newton mentions the irregularities – deviations from ellipses – that troubled his predecessors and which his theory of gravity explained. Yet he himself was not clear about it. Until a critical rationalist theory says, new theories overcome troubles of their predecessors, this oversight is not too surprising. But what made it difficult to admit this? Here Wettersten has something amazing to tell: the public relations of science make it advisable to stress not problems but solutions. Science must look good. What philosophy is this? How does it get established?

### 4. *Intellectual History as a History of Critical Thinking*

Our concern here is the history of philosophy. The writing of critical histories of the physical sciences is the easiest to offer criteria for, at least to begin with; the critical history of philosophy is perhaps the hardest. The role of history is to explain the past; appreciation of a scientific theory is explained by reference to its past success as still recognized by present science. The innovation of critical rationalism is the idea that the past appreciation of a theory should never be forgotten but explained. Philosophers of science, let me report my empirical finding, are almost unanimous in their view that a rejected theory is and should be viewed with contempt. This view is highly confusing, as it leads to the silly conclusion that Newton's valuable theory is true and thus to many lengthy, pointless reflections on the nature of truth.

Uncritical and critical historians share the readiness to impose structures on factual information. This always raises the simple serious question: what happens when facts refuse to enter the selected mold? In the present case, what happens when a new theory does not explain known facts as well as the old one? The imposition of a structure on given material seems above criticism, as it is the imposition of criteria of importance. Are the criteria of critical rationalism not contestable? First, they can and should be contested; second, they are not.

That critical rationalism can and should be contested is evident from the usefulness for historians of Popper's extension of his theory from empirical refutations to any form of criticism. This invites the search for non-empirical criticisms in the rational portion of the history of philosophy, such as those leveled by Aristotle against Plato, such as the criticism leveled by Aristotle's commentators against his views, such as the Modern critique of mediaeval philosophy, and so on. Much philosophy is not open to such critique, and it may be judged poor; in fact we will find that we have judged it poor anyway; this may make us go the other way too, and find ourselves more open to the appreciation of some old, defunct philosophical ideas. (Writing the history of western philosophy, Russell reports in his autobiography, he read some mediaeval texts and was surprised to be quite favorably impressed by its acuity.) So much change should not be taken for granted unexamined.

Second, today no one contests the critical rationalist view of science, so that we have to go to history to find important alternatives to it. Admittedly, philosophers reject the critical terminology (Thomas S. Kuhn said explicitly that his main disagreement with critical rationalists is terminological). They also taboo certain discussions. In particular, they hardly ever discuss the reason for the past popularity of a scientific theory now defunct, because they lack the courage to admit the facts. Many of them even deny these facts: scientific theories fall out of fashion, but, they insinuate, scientific theories are never defunct. Why are they out of fashion, then? Allegedly not because they are defunct but because better ones are available. How come better ones are available, then? They are invented. Why are they invented, then? Because the old ones are not good enough. This is what makes them defunct, then. Still, most philosophers of science will not use the critical terminology, perhaps because it is too direct. So they still speak of scientific success as if it were inductive or pragmatic. This is not clear at all, but presumably it means that they declare a scientific theory successful when it rests inductively on data, and/or when it works. Yet what this means is not transparent, since notoriously no theory rests inductively on data and since many scientific ideas are useless and many useful ideas are barely scientific, not to mention the possibility that the same idea works one way and fails in another. (Usually 'work' means yields sufficiently accurate predictions to be applicable. But work may have different roles, such as scientific education and popularization, or clarification, or even politics).

When philosophers talk so vaguely it may be best simply to ignore what they say as not too serious. This is particularly the case when it turns out, upon careful analysis, that what they mean when they say that a theory rests on data or that it works is not meant to differ from the criterion here posited. (This reading of the fallibilist criterion has an inductivist name: it is "induction to the best explanation." It is not quite the same as the criterion here posed, though, as it unwittingly assumes the obligation to choose one alternative and that the alternative is always unique. When more than one best explanation exists, the advice is to go for a crucial experiment; but putting things this way is too fallibilist for the popular taste). By tradition scientific proof is very important, and then the only proofs that were ever posited were logical or mathematical. So different items were marshaled as proof: deep conviction (of truth or of some other quality that allegedly guarantees truth, usually clarity) or some empirical basis.

Duhem raised the question, since inductivism is false, how come Newton adopted it? He gave this a vague explanation and generally dismissed the views of scientists about science. This was important for him, since he declared that science progressed in the Middle Ages, during which the idea of progress was discarded. The style of mediaeval writings greatly differs from Duhem's. His reconstruction is extremely interesting and even exciting; but he distorted more than historians of architecture would had they dared describe Gothic cathedrals while concealing their ornamentation as if they were Romanesque or Renaissance.

Like political history, intellectual history can present Newton's attitude without endorsing it: he was uneasy about the logical relations between his own theory and those of his illustrious predecessors. The result of such intellectual history is far more interesting and human.

This is not to say that the critical reading is a vindication of critical rationalism. The critical attitude foregoes all vindication anyhow. All my discussion may do is illustrate a method of historical research that may be rewarding. The question to examine next is, is this also good for the study of the history of philosophy?

This cannot be the end of the story. The scientific ethos almost always included objectionable items, and some rules about the way to consider them are needed. When I said the history of science considers historical items pertinent to science, I opened the road to discussing both positive and negative evidence. But then both kinds of evidence are positive factors, contributions to scientific progress; what should historians of science do with negative factors, with items that were obstacles to scientific progress? Answer: if the impediments are from the non-scientific community then they may be discussed. This rule borders on the permission to perpetrate lies.

It is a distortion of the history of science to present its heroes as if they were critical rationalists. Political historians have learned the hard way to highlight the differences of opinions they have with their heroes. How else can a liberal write about a tyrant? Particularly when liberal historians wish to praise some old progressive tyrants, they have to explain that their praise is offered from a viewpoint not acceptable to the object of their praise. The praise is not a reward but a part of our task of coming to terms with our past, which task is one of the important challenges for political historians at all times. Why is it so difficult to transfer this standard item from the historiography of politics to that of science? I asked this question in my earliest publication on the topic, but I did not answer it.

Nor is it my claim to be the first to ask this question. The most currently famous legal philosopher in the English-speaking world is perhaps Ronald Dworkin. He declares each and every legal system to respect human rights even though most of them condone slavery and torture. He insists that they also say the opposite. This is no attempt to vindicate him. Yet this is not to censure him either. Whereas historians of science revise history without properly warning their readers, he does not have to do so: he cannot possibly mislead his readers the way historians of science do. Perhaps this is too hard on the historians of science. When Pierre Duhem invented the history of mediaeval science it was clear that at least to some extent he was engaged in reconstruction, not in straight reporting: just as readers of a history of mathematics full of modern mathematical notations are scarcely deceived by it or by the clarity that it introduced into a more murky historical picture, so Duhem's picture of mediaeval science did not mislead his readers to assume that there are mediaeval science texts akin to ours. Yet it is not known that he picked his raisins from texts that were not only unscientific but also mostly superstitious, and this way he did mislead his readers.

What is here at issue is not what is written but the intent to deceive, and deception unintended and easily avoidable. It is not too much to ask historians of science to report how they oversimplify and why; yet they will not heed this: they habitually deceive. (Historiographers of science deceive more; they couple my views with those of Herbert Butterfield, which I reject, not with those of Popper, which I constantly refer to as my inspiration.)

Why, then, is the matter more difficult in the study of the history of philosophical ideas than of scientific ones, and more there than in the study of the history of politics or of art? The answer is rather obvious. It is easy to assess the same political fact in different political viewpoints; it is easy to assess the same scientific fact in different viewpoints about science; but the same is not at all easy when it comes to philosophy, including the philosophy of science. Let me mention the ease of finding dogmatic criteria for the historiography of philosophy and the difficulties of finding fallibilist criteria for the historiography of philosophy akin to those in the historiography of science or of politics. The ease is easy to observe: authors simply praise passages that they find in their texts and agree with, and they ignore or light passages that they find there and disagree with.

This may raise the question, is a dogmatic history ever useful? This question is context-dependent. Historians are delighted in interesting source material and simply write off the dogmatism of its author. As all good critical historians know, dogmatic history is blindfolded, but if they are able critical historians then they know that all historians are blindfolded to this or that extent; they should all the more appreciate the critical works of dogmatic historians who record items that they would really rather overlook but for their respect for the truth. Also, they can also try to fill gaps due to the blindness of recorders, whether they were more dogmatic or less, more truthful or less. The more interesting point here is Popper's: the history of ideas known to us is largely due to the very invention of dissent and criticism; it is this invention that made members of the Ionian school of thought, the earliest *physiologoi*, present ideas of their different leaders without effort to compromise between them, without smoothing over opposition as is so common among dogmatists. It is easy to record different opinions as they are known to us, or to copy old records. How do we assess them? How do we approach them in order to decide how important they are for us when we study this or that historical episode?

## 5. *A Critical History of Philosophy*

To some extent uncritical (descriptive) histories of ideas have contributed to political history in this respect: when political figures were influenced by some philosophy, especially religious one, to act this way or that, then this must enter the record of their political conduct. Of course, we do not know if and how a philosophy held by some given persons influences their conduct. It is an open question whether fatalism moves one to act or not, for a conspicuous example, even though clear cases exist of different kind of influence that fatalism has effected, for example, ruthlessness and passivity. If we refuse to accept these instances, as logical positivists should, for example, then we may wish to ask, how is the appearance of these examples to be explained away and what should the explanation by reference to philosophy be replaced with? If we do accept these instances, we may want to ask, when does fatalism influence this or that action? We may ask these questions in general, as a part of our social science, and we may ask them in a historical context as historians.

Classical rationalists took rationality to be the same as proof. They had no proof theory. The need for a proof theory was not felt until recently. Perhaps it is still very unclear what proof in mathematics amounts to: Lakatos had a terrific idea about it, yet one that runs contrary to the generally received view, and one that invites much clarification and elaboration before it will be taken as canonical. But this is not important: what is very important is to realize that in earlier generations the idea of proof was much more vague than what is available today. What does one do in such cases? Take concrete examples of increased understanding: today we do not use old names for diseases as they are too vague; so historians must remind their readers that the old concepts are vague and they may try to improve descriptions of the case histories which they study. The same holds for any history that uses old, vague concepts, political or military as the case may be. The same holds for proof. What Newton or Marx understood by proof is vague by current standards. Popper has analyzed Marx's alleged proofs and showed they are unacceptable. He left open the question, why were they accepted by his followers? Perhaps because they were not academics and not very critically oriented. This is no answer, since the fact that the academics then ignored Marx until they could criticize him. Standards of proof have to be reconstructed that render his proofs admissible. The same goes for Newton's proofs. Lakatos tried to present Newton's idea of proof in this way in his doctoral dissertation. (He did not follow this vigorously and then fell back on much less promising ideas).

This is again an imposition of an idea on history. The idea imposed is that of critical rationalism: some old ideas, Newton's, Marx's, are false but valuable, and their value has to be explained historically in a manner that reflects but does not endorse their own views. This idea, so common in political history, is still uncommon in the history of science and of philosophy. Wettersten's admiration of and dissent from Whewell's views is in this framework. This is first very interesting and second problematic: is it not too dogmatic?

I do not know. Suffice it that it is clear and open to critical examination. We may start by asking, is his judgment not too severe? For, to repeat, whenever it is possible not to ascribe deceit we should try it. To be clear about it we should have more studies of the intent of individuals who were very influential in the commonwealth of learning. Thus, for example, when Spinoza claimed that his philosophy is as certain as Euclidean geometry, it is clear that he was in error but no one will seriously doubt his good faith. In particular it is amusing to read his claim that he was as self-critical as he could be, which is no doubt true, and which indicates that he thought this is a kind of proof. This is very different from Robert Boyle's suggestion that criticism should not be explicit as it will hurt the individuals whose views are criticized. Whatever we think of this, its honesty and good sense are above question, as is the need today to render old implicit criticism explicit. Matters are less clear in the case of Claude Bernard's Whewellian text: was he aware of Whewell's studies? This question invites more study. Wettersten shows clearly enough, however, that Whewell's contemporaries merit much less the benefit of the doubt. Why then did they pretend? For the greater glory of science, of course.

To repeat, Whewell's ideas suffered from being critical but not fallibilist. This was clear at least to Charles Sanders Peirce, who invented the term. But he too was more ready to mention Whewell in his lectures than in his publications. Why? Was pressure put on him? Surely he was a person who stood up to pressure well. What then is his restraint in acknowledgment to Whewell? Again this invites more study. Yet one item is clear: Whewell offered more challenge than comfort. He may be compared to the rare, vulnerable species that has left one econiche but has not yet arrived at the next safe one. This should teach us not to disregard half-baked ideas, and we can do so by attending to the problems they come to solve, which often are the criticisms of extant ideas.

*Tel-Aviv University* and *York University*
37, Legi Eshkol Str.
Herzalia 46745
Israel
agass@post.tau.ac.il

**Ron Curtis**

# THE THEOLOGICAL DEDUCTION

ABSTRACT. Any appraisal of the reception of Whewell's philosophy must take into account the reaction against his opposition to Darwin's theory and the fact that he has been incorrectly cast by inductivist historians as a special creationist. Although a vigorous opponent of inductivism, Wettersten does defer to the tradition to the extent that he is embarrassed by Whewell's theology and, in an attempt to rescue Whewell, mistakenly denies that it plays a fundamental role in Whewell's philosophy of science.

Whewell has suffered severely at the hands of inductivist historians for his vigorous, lifelong opposition to evolutionism, and any appraisal of his reception over the last 150 years must take this into account. Inductivist historians, who work, as Agassi has explained (1963; 1995), in the service of scientific orthodoxy, are always too wise after the event and readily classify all ideas in the history of science as either scientific or unscientific, good or bad, white or black, depending on whether they are true or false according to the most up-to-date science textbook. Evolutionism is, on this standard, of course, all white: "Evolution is fact, not theory," says Richard Lewontin, Agassiz Professor of Zoology at Harvard (quoted in Bethell 1985, p. 61). Inductivist historians confirm evolutionism, show that it is fact, by telling the story of its triumph over a black idea – creationism – that is treated as the only alternative. They read nineteenth-century biology in the light of intellectually insignificant skirmishes between late twentieth-century evolutionists and "scientific creationists" and construct nineteenth-century versions of contemporary biblical literalists such as Henry Morris, author of *Scientific Creationism*, or Duane T. Gish, who thinks that species were created by special acts of divine intervention in the course of nature (1988, p. 226). It has been Whewell's misfortune to have been assigned the role of one of the nineteenth-century counterparts to Morris and Gish.

In: John Wettersten, *Whewell's Critics: Have They Prevented Him from Doing Good?*, edited by James A. Bell (*Poznań Studies in the Philosophy of the Sciences and the Humanities*, vol. 85), pp. 331-339. Amsterdam/New York, NY: Rodopi, 2004.

Even though Whewell often insisted that the Bible cannot teach us science (it is intended only for our moral education), and even though Darwin quotes Whewell in the epigraph to the *Origin* in support of his attempt to provide a law-like explanation of the origin of species,[1] Whewell is nevertheless treated as a special creationist who thought that God intervenes at will in individual cases to create species nicely adapted to their environments.

By subsuming the diversity of nineteenth-century opposition to evolution, or transmutation, under the fiction of special creationism, inductivist historians have constructed a simple proof of evolution that lies just below the surface of their narrative. Lewontin states it openly: "Either you must say God created birds, say, when his finger went out and touched some primordial muck, or you must admit they arose by natural phenomena. If they arose by natural phenomena, they had to evolve" (Bethel 1985, p. 61). This is induction by elimination: there are said to be only two alternatives, and if you reject the first – and you must, since it is obviously quite absurd – then you must accept the second, which is thus "fact, not theory." Lewontin is actually simply restating a dilemma originally presented by T.H. Huxley, arch-propagandist for Darwinism. Huxley mentions Whewell as a supporter of the first alternative "who decides in favour of acts of creation and extinction we may properly call miraculous" (Huxley 1887, p. 548).

Michael Ruse, who labours as hard as anyone in the service of scientific orthodoxy, uses the Whewellian special-creationist persona in many of his writings. In "William Whewell and the Argument from Design" he says, for example, that in moving from Whewell to Darwin, we see a "change of vision from an organic world which is due to special Divine miraculous interventions to Darwin's organic world which is caused by [law-like] natural selection" (Ruse 1977, p. 258). And in *Darwinism Defended* he likens Whewell to contemporary scientific creationists who "believe the world started miraculously" (Ruse 1982, p. 322). The only good thing that can be said of Whewell and his special creationism is that "at least he was candid" (Ruse 1982, p. 323).

I have refuted the legend of Whewell the special creationist in detail in my (1990), where I show that Whewell was in fact highly critical of Charles Lyell, for example, for including special creations in his *Principles of Geology*. Whewell complained that Lyell had not consistently used lawlike natural causes in his book: rather, in order to explain certain facts about species distribution, Lyell had suggested that

[1] Whewell says in the quoted passage from his *Bridgewater Treatise* that "with regard to the material world at least, we can perceive that events are brought about not by insulated interpositions of divine power, but by the establishment of general laws" (Darwin 1959, p. 40).

"when the Author of Nature creates an animal or plant, all the possible circumstances in which its descendants are destined to live are foreseen" (Lyell 1830-33, vol. 2, p. 23). Whewell objected because in supposing the Author of Nature, in distributing organisms, "to study attentively the physical conditions of each spot and to use various precautions," Lyell was using the idea of the direct intervention of the First Cause – which had no proper place in science (Whewell 1837b, vol. 3, pp. 489-490).

The legend of Whewell the special creationist is unlikely to die easily, however, since it is central to the narrative proof of evolution. Whewell can be presented, moreover, as a formidable opponent of Darwin – even though his alleged special creationism is itself of little value and, one must suppose, therefore, easily vanquished – because, unlike his twentieth-century counterparts, he was at the center of the scientific institutions of his day.

The disrepute surrounding Whewell the special creationist in biology has spilled over onto Whewell the philosopher of science. Many commentators find that a methodologically disreputable special creationism is central not only to his antievolutionism but also to his attempt to save his anti-empiricist philosophy. Whewell's "theological deduction," in particular, is regarded as a desperate, ad hoc auxiliary hypothesis brought in at the last minute to save a system that had run into serious trouble. Robert Butts thinks that *only after* 1841, allegedly as a result of Herschel's complaint that the principles of science were for Whewell mere matters of convention or imagination, was Whewell bothered "by the realisation that his philosophy of science had not been given the kind of metaphysical justification he himself recognised was required" (Butts 1965, p. 175). Only after 1841 did Whewell begin to confront the conventionalism/realism problem, which he eventually solved in 1853 in the *Plurality of Worlds* with a theological deduction of the Fundamental Ideas. Butts quotes from an unpublished passage of the book: "Why does the constitution of the world as an object correspond with the constitution of mind as mind? First . . . because they are the works of the same Maker . . . further, because the constitution of the world is marked with the thoughts of the Divine mind, and the human mind is in part a sharer in the thoughts of the Divine mind" (p. 179).

In Butts' narrative, as Whewell's *Philosophy* founders, his special creationism comes in to save it, in a desperate return to classical modern metaphysics: "In his youth Whewell had proclaimed the inductive philosophy the 'true faith,' only to discover in his mature years that he needed precisely that metaphysics that had been regarded as the arch-enemy of the true faith to save [his] faith in induction" (pp. 176-7). Whewell was "perceptive enough" by then, says Butts, "to realize that there must be some principle of unity that subsumes [ideas and things] . . .

some 'third man' . . . the third man that Whewell finally discovered (though perhaps 'discovered' must here be regarded as a euphemism) was God" (p. 179). Though more restrained and scholarly, Butts' tone resembles that of Ruse. The term 'third man' is Butts' term, not Whewell's, and it is, of course, a picturesque allusion to Whewell's legendary special creationism.

Stoll, in a passage quoted by Wettersten in this volume (§ 4.3), treats the theological deduction as a similarly desperate attempt to save the *Philosophy*: "his conviction that a Beneficent Designer must be responsible for the perfect harmony of the a priori mental order and the physical order . . . represented the only answer Whewell could propose in an attempt to save his rationalism" (Stoll 1929, p. 63). Wettersten replies to Stoll that she fails to understand what is the proper order of inquiry for Whewell. But it seems to me that on this point- and on this point alone, perhaps – Wettersten fails too, as I explain below.

Wettersten is, of course, a vigorous opponent of inductivist historiography. But he is, it seems to me, nevertheless *embarrassed* by Whewell's theology – his alleged special creationism – and he tries to downplay the importance of Whewell's theology throughout this book, to treat it as a relatively unimportant implication of Whewell's position, one that emerges late in the game and one that Whewell's readers may accept or reject as they choose. In this way he does defer to the inductivist tradition. At one point he even defers to the notion that Whewell's *Astronomy* is a bad book (Ian Jarvie once suggested to me that Whewell's reputation would never recover from the fact that he had contributed to a series *The Bridgewater Treatises* that was intended, in the words of its sponsor, to demonstrate "the power, wisdom, and goodness of God as manifested in the creation") and that the theological deduction is the sort of ploy one might expect from the man who wrote it: "should we really be surprised that the author of the *Bridgewater Treatise,* who explained such facts as the distance of the Earth from the Sun as signs of God's plan, should now explain the possibility of knowledge with a theology?" – says Wettersten (this volume, p. 191).

I find Wettersten's argument that Whewell does not need the theological deduction – the disreputable part of Whewell's philosophy – to be the one Wettersten has the most trouble with and the one he himself seems to be most uneasy with, since he returns to it again and again throughout his book, using almost the same arguments each time. Arguing that Stoll has not understood the proper Whewellian order of inquiry, Wettersten says that in his view Whewell "clearly intended these speculations [in the theological deduction] to be a mere analysis of the truth of ideas which he had independently established" (§ 4.3). Wettersten's Whewell is remarkably lenient in matters of theology:

"theology is not the cornerstone of his philosophy . . . some may prefer to accept his theological deduction," while others who "do not believe in a God may choose not to . . . Whewell chooses merely to spell out the consequences of his own view" (§ 4.3). These consequences can be "accepted or not" (§ 4.4.4), and neither the theological deduction "nor any other explanation was needed to carry out the task Whewell set himself when he sought to explain the growth of knowledge. "It was an addition open to those who believed. One was free to reject it as one chose" (§ 8.2.5). Wettersten conflates two questions: Did Whewell think he needed the theological deduction to justify his philosophy? and Did he need it, in fact? Wettersten's answer to both questions is "no"; my answer to both is "yes."

Wettersten has some difficulty finding textual evidence to support his answer to the first question: some of the quotations he does give go directly against it. For example, when Whewell says "*supposing* such a reference [to God, in the theological deduction] to be admitted, let us see what aspect our doctrines give it." Wettersten italicizes 'supposing' (§ 7.4.4) and squeezes out of it the meaning that "you may accept or reject my position, as you wish." To me this passage says only, "Let us trace the consequences of my view." Whewell in fact makes very clear in another passage quoted by Wettersten that to reject his theological deduction is to reject the core of his philosophy, his anti-empiricism and apriorism: "those who do not believe would reply that man's thought and ideas agree with the world [because] they are borrowed from the world and that the persuasion that Ideas and truths derived from them have any origin except the world without is an illusion" (quoted in § 7.4.4).

Even if Whewell did think he needed the theological deduction, Wettersten would reply that he was mistaken. Wettersten argues in several places that "his philosophy was justified because it . . . explained the history of science" (§ 8.2.7) : "his philosophy of science is itself a science – a science of science – and, as such, is justified by the fact that it can account for the history of science as the growth of knowledge and no other theory can" (§ 8.3.6). Wettersten does not give any direct quotations to show that Whewell shared his view that the explanatory success of his philosophy provided the *only* justification it needed. And since, in fact, the philosophy of science was, as Whewell said, a science like any other, precisely the same problem of justification arises for it as for the other sciences. The philosophy of science, like other sciences, is developed using ideas arising not out of the facts in its domain – in this case, arising not out of the history of science – but having a separate and independent mental origin: the notions of a fundamental antithesis, of Fundamental Ideas, of thoughts and things, of the colligation of facts, and so on. Why, therefore – to repeat once again the problem of justification that arises for all the other sciences – have the ideas used in the science of science led to the truth of the matter? Why does the constitution of the sciences as object correspond with the constitution of these mental conceptions as mental conceptions? And have our ideas indeed led us to the truth of the matter, or have we somehow been deceived in thinking that, say, the educated intuition of scientists whose ideas are clear and distinct is not misguided, or in thinking that explanatory success is a mark of truth, or in thinking that Whewell's theory and no other can account for the history of scientific success? For these and any other ideas Whewell uses to justify his theory Whewell still needs the guarantee that God is no deceiver; he still needs his theory of Divine preattunement.

While Butts and many others are right to argue that Whewell needs his theology, the theological deduction was not, as everyone seems to think, a late, epicyclic addition. On this point, Whewell might say, everyone, including Wettersten, misses the proper order of inquiry. A version of the theological deduction was part of Whewell's thinking very early on, and it developed as Whewell's own thinking as the philosophy of science developed. This is as it should be on Whewell's own theory of science. Ideas – in this case the Idea or Conception of Design or First Cause – guide and order inquiry from the beginning. Though at first perhaps only dimly perceived, they become clear and distinct as inquiry progresses. It would not, I think, be too fanciful to write something like a Whewellian history of the development of Whewell's Idea of Design.

Whewell used that Idea as early as his (1825), when he confronted the conventionalism/realism problem with a version of the theological deduction that he allegedly used only after 1841. "Everyone who has studied the laws of nature feels," he said in 1825, "an irresistible conviction that they . . . are really the rules to which nature is subjected."[*] They really do "govern and determine" planetary motions, for example. But why should this conviction arise? We have found rules by which events are produced, but "except in the mind of an intellectual spectator, such a rule is nowhere and nothing." An analysis of the laws of nature teaches us that they have a subjective ideal element: it "seems that laws exist only in our minds and operate not upon matter but upon our way of considering it." Yet we remain convinced that they give us realistic, certain knowledge of the world. They are not "arbitrary and conventional . . . not mere classifications of the phenomena." The only solution to the apparent paradox, the only way we can hold both that the universe really is governed by the laws we have discovered and that these laws exist not only as mental entities is to accept that "these general laws must have

---

[*] Editor's note: All the quotations in this paragraph are drawn from Whewell (1825), pp. 362-365.

prevailed in the mind of an intellectual being by whom the material world is governed. There is no other manner in which they could exist." We have been able to discover some natural laws "because God has created us with an intellectual nature that so far resembles His . . . that something of what He does we can understand. We can consider nature until the material shell falls away and we see the Divine ideas almost before they are embedded in matter."

The Idea of Design that is implicit in this early version of the theological deduction was developed further over the years, rendered clearer and more distinct; one might say, as Whewell conducted his researches about science. He finds in his *Bridgewater Treatise* that a study of the progress of science does not weaken but rather strengthens belief in God. Scientific discovery seems, again, to resemble "the mode in which one intelligent being understands the conception of another" (Whewell 1837a, p. 307). As for the place of the Idea of Design in his thinking, it is not merely an implication of his views: since "purpose and design cannot be inferred from the arrangements we see around us . . . it is not at the end but at the beginning of our syllogisms that we must place it" (Whewell 1837a, p. 343).

By 1840, in the *Philosophy*, again, he says that the "Conception or Idea of a First Cause is not gathered from the effects we see around us; it is not extracted from the phenomena but assumed that the phenomena may be intelligible" (Whewell 1840, p. 706), although, like any other Idea, "it depends on observation for its development." By now, as a result of his extensive study of the sciences, the Idea is so clear and distinct in Whewell's mind that he "cannot doubt" that the mental world, including "language and all men's other faculties" (p. 707) and also "Facts and Ideas . . . the materials of knowledge . . . the stones of our building . . . which form part of the lofty temple of Truth [were] intended for this purpose by the Great Architect" (p. 708). By 1840, in other words, Whewell's science of science has rendered the Idea of Design so clear and distinct that he cannot doubt the theological deduction, which had guided his researches from the beginning.

My view of the place of the theological deduction in Whewell's thinking is therefore as follows: as Whewell develops his science of science he also justifies, or confirms, the theological deduction, which itself returns in a circle to justify his science of science. Whewell's contemporaries certainly understood it to be an integral part of the *Philosophy*. Herschel (1841, p. 182), for example, describes the Fundamental Ideas as "endowments of the creator [which are] in harmony with universal nature . . . making us capable of attaining to real knowledge."

Whewell is in my opinion much more the hard-nosed theologian than Wettersten would like to admit. But Whewell's admirers need not be embarrassed by the theological deduction, for it is not a special, creationist's hypothesis. Whewell did not try to explain *how* the Fundamental Ideas came to be so nicely adapted to the world: he did not, in particular, say that they were the result of insulated interpositions of a "third man." Whewell always maintained that the action of the Deity in so far as we can know it is lawlike, that scientific explanations must be lawlike. But he himself did not attempt a scientific explanation of the origin of Fundamental Ideas. To do so would be to venture into a science of the science of science. Things become increasingly difficult as we ascend to this lofty level. I think that at times Whewell thought that a science of this sort was probably not possible and that at times he believed it was impossible. How far one could go in this direction he did not know (see Curtis 1990).

Whewell was convinced, however, that transmutationists could not solve the problem. They could not explain the adaptations of organs of extreme perfection, such as the eye (Whewell used the example of the eye in his argument against Darwin's theory in the preface to the seventh edition of his *Bridgewater Treatise*) or the human mind with its Ideas. If species changed to the extent required by the transmutationists, they would lose their perfect adaptations, and if humans in particular "had originated through a transmutation of some species of the ape or baboon tribe" said Whewell (1837b, vol. 3, p. 480), scientific knowledge as we know it would be impossible, since the a priori mental order could not be perfectly adapted to the physical world (see Curtis 1987; see Wettersten on Hull, this volume, p. 239*ff*).

But though Whewell doubted that a science of his science of science was possible and though he thought that if it was possible it could not be transmutationist, he did formulate the problem in such a way that the young Darwin tried to solve it in his notebooks on transmutation in the late 1830s. Darwin tried to provide a transmutationist explanation of the origin of Fundamental Ideas that would stand up to Whewell's criticisms (Curtis 1987). This is one reason why inductivist historiography is wrong to paint Whewell's theory black. By formulating problems and taking critical positions to which Darwin had to try to respond, Whewell ensured that Darwin's scientific reduction was not too easy. He, along with many other opponents of transmutationism, thus forced Darwin to produce the extended argument against them that he published in the *Origin*.

RR1
Sydenham, Ontario K0H 2T0
Canada
rcurtis@kingston.net

## REFERENCES

Agassi, J. (1963). *Towards an Historiography of Science*. Beiheft 2 to *History and Theory*. Mouton.

Agassi, J. (1995). Contemporary Philosophy of Science as Thinly Veiled Antidemocratic Apologetics. In K. Gavroglu, J. Stachel and M. Wartofsky (eds.), *Physics, Philosophy, and the Scientific Community*, pp. 153-170. Amsterdam: Kluwer.

Bethell, T. (1985). Agnostic Evolutionists. *Harper's Magazine* **261**, 49-61.

Butts, R.E. (1965). Necessary Truth in Whewell's Theory of Science. *American Philosophical Quarterly* **2**, 161-181.

Curtis, R. (1987). Darwin as an Epistemologist. *Annals of Science* **44**, 379-408.

Curtis, R. (1990). Baden Powell and the Whewell Legend. *Annals of Science* **47**, 301-312.

Darwin, Ch. (1959). *The Origin of Species*. Edited by Morse Peckham. Philadelphia: University of Pennsylvania Press.

Gish, D.T. (1988). Creation, Evolution, and the Historical Evidence. In M. Ruse (ed.), *But Is It Science?*, pp. 266-282. New York: Prometheus.

Herschel, J.F.W. (1841). Review of the History and Philosophy of the Inductive Sciences by the Reverend William Whewell. *Quarterly Review* **68**, 177-238.

Huxley, T.H. (1887). On the Reception of the *Origin of Species*. In: F. Darwin (ed.), *The Life and Letters of Charles Darwin*, vol. 1, pp. 533-580. New York: Appleton.

Lyell, Ch. (1830-33). *Principles of Geology*. 3 vols. London.

Ruse, M. (1982). *Darwinism Defended*. Reading, Mass.: Addison-Wesley.

Ruse, M. (1977). William Whewell and the Argument from Design. *The Monist* **60**, 244-261.

Stoll, M.R. (1929). *Whewell's Philosophy of Induction*. Lancaster, PA.

Todhunter, I. (1876). *William Whewell, D.D. An Account of His Writings*. London: Macmillan.

Whewell, W. (1825). Reflexions of God. In: Todhunter (1876), pp. 360-365.

Whewell, W. (1837a). *Astronomy and General Physics Considered with Reference to Natural Theology*. London: J.W. Parker.

Whewell, W. (1837b). *History of the Inductive Sciences*. 3 vols. London: J.W. Parker.

Whewell, W. (1840). *The Philosophy of the Inductive Sciences*. 2 vols. London: J.W. Parker.

**Maurice A. Finocchiaro**

# WAS WHEWELL AN INDUCTIVIST?

ABSTRACT. The question of Whewell's inductivism is clarified by distinguishing several aspects of his work and several meanings of the concept. I focus on the question whether his historical account of the Copernican Revolution is inductivist in the empiricist sense and in Agassi's sense of evaluative moralism. My analysis shows that his historical interpretations are not empiricist but that they are not really apriorist either, and that his explanationism is a way of combining observation and theory. I also show that his historical interpretations exhibit a subtle awareness of the interaction between truth and error.

## I

In his interesting monograph, Wettersten has established to my satisfaction that Whewell constitutes an instructive case in modern philosophical history. Whewell certainly had insightful ideas about the nature of science and the relationship between the history and the philosophy of science, ideas which in this century were re-discovered and elaborated by Karl Popper and his followers. And Wettersten makes a convincing case that Whewell's ideas have been often misunderstood and improperly criticized.

However, after reading Wettersten's book, I am still unclear as to why Whewell's views should have been so frequently misinterpreted and so unfairly criticized. Moreover, my impression is that Wettersten goes too far in his charges of misinterpretation and in his criticism of the criticism. In particular, he criticizes too indiscriminately those interpretations that lead to favorable evaluations of Whewell as contrasted with those interpretations that lead to unfavorable evaluations. I believe that the two types of account ought to be distinguished; that they represent different hermeneutical situations; and that the favorable-oriented interpretations cannot be dismissed as easily as the unfavorable-oriented ones because the former do not need as strong a documentation as the latter, and thus in

In: John Wettersten, *Whewell's Critics: Have They Prevented Him from Doing Good?*, edited by James A. Bell (*Poznań Studies in the Philosophy of the Sciences and the Humanities*, vol. 85), pp. 341-356. Amsterdam/New York, NY: Rodopi, 2004.

turn Wettersten's criticism of the former should be more well-documented than his criticism of the latter.[1]

Another difficulty I have is that, although some of Wettersten's critiques are plausible and sound insofar as they go, they do not go very far; the reason is that they raise further important and interesting issues which he does not explore.

For example, Wettersten discusses Laudan's (1971) account of Whewell's principle of the "consilience of inductions" and criticizes it in several ways. According to Whewell (1847, vol. 2, p. 469), "the consilience of inductions takes place when an induction, obtained from one class of facts, coincides with an induction, obtained from another different class." Wettersten objects that Laudan fails to understand Whewell's justification of the value of this principle. Whewell's justification is that this principle can account for the facts of the history of science, and that no other principle can. The most relevant fact is that, in Whewell's own words, "no example can be pointed out, so far as I am aware, in which this consilience of inductions has given testimony in favor of an hypothesis afterwards discovered to be false" (Whewell 1847, vol. 2, p. 67). As Wettersten stresses, for Whewell such a justification is a scientific one by Whewell's own criteria, applied to his own field of investigation. However, while Laudan may have misunderstood Whewell's supporting argument, one would like to know whether the alleged fact mentioned by Whewell is really a fact and whether the principle of consilience really accounts for it. Wettersten does not discuss these questions.

## II

More generally, Wettersten interprets Whewell as an anti-inductivist and evaluates him favorably for being so. Wettersten also defends Whewell from his many inductivist critics and criticizes those who interpret him as an inductivist. In this regard, I wish Wettersten had clarified the issue by distinguishing several aspects of Whewell and several meanings of inductivism.

To begin with, because Whewell's work has two main components, historical and philosophical, we need to make a corresponding distinction. That is, the question of Whewell's inductivism could refer to whether or not in his historical writings Whewell portrays particular scientists as inductivists; and it could refer to whether or not in his philosophical writings he advances general epistemological claims to the effect that scientific knowledge is inductive.

---

[1] For more details about this asymmetry, see Finocchiaro (1979; 1980, pp. 332-341; 1981, pp. 17-18; 1988, pp. 28-29, 121-122, 141, 245-248; 1992).

Moreover, there is also the question of whether or not Whewell's procedure in his own inquiries is inductive. This question in turn splits into two, depending on whether procedure refers to the context of discovery or the context of justification. That is, we have the question of whether or not Whewell arrived at his historical interpretations and epistemological claims by following an inductive approach; and we have the question of whether or not the arguments he used to justify them were inductive.

Thus, there are at least four ways in which Whewell may or may not be an inductivist. They refer, respectively, to the content of his historical interpretations, to the content of his epistemological claims, to how he first arrived at his conclusions, and to how he argued when he tried to justify them.

If these distinctions are necessary to clarify what is meant by 'Whewell' when we ask whether he was an inductivist, others are equally necessary to clarify what is meant by 'inductivism'. In this context, we should begin by stressing that Whewell described himself as an inductivist; indeed, he advanced an original theory on the nature of induction. This is immediately apparent from the fact that the term is used in the title of his major works, *History of the Inductive Sciences* and *The Philosophy of the Inductive Sciences*. And this initial impression can be substantiated from the actual content of these multi-volume works. For example, the key notion of the *History* is that of an "inductive epoch," such as the inductive epoch of Copernicus, that of Kepler, that of Galileo, and that of Newton (see, respectively, Whewell 1857, vol. 1, pp. 274-290 and 317-331; and vol. 2, pp. 19-41 and 117-143). And the heart of the *Philosophy* is a discussion of "Certain Characteristics of Scientific Induction" and of the "Logic of Induction" (Whewell 1847, vol. 2, pp. 46-94). To summarize a very long story which here I have no space to elaborate, I would say that Whewell's notion of induction essentially reduces to a kind of inference to the best explanation (Whewell 1847, vol. 2, pp. 88-90). I shall call Whewell's conception the *explanationist* sense of induction.

One confusing aspect of Wettersten's monograph is that he does not make it sufficiently clear that the meaning he himself attaches to the term is different from the one Whewell attached to it. When Wettersten says that for Whewell science is not inductive, Wettersten is referring partly to induction as inference from particulars to a generalization of which they are instances. Of course, in so doing Wettersten is following a tradition which may be said to go back to Aristotle and to have continued after Whewell, with such philosophers as Augustus de Morgan.

However, despite the weight of such a tradition, it is unclear to me why a student of Whewell would want to concede this particular point to his critics, namely the point about the proper meaning and interpretation of 'induction'. This is all the more puzzling in light of the fact that nowadays the Aristotle-de Morgan conception of induction is not even the most up-to-date definition, which is rather a *probabilistic* one (Salmon 1984, pp. 14-18; Skyrms 1966, pp. 6-16). That is, the most common current definition is that of induction as inference such that it is improbable (though not impossible) for the conclusion to be false while the premises are true; in other words, induction as inference where the premises render the conclusion true with probability rather than with certainty; in short, induction as probable inference.

Another confusing element that Wettersten adds to his notion of induction is empiricism. This is reflected in his claim that according to inductivism, all knowledge is reducible to observation. In other words, the particulars from which a generalization is inferred in the process of induction are supposedly observational, sensory, or experiential. For this reason, I shall call Wettersten's sense of induction the *empiricist* conception.

Wettersten wants to attribute to Whewell an anti-inductivist epistemology, according to which scientific theories are not inferred by generalization from the facts, but are rather explanations from which the facts are deducible; they not reducible to observations but rather have an inherently hypothetical element; and, indeed, all observation is theory-laden.

A fourth and final sense of inductivism stems from Joseph Agassi (1963; 1975; 1981). It is important to consider it in more detail because Agassi is both an admirer and a critic of Whewell, and one whose views are discussed by Wettersten in great detail. Now, to some extent, Agassi's views overlap with Wettersten's. However, there is a distinct element, and this is what I wish to stress here. It involves Whewell's allegedly inductivist historiography. In the historiographical context, besides portraying scientists as inductivists in the empiricist sense of the word, inductivism uses an excessively polarized principle of classification and evaluation, namely the latest textbook. That is, history thus becomes primarily a chronological re-arrangement of the latest textbook, accompanied by accounts of who made which discovery and how; figures and ideas of which there is no trace in the latest textbook are left out; correctness and incorrectness are defined as correspondence or lack of correspondence with that textbook; thus, errors and even the gray area between right and wrong are ignored. I shall use the label of *evaluative moralism* (or just *moralism*, for short) to refer to this evaluative stance which ignores or downplays the existence or importance of scientific errors as well as the interactions between science and nonscience; correspondingly I shall speak of the *moralist* sense of inductivism.

The four aspects of Whewell's work and the four senses of inductivism generate sixteen abstract possibilities, and so the question whether he is an inductivist splits into sixteen parts. Some of these lack interest in the present context; others are easily resolved; so I shall focus on only some of them which happen to be interrelated into a cluster. For example, I do not think that the probabilistic concept of induction has much relevance here.

On the other hand, the question of whether Whewell's history is inductivist in Agassi's sense is of great import.[2] One reason for this is that Wettersten advances an interesting, subtle, and elegant criticism of Agassi's claim that Whewell's historiography is inductivist. Wettersten objects that "to classify Whewell in this fashion is to see the historiography of science as black or white and to neglect the importance, originality and uniqueness of gray figures such as Whewell" (pp. 247-248). In short, Wettersten is charging Agassi with inductivism in the field of the history of the historiography of science, namely with treating Whewell the way Whewell allegedly (according to Agassi himself) treats scientists. This is an elegant and essentially correct and fair criticism. It is fair in the sense that when Agassi claims that Whewell was an inductivist historian, this is used to point out the limitations and inadequacy of his history. However, in his criticism, Wettersten does not discuss enough details of Whewell's historical writings to prove the point that they are not inductivist (in the moralist sense). We will want to examine whether this is so.

Another reason why it is important to undertake this examination is that Agassi's notion of inductivism refers to both the portrayal of scientists as inductivists and certain features of the historian's own procedure. Now, the examination of Whewell's historiographical procedures is important because his epistemology is undeniably explanationist and anti-inductivist (in the empiricist sense of inductivism). The connection is that Whewell's explanationism is meant to apply not only in the domain of natural science, but also in his own field of investigation, namely the science of science. This means that his primary justification for his epistemological principles is that they account for the facts of the history of science, and no other principles can. Thus, Whewell's epistemology is intimately related to his history.

In short, an examination of Whewell's history will give us information not only about the character and adequacy of this important aspect of his work, but also about the adequacy of his epistemology (whether it can be justified by his history), as well as about the character of the procedures he follows (whether indeed he justifies it in this manner).

---

[2] For other appreciations and criticisms of Agassi (1963), see Finocchiaro (1973; 1987).

It is beyond the scope of this brief paper, as well as beyond my competence, to examine all of Whewell's historical writings. As is well known, they cover not only all historical periods since antiquity, but also all disciplines from astronomy to zoology. However, there is one area that I happen to know is some depth and detail, and so I shall focus on it; that is Whewell's account of the Copernican Revolution, Galileo, and related aspects of the history of astronomy and of mechanics. This is found primarily in books V and VI of the *History* (Whewell 1857, vol. 1, pp. 271-328). To this I now turn.

## III

Whewell begins his account of the "inductive epoch of Copernicus" with an explanation of why, before Copernicus, geocentrism prevailed over heliocentrism. He argues that the explanation lies primarily in the fact that, although both theories could account for such basic observations as diurnal and annual motions, only geocentrism accounted for the inequalities of planetary motions; it did so, of course, by the devices of eccentrics, deferents, and epicycles. On the other hand, a heliocentric explanation of planetary inequalities simply did not exit, because no one had worked it out. In describing this key advantage of geocentrism, we may speak of its superiority in regard to the criterion of *explanatory accuracy* since Whewell himself uses this notion when he asserts that "the heliocentric theory . . . would not approach the Ptolemaic, in the accurate explanation of facts" (Whewell 1857, vol. 1, p. 275).

However, Whewell's explanatory accuracy is not the same thing as mere empirical accuracy, in the simple sense of correspondence to direct observation. For although Whewell admitted that the criterion of direct sense-experience favored geocentrism, he thought this argument was inconclusive, as inconclusive as the simplicity argument which favored heliocentrism; he seemed to think that these two arguments counter-balanced each other. Clearly, Whewell was no empiricist: "on the heliocentric side we have the advantage of simplicity: – true; but we have, on the other, the testimony of our senses . . . Both these arguments . . . are . . . indecisive. We cannot establish any strong preponderance of probability . . ." (Whewell 1857, vol. 1, p. 275).

In regard to considerations that may be called cosmological, Whewell again seemed to think that neither alternative was favored: "Thus, as it had been urged in favour of the geocentric doctrine, that the heaviest body must be in the centre, it was maintained, as a leading recommendation of the opposite opinion, that it placed the fire, the noblest element, in the centre of the universe" (Whewell 1857, vol. 1, p. 278).

Finally, Whewell expressed the same judgment in regard to physical considerations. Two examples are mentioned. One is the geokinetic argument that it would be impossible for so many large bodies to revolve around the earth at such great speeds, as would be required by the diurnal motion of the heavenly bodies; the other is the geostatic argument that on a rotating earth it would be impossible for terrestrial bodies to fall and move as they do. Although Whewell admits that the situation changed after Copernicus, he claims that "the force of such arguments was at his time almost balanced" (Whewell 1857, vol. 1, p. 279).

I do not know why Whewell does not mention, for example, that the ether composition of heavenly bodies provided an obvious answer to the former argument. Nor am I sure that there was an equally ready answer for the latter argument. However, the oversight is in accordance with his intellectualist and explanationist portrayal of pre-Copernican astronomy.

Moreover, regarding the issue of evaluative moralism, far from being moralistic, this part of Whewell's writings contains some excellent examples of the opposite stance. In fact, in this discussion his central aim is to determine whether it is possible to avoid being wise after the event. Whewell is clear that "the astronomers who, before Copernicus, held the heliocentric opinion, cannot on any good grounds, be considered as much more enlightened than their opponents" (Whewell 1857, vol. 1, p. 276). Indeed, he goes so far as to suggest that perhaps the Ptolemaic theory was a necessary error on the path to the truth (Whewell 1857, vol. 1, p. 278).

## IV

The next step in Whewell's account of the "inductive epoch of Copernicus" is an account of Copernicus's own contribution. It should come as no surprise that Whewell interprets it in terms of explanationism rather than empiricism. Whewell does not discuss the substantive details of the Copernican achievement, but we do find a good statement of explanationism which clarifies several things:

> In this, as in other cases of discovery, a clear and steady possession of abstract ideas, and an aptitude in comprehending real facts under these general conceptions, must have been leading characters in the discoverer's mind . . . He must have seen, with peculiar distinctness, the consequences which flowed from his suppositions as to the relations of space and time, – the apparent motions which resulted from the assumed real ones; and he must also have known well all the irregularities of the apparent motions for which he had to account. (Whewell 1857, vol. 1, p. 282)

This statement makes clear that, while explanationism is different from empiricist inductivism, it is not per se unempirical or intellectualist; it is rather a manner of combining intellectual and empirical elements.

On the other hand, we have seen that for Whewell explanationism is a framework that can also be applied to geostatic astronomy. It does not by itself provide reasons for articulating or asserting heliocentrism. For Whewell, Copernicus's reasons involved not only simplicity (which was an old consideration) and explanatory accuracy (which he had himself proved for the first time), but also something which Whewell calls the "symmetry of the system" (Whewell 1857, vol. 1, p. 282). This phrase is meant to refer to the fact that the various elements of the system "are so connected in order and magnitude, that no one part can be transposed without disturbing the rest" (Whewell 1857, vol. 1, p. 282).[3] This criterion could also be called the explanatory coherence of the theory.[4]

In regard to the question of moralism, Whewell does not hesitate to discuss Copernicus's errors. One of these is the "third motion" he attributed to the earth, in an attempt to account for the fact that the axis of terrestrial rotation remains always parallel to itself. However, Whewell goes further than merely admitting such an error, which seems to have no special significance. He also mentions Copernicus's principles of circularity and uniformity. And these are significant errors insofar as they result from carrying too far the ideal of explanatory coherence, which up to a certain point is sound and beneficial. In Whewell's own eloquent words:

> The tendencies of our speculative nature, which carry us onwards in pursuit of symmetry and rule, and which thus produced the theory of Copernicus, as they produce all theories, perpetually show their vigour by overshooting their mark. They obtain something by aiming at much more. They detect the order and connexion which exist, by imagining relations of order and connection which have no existence. Real discoveries are thus mixed with baseless assumptions; profound sagacity is combined with fanciful conjecture; not rarely, or in peculiar instances, but commonly, and in most cases. (Whewell 1857, vol. 1, p. 284)

Finally, Whewell attributes to Copernicus another intellectual trait which, while not directly related to the issue of whether his historiography is inductivist, is indirectly related to it as well as intrinsically interesting. The trait relates directly to the question of the methodological role of boldness, independent-mindedness, and respect for tradition and authorities. Whewell says that, on the one hand, "Copernicus must have had great intellectual boldness and vigour, to conceive and fully develop a theory so different as his was, from all received doctrines" (Whewell 1857, vol. 1, p. 286). On the other hand, Whewell also quotes approvingly Rheticus's remark that Copernicus had great respect for Ptolemy, so much so that in a sense Copernicus's intention was "to aim at the same mark at which Ptolemy had aimed, though with a bow and with shafts of a very different material from his" (Whewell 1857, vol. 1, p. 286). The general point is that "this deference for the great men of the past, joined with the

[3] Quoted by Whewell from Copernicus's own Preface to *De Revolutionibus*.

[4] For more details on explanatory coherence, see Finocchiaro (1997, pp. 28-31); see also other authors who use different terminology for what is essentially the same concept, for example, Lakatos and Zahar (1975, pp. 368-381), Millman (1976), and Thomason (1992).

talent of seizing the spirit of their methods when the letter of their theories is no longer tenable, is the true mental constitution of discoverers" (Whewell 1857, vol. 1, p. 286). The indirect connection with the question of inductivism is that Whewell is here extolling the anti-holistic attitude that it is important to learn to distinguish the right from the wrong in the ideas of great predecessors, rather than regarding them to be totally right or totally wrong.

## V

Highly revealing and relevant is also Whewell's account of the effect of Galileo's telescopic discoveries on Copernicanism. This account tends to downplay the importance of these empirical discoveries. Like his account of pre-Copernican astronomy, this de-emphasis does clearly show that Whewell is not inductivist in the empiricist sense, but it also indicates his penchant toward intellectualism.

To be sure, the title Whewell gives to this section is "The Heliocentric Theory confirmed by Facts. Galileo's Astronomical Discoveries" (Whewell 1857, vol. 1, p. 299). However, the main claim with which he concludes the section is that "these events are a remarkable instance of the way in which a discovery in art . . . may influence the progress of science" (Whewell 1857, vol. 1, p. 303).

More importantly, when Whewell discusses the concrete details of the claimed "confirmation," his account is somewhat superficial and does not do justice to the situation. Consider for example, the satellites of Jupiter. Whewell does state that this discovery

> showed that the real universe was very different from that which ancient philosophers had imagined, and suggested at once the thought that it contained mechanism more various and more vast than had yet been conjectured. And when the system of the planet Jupiter thus offered to the bodily eye a model or image of the solar system according to the views of Copernicus, it supported the belief of such an arrangement of the planets, by an analogy all but irresistible. (Whewell 1857, vol. 1, p. 301)

However, he explains neither the specific difference, nor the mechanism, nor the analogy.

Now, in regard to the difference, Jupiter's satellites provided conclusive empirical evidence that some heavenly bodies do not revolve around the center of the universe, thus disconfirming the Aristotelian laws of natural motion as well as the earth-heaven dichotomy.

In regard to the talk of mechanism, it is unclear what Whewell has in mind; perhaps it is the following. One of the arguments against Copernicanism involved the lunar orbit. The objection was that, since it is undeniable that the moon revolves in a monthly geocentric orbit, if the earth were revolving around the sun, then the moon would be left behind;

the usually unstated reason why the moon would be left behind was that no mechanism was known whereby one body could revolve around a second while the second is in motion around a third. Jupiter's satellites enabled Galileo to answer this objection by pointing out that, since both sides of the controversy agreed that Jupiter revolves in an orbit, its satellites are obviously not being left behind (see Galilei 1890-1909, vol. 3, p. 95; and vol. 7, p. 362; and Finocchiaro 1997, pp. 237-244); thus, they provide clear empirical evidence that it is possible for one body to orbit a second while the latter orbits a third. In other words, whatever (unknown) mechanism allows Jupiter's satellites to keep up with the planet, the same mechanism would allow the moon to follow the earth if Copernicanism were true. Galileo's reply is in part an argument from analogy, and it does strengthen Copernicanism by removing an objection against it. While the analogy is a good one, it is far from being "irresistible," and it involves a comparison between the Jovian system and the earth-moon system rather than between the former and the solar system.

Whewell's talk of an irresistible analogy between the Jovian system and the solar system may be a glimpse at another argument which is almost irresistible but is not an argument from analogy. Jupiter's satellites could be observed to revolve around the planet in periods which increased as the distance from Jupiter increased. Given this aspect of their motion, it was almost irresistible to compare it with the similar pattern known to exist for the five agreed-upon planets (Mercury, Venus, Mars, Jupiter, and Saturn), regardless of whether the center of planetary revolutions was the earth or the sun. This comparison suggested that the generalization previously known to hold for the planets (by simple and exhaustive enumerative induction) was a universally valid law of nature; the law would state that, whenever several bodies are revolving around a common center in orbits of different sizes, the periods of revolutions become longer as the orbits become bigger. This law is of course an approximation to Kepler's third law. In the Galilean context (Galilei 1890-1909, vol. 7, pp. 144-145; Finocchiaro 1997, pp. 134-36), the more rudimentary law provides a powerful argument for the earth's motion: the argument is that the Copernican system is in accordance with the law, whereas any geostatic system violates it insofar as the diurnal rotation of the stellar sphere involves the largest orbit but the shortest period.

The upshot of these considerations is that Whewell does not seem to show the proper appreciation for the empirical discoveries made possible by the telescope. This does not mean that either he or his system lack the resources for such a proper appreciation. For example, his own notion of the consilience of inductions seems readily applicable to the just mentioned example of Jupiter's satellites and the Galilean law of revolution. Nevertheless, Whewell's superficial account of the import of Galileo's empirical discoveries does suggest an intellectualist or anti-empirical bias, even as it also shows that he is no inductivist in the empiricist sense of the term.

## VI

As a "prelude to Kepler's astronomical discoveries" (Whewell 1857, vol. 1, p. 312), Whewell discusses the status of the heliocentric theory from the point of view of physical considerations. By physical Whewell means pertaining to the causes of motion, and so it corresponds to what we might call mechanical or dynamical. He discusses two of the traditional objections to the earth's motion, the argument from natural motion and the argument from vertical fall. Two conclusions are especially relevant to our subject.

The first is that "the loose and inaccurate notions which prevailed respecting the nature and laws of force, prevented, for some time, all distinct reasoning on this subject, and gave truth little advantage over errour" (Whewell 1857, vol. 1, p. 312). Here Whewell seems to admit that for a while physics did not favor Copernicanism. This admission is perhaps too weak insofar as the available physics at the time actually disconfirmed the Copernican theory. Nevertheless, to judge that in a given situation truth had little advantage over error is certainly a sign of an anti-moralistic attitude, and so of an anti-inductivist attitude in that sense of the term. The same is the impression conveyed by Whewell's frank admission of Kepler's errors, claiming that "Kepler was right no further than in his suspicion of a connexion between the cause of motion and the distance from the center; not only was his knowledge imperfect in all particulars, but his most general conception of the mode of action of a cause of motion was erroneous" (Whewell 1857, vol. 1, p. 316).

In fact, the case of Kepler offers Whewell a great opportunity to show that his own view of scientific progress is far from being moralistic. He begins his discussion of the "intellectual character of Kepler" (Whewell 1857, vol. 1, p. 317) by referring to several authors who have expressed dismay at the fact that he made his discoveries on the basis of very fanciful conjectures. Whewell argues that such a process is the rule rather than the exception; what is unusual in Kepler's case is the fact that he wrote down a detailed account of his procedures and that he was somewhat slow at finding the flaws in his conjectures. Indeed, "all who discover truths must have reasoned upon many errors, to obtain each truth . . . Discovery is not a 'cautious' or 'rigorous' process" (Whewell 1857, vol. 1, p. 318). According to Whewell, Kepler's virtues were fertility of imagination and perseverance in testing his conjectures; his weaknesses were mathematical imprecision and slowness in rejecting errors.

## VII

Let us now turn to what Whewell calls the "Inductive Epoch of Galileo. – Discovery of the Laws of Motion in Simple Cases" (Whewell 1857, vol. 2, p. 19). This development provides additional excellent test cases for exploring the question of moralism. The two cases are the figures of Descartes and Galileo. Because the former was mostly wrong and the latter mostly right in their respective scientific contributions, the inductivist-moralist historian can hardly resist the temptation to ignore Descartes completely and, for the case of Galileo, to report only his successes. Clearly, this is not Whewell's procedure.

On the one hand, Whewell does reach the general conclusion that "of the mechanical truths which were easily attainable in the beginning of the seventeenth century, Galileo took hold of as many, and Descartes of as few, as was possible for a man of genius" (Whewell 1857, vol. 2, p. 39). However, Whewell does not paint a completely negative picture of Descartes. For example, he agrees that, in regard to the first law of motion, "Descartes was the first to state this generally" (Whewell 1857, vol. 2, p. 20). And he does not hide Galileo's limitations. For example, in regard to the same law, Whewell claims that although he eventually got it right (in *Two New Sciences*), in the *Dialogue* "he asserts circular motion alone to be naturally uniform, and retains the distinction between natural and violent motion" (Whewell 1857, vol. 2, p. 20).[5] Whewell also criticizes as "somewhat insecure" (1857, vol. 2, p. 24). Galileo's simplicity argument for the principle that the velocity of falling bodies increases proportionally to the time elapsed, an argument found in the *Two New Sciences*. Finally, on another occasion, although Whewell does not really criticize Galileo, he objects to some of his admirers for regarding his principle of virtual velocities as one of his great contributions to mechanics. This principle states that "in any machine, the weights which balance each other, are reciprocally to each other as their virtual velocities" (Whewell 1857, vol. 2, p. 32). Whewell has reservations about Galileo's proof of this principle: "instead of offering any reasons which could independently establish this principle, he gives his readers a number of analogies and illustrations, many of them very loose ones . . . Such images may please the fancy, but they cannot be accepted as mechanical reasons" (Whewell 1857, vol. 2, p. 33).

[5] This thesis by Whewell is not exactly right, but the discussion of the relevant issues is beyond the scope of this paper; for some details, see Finocchiaro (1980, pp. 33-34, 87-92, 349-353).

In regard to the issue of empiricism, there are several relevant passages. After reporting Descartes' theological justification of the first law of motion, Whewell evaluates it as follows: "Reasoning of this abstract and a priori kind, though it may be urged in favour of true opinions after they have been inductively established, is almost equally capable of being called in on the side of error" (Whewell 1857, vol. 2, pp. 20-21). Such an objection to Cartesian apriorism may be taken an indication of some kind of empiricism on Whewell's part.

The same impression is conveyed by Whewell's remarks about the Aristotelian principle that heavier bodies fall faster. Whewell says that several people at the time of Galileo questioned this principle. He even admits that the refutation of the Aristotelian principle "followed by very obvious reasoning; for ten bricks, in contact with each other, side by side, would obviously fall in the same time as one; and these might be conceived to form a body ten times as large as one of them" (Whewell 1857, vol. 2, p. 36). A version of this refutation had been given by Galileo himself in *Two New Sciences* (Galilei 1974, pp. 66-67). However, Whewell also asserts that "the merit of proving this by experiment, and thus refuting the Aristotelian dogma, is usually ascribed to Galileo, who made his experiment from the famous leaning tower of Pisa, about 1590" (Whewell 1857, vol. 2, p. 36). And this empiricist myth is repeated in Whewell's *Philosophy of the Inductive Sciences* (1847, vol. 2, p. 217), in a context where he does not even mention the possibility of a theoretical refutation of the principle.

Of course, it is unclear whether this assertion indicates a deeply held belief on Whewell's part, or whether it is merely an example of uncritical acceptance of sources. Nevertheless, even if it is the latter, the uncritical acceptance of an empiricist source may be itself a sign of empiricism.

However, such empiricist leanings in Whewell's discussion of the inductive epoch of Galileo must be taken in context. For one thing, they should be counterbalanced with his anti-empiricist account of pre-Copernican astronomy and of post-Copernican telescopic discoveries discussed above. For another, even within this inductive epoch there are other things he says which indicate a more complex philosophy than simple empiricism. A good example is what he says about the justification of the first law of motion:

> The induction by which the first law of motion is established, consists, as induction consists in all cases, in conceiving clearly the law, and in perceiving the subordination of facts to it. But the law speaks of bodies not acted upon by any external force, – a case which never occurs in fact; and the difficulty of the step consisted in bringing all the common cases in which motion is gradually extinguished, under the notion of the action of a retarding force. In order to do this, Hooke and others showed that, by diminishing the obvious resistances, the retardation also became less; and men were gradually led to a distinct appreciation of the resistance, friction, &c., which in all

> terrestrial motions, prevent the law from being evident; and thus they at last established by experiment a law which cannot be experimentally exemplified. The natural uniformity of motion was proved by examining all kinds of cases in which motion was not uniform. Men culled the abstract rule out of the concrete experiment; although the rule was in every case, mixed with other rules, and each rule could be collected from the experiment only by supposing the others known. (Whewell 1857, vol. 2, p. 21)

This is a very revealing passage. It reveals that, as already mentioned, Whewell was an inductivist of sorts, although he had an original conception of induction. It reveals that that he conceived induction in an explanationist, not in an empiricist, sense. It reveals that his explanationism is not apriorist or anti-empirical but rather gives empirical considerations an important role to play. There is even a hint (in the last sentence) that some kind of holism is needed in methodology, although (as far as I know) this is not a point generally stressed by Whewell, and it was left to Duhem and others to elaborate the problem.

## VIII

In conclusion, I began by clarifying the question of whether Whewell was an inductivist. I have examined this question by focusing on Whewell's historical writings on the Copernican revolution and Galileo. Focusing on such writings is important not only because the history of science is such a large part of Whewell's work, but also because of his view of its relationship to the philosophy of science.

These writings show clearly and conclusively that he is not an inductivist in the sense of evaluative moralism. On the contrary, his interpretations and procedures exhibit a sound appreciation for the inescapability and significance of error in scientific inquiry. Indeed, he is so insistent and eloquent about this that reading what he says about it feels like one is reading some of what Agassi (1973) has written on the topic.

Whewell's historical writings also show that he is not an inductivist in the empiricist sense of the word. If anything there are indications that he has apriorist, intellectualist, or anti-empirical leanings. However, such leanings should be viewed in context, that is in the context of the fact that sometimes he seems exhibits empiricist leanings, and more generally in the context of his explanationism. As the last quotation vividly illustrates, explanationism (at least in Whewell's version) is an interesting and viable way of taking into account both empirical and intellectual considerations. Thus, the opposite tendencies detectable in Whewell's work do not really amount to a case of incoherence or inconsistency.

Finally, it cannot be ignored that Whewell's version of explanationism is also a version of inductivism, in his sense of the term induction. This is a problem which I have not solved, but merely formulated.

*University of Nevada, Las Vegas*
Department of Philosophy
Las Vegas, NV 89154-5028, USA
mauricef@nevada.edu

## REFERENCES

Agassi, J. (1963). *Towards an Historiography of Science. History and Theory*, Beiheft **2**. The Hague: Mouton.

Agassi, J. (1975). *Science in Flux*. Dordrecht: Reidel.

Agassi, J. (1981). *Science and Society*. Dordrecht: Reidel.

Finocchiaro, M.A. (1973). *History of Science as Explanation*. Detroit: Wayne State University Press.

Finocchiaro, M.A. (1979). Methodological Criticism and Critical Methodology. *Zeitschrift für allgemeine Wissenschaftstheorie* **10**, 363-374.

Finocchiaro, M.A. (1980). *Galileo and the Art of Reasoning*. Dordrecht: Reidel.

Finocchiaro, M.A. (1981). Fallacies and the Evaluation of Reasoning. *American Philosophical Quarterly* **18**, 13-22.

Finocchiaro, M.A. (1987). The Methodology and Philosophy of History of Science: Recent Issues and Developments. *Organon* (Warsaw), Nos. 22/23, 1986/1987, 99-114.

Finocchiaro, M.A. (1988). *Gramsci and the History of Dialectical Thought*. Cambridge: Cambridge University Press.

Finocchiaro, M.A. (1992). Asymmetries in Argumentation and Evaluation. In: F.H. van Eemeren, R. Grootendorst, A. Blair and C. Willard (eds.), *Argumentation Illuminated,* pp. 62-72. Amsterdam: Stichting International Centre for the Study of Argumentation.

Finocchiaro, M.A., trans., ed. (1997). *Galileo on the World Systems*. Berkeley: University of California Press.

Galilei, G. (1890-1909). *Le Opere di Galileo Galilei*. 20 vols. National Edition by A. Favaro. Florence: Barbera.

Galilei, G. (1974). *Two New Sciences*. Translated by S. Drake. Madison, WI: University of Wisconsin Press.

Lakatos, I. and E. Zahar (1975). Why Did Copernicus' Research Program Supersede Ptolemy's? In: R.S. Westman (ed.), *The Copernican Achievement,* pp. 354-83. Berkeley: University of California Press.

Laudan, L. (1971). William Whewell on the Consilience of Inductions. *The Monist* **55**, 368-391.

Millman, A.B. (1976). The Plausibility of Research Programs. In: F. Suppe and P.D. Asquith (eds.), *PSA* 1976*: Proceedings of the* 1976 *Biennial Meeting of the Philosophy of Science Association*, vol. 1, pp. 140-148. East Lansing, MI: Philosophy of Science Association.

Salmon, W.C. (1984). *Logic*. Third edition. Englewood Cliffs, NJ: Prentice-Hall.

Skyrms, B. (1966). *Choice and Chance*. Belmont, CA: Dickenson.

Thomason, N. (1992). Could Lakatos, Even With Zahar's Criterion for Novel Fact, Evaluate the Copernican Research Programme? *British Journal for the Philosophy of Science* **43**, 161-200.

Whewell, W. (1847). *Philosophy of the Inductive Sciences Founded upon their History*. Second edition. 2 vols. London: John W. Parker, 1847. Reprinted by New York: Johnson Reprint Corporation, 1967.

Whewell, W. (1857). *History of the Inductive Sciences*. Third edition. 3 vols. London: John W. Parker and Son, 1857. Reprinted by Frank Cass & Co. Ltd., 1967.

**Godfrey Guillaumin J.**

# WILLIAM WHEWELL'S IDEA OF HISTORICAL CAUSATION. SOME METHODOLOGICAL AND EPISTEMOLOGICAL DIFFERENCES WITH HERSCHEL

ABSTRACT. This article explores the Whewellian notion of Palætiological sciences and some differences with Herschel's idea of science. Whewell maintained, as the core of his idea of scientific enterprise, a difference between mechanical and historical causation. According to Whewell, while mechanical sciences have in fact arrived at the methodological and epistemological maturity, historical ones have a long way to go before they can be reliable. Whewell thought that Lyell's geological theory was an historical science, which he called palætiological science. Herschel, on the other hand, considered Lyell's theory a good one mainly because it complied with his methodological standards of *vera causa*. Whewell and Herschel disagreed on a fundamental and totally new issue-namely, whether geology has the same scientific status as the successful sciences. This episode was crucial in the history of methodology because, among other reasons, it implicitly revealed fundamental differences with respect to the leading science of that time: Newtonian mechanics.

## 1. *Introduction*

Two of the more important epistemological evaluations of Lyell's geological theory were those of Herschel and Whewell. While, roughly speaking, Herschel considered Lyell's theory to be a good one, particularly Lyell's climate theory, Whewell thought that the appraisal of Lyell's work hangs on getting greater clarity about some key methodological issues. According to some scholars (Ruse 1975; 1979) such discrepancies between Herschel and Whewell depended upon what each of them took as a true cause (*vera causa*).

In the present paper, I focus on Whewell's evaluation of Lyell's theory and I emphasize and analyze the role of the Whewellian notion of *Palætiological* sciences; in fact, there is no other way adequately to

In: John Wettersten, *Whewell's Critics: Have They Prevented Him from Doing Good?*, edited by James A. Bell (*Poznań Studies in the Philosophy of the Sciences and the Humanities*, vol. 85), pp. 357-375. Amsterdam/New York, NY: Rodopi, 2004.

understand Whewell's thoughts about geology. I shall maintain the thesis that Herschel's and Whewell's differences on Lyell's theory are not only about their understanding of the *vera causa* notion and their general epistemic position about scientific knowledge (rationalism and empiricism), but they also depended upon a different set of problems which Whewell found in historical causation, for example, some epistemological difficulties detected by Whewell concerning how to understand and to justify historical causation.

My intention here is to answer, if partially, a series of important questions that John Wettersten has outlined in the present anthology-namely, were there significant methodological as well as philosophical differences between Herschel and Whewell? If so, which are they? Especially with respect to the way in which historical causation is integrated in scientific explanation, I think there are significant methodological and philosophical differences between Herschel and Whewell. This topic has not been approached and my purpose in this article is to bridge this gap.

The strategy I am going to follow is to expound very briefly some of the philosophical and empirical problems Lyell faced. Then I shall analyze Herschel's methodological appraisal of Lyell's theory and the relation between natural laws and causality that Herschel maintained. Next, I shall examine the Whewellian notion of palætiological sciences in order to stress its philosophical and epistemological relevance. Finally, I shall conclude by accenting the fundamental differences between Herschel and Whewell about this issue.

## 2. *Lyell's Geological Theory as a Theory of* Vera Causa

Between the late eighteenth century and the early nineteenth century, the problem of how to explain phenomena that involved an historical dimension arose with some force in different disciplines. German linguistics, geology, embryology, human history, etc., faced difficulties in describing and classifying the development of phenomena. De Candolle, Hutton, Werner, and Lyell are some of the scientists involved with that kind of problem.

Lyell's masterpiece on geology, *Principles of Geology* first published in 1830-33, appeared as a particular case among that problematic class of investigations which had historical explanation as its core. Faced with different general geological theories, like Werner's or Cuvier's, which postulated unknown phenomena – huge catastrophes, the hot molten center of the earth, etc. – Lyell's goal was to create a geological theory based on Hutton's geological principles (Hutton 1788), on the basis of "true causes" and rejection of the hypothetical character of most of the geological theories of his day. To Lyell, a fundamental epistemological problem was how to know whether geological phenomena postulated by rival theories really existed.

The *vera causa* principle was a methodological inheritance from Newton based on his first *regula philosophandi* which says that we are to admit no more causes of natural things than such as are both true and sufficient to explain their appearances. Lyell articulated this principle under the assumption of an uniform development of the geological history of the Earth. He thought that "the laws of nature have not changed over time" (Laudan 1987, p. 205) and only phenomena like earthquakes, volcanic activity, sedimentation deposits, the movement of the ocean, etc., were the unique acceptable geological causes in his non-hypothetical geology. At the beginning of his third volume of *Principles*, Lyell said that "in our attempt to unravel these questions (geological ones), we shall adopt a different course, restricting ourselves to the known or possible operations of existing causes . . ." (Lyell 1830-33, vol. 3, p. 6). Likewise, he maintained that such causes or forces currently operating are of the *same* kind that had been operating through past time up until the present, and that its magnitude had been equal in all times. These two basic characteristics of his theory, Whewell called *uniformitarianism* and *actualism* respectively.[1] Lyell expressed the idea of actualism in the subtitle of his *Principles* as "an attempt to explain the former changes of the Earth's surface, by reference to causes now in operation."

Lyell's methodological strategy successfully explained climate change, but he found some difficulties to explain the elevation of continents by a *vera causa* theory. One of these difficulties was that Lyell could not find any true cause (actually observed) responsible for the elevation of continents; in order to explain that, he would have had to hypothesize an unknown cause. Faced with those difficulties, however, he did not resort to hypotheses because "the method of hypothesis, Lyell thought, was unreliable; consequently, he eschewed hypotheses about the causes of volcanoes and earthquakes as far as he could" (Laudan 1987, p. 217). In fact, Lyell adhered to *vera causa* methodology and his basic epistemic principle: only causes now in operation were to be admissible as agents in explaining the past. He rejected the hypothesis method arguing that the right method of analysis consists in inferring the past from the present; the main thesis here is that "all hypotheses about unobserved causes or effects must be founded squarely on what we have observed" (Laudan 1987, p. 204). In the following passage of the *Principles*, Lyell clarified that issue:

---

[1] The *uniformitarianism* notion is too complex to explain here. Some scholars have found that Lyell used it in three different ways. See Rachel Laudan (1987), p. 205. For my purposes, however, it is enough to note Lyell's removal from hypothesis method.

> Instead of vague conjectures as to what might have been state of the planet at the era of its creation we fix our thoughts steadily on the connection at present between climate and the distribution of land and sea . . . we may perhaps approximate to a true theory. If doubt still remains, it should be ascribed to our ignorance of the laws of Nature, not to revolutions in her economy – it should stimulate us to further research, not tempt us to indulge our fancies in framing imaginary systems for the government of infant worlds (Lyell 1830-33, vol. 1, p. 105).

It is crucial to notice that *vera causa* method, according to Lyell, was a means to obtain true knowledge about geological matters only under the assumption that "the past history of the Earth had never been fundamentally different from what we can observe about it in the present" (Rudwick 1992, p. 65). The present is not only the key for the past but also the key for the future, because natural geological laws have always been and will always be of the same kind; they govern and will continue to govern geological matters in the same way as in the past. Indeed, Lyell was not interested in the future, but it is likewise clear that his uniformitarian principle was supposed to be extended to the future mainly because laws of nature are invariable and universal. According to Lyell, geological change is continuous and the Earth is in balance, i.e., dynamic steady state. Gould (1965; 1987) and Rudwick (1972) have pointed out that Lyell used his uniformitarian principle in four different ways. I am not going to criticize their interpretation here but I should mention one of them in order to throw some light on the point I am stressing.

The sense of uniformity in which I am interested states that "natural laws are constant in space and time. Philosophers have long recognized . . . that assumptions about the invariance of natural law serve as a necessary warrant for extending inductive inference into an unobservable past" (Gould 1987, p. 119). According to Lyell, geological phenomena are governed by natural laws, but for Lyell geology is an historical science. He said that "Geology is the science which investigates the successive changes that have taken place in the organic and inorganic kingdoms of nature" (Lyell 1830-33, vol. 1, p. 1). In spite of geology's concerns with natural invariable laws, some of the epistemic functions of those laws are not present in geology, for example, prediction. Lyell was clearly aware of the difference with mechanical science and particularly with the distinctive character of historical inquiry – namely, the explanation of present phenomena as the contingent results of a past that might have been different, not as predictable products of nature's laws (Gould 1987). Thus, the uniformitarian principle in geology, insofar as it deals with contingent events, has not the same epistemic character as the physical sciences, even though it involves nature's laws. Lyell traced the difference in the following words:

> By the geometer were measure the regions of space, and the relative distances of the heavenly bodies – by the geologist myriads of ages were reckoned, not by arithmetical computation, but by a train of physical events – a succession of phenomena in the animate and inanimate worlds-sings which convey to our minds more definite ideas than figures can do, of the immensity of time. (Gould 1987, p. 154)

Lyell realized that history must be a succession of phenomena, where every singular phenomenon obeyed nature's laws but the contingent sequence of all of them constitutes the historical character of geological phenomena. Thus, geology has an empirical character insofar as it works with true causes and nature's laws. At the same time, it has an historical character because geological explanations are causal narratives which incorporate successive changes that have taken place in the past. Herschel appreciated Lyell's theory because it followed *vera causa* principles, but mainly because it was adequate to a certain idea of the relation between causes and laws which Herschel maintained.

## 3. *Herschel's Methodological Position on Geological Causality*

In the famous article that Herschel wrote in the late 1824s for the *Encyclopedia Metropolitana*, titled "Light," he had already developed the general outline that he was to articulate further in his *Preliminary Discourse*. In fact, "it was here [in the article 'Light'] that Herschel explicitly used the term *vera causa* for the first time . . . he used it in his discussion of Fresnel's undulatory explanation of the colors produced by polarized light in crystals" (Good 1987, p. 37). This close relation between his optical works[2] and some of his methodological ideas was an important factor in developing a certain idea about natural laws, as well as about their methodological role in the scientific explanations. To Herschel, as we shall see, there is no fundamental distinction in the methodological function of optical laws and geological ones.

About the origin and some characteristics of true causes, Herschel said that

> experience having shown us the manner in which one phenomenon depends on another in a great variety of cases, we find ourselves provided, as science extends, with a continually increasing stock of such antecedent phenomena, or causes (meaning at present merely proximate causes), competent, under different modifications, to the production of a great multitude of effects, besides those which originally led to a knowledge of them. To such causes Newton has applied the term *verae causae*; that is, causes recognized as having a real existence in nature, and not being mere hypotheses or figments of the mind. (Herschel 1987, pp. 144-145)

---

[2] Herschel's most important and original research in optics was done between 1819 and 1820. It was dedicated to discover the axioms or physical Laws from which one can deduce the particular color patterns produced when polarized light is transmitted through crystals of one or two optic axes. See Buchwald (1989), p. 291.

Even though the main conditions for recognizing a true cause is that it really exists (viz., it is not a figment of the mind), Herschel's central methodological motivation, like Newton's and Lyell's, was to avoid hypotheses in Natural Philosophy. Some scholars of Herschel's methodology have tried to clarify the demands that he imposed on a *vera causa* – real existence of the cause – and they have underscored that this condition is sufficient and necessary to identify *verae causae*. Herschel's favorable evaluation of Lyell's geological theory arises not only because Lyell referred to causes that were sufficient and adequate, but mainly because Lyell relied on empirical positive support; unlike rival geological theories which involved speculative hypotheses. Empirical positive support shall be understood as a necessary condition to consider a cause as a *vera causa*, and it means the existence of known facts which support a theory. Considering Lyell's theory, Herschel said that "We have here, therefore, an evident real cause, of sufficient universality, and acting in the right direction, to account for the phenomena. Its adequacy is another consideration" (Herschel 1987, p. 148).

There are other necessary conditions, according to Herschel, in order to consider a cause as *vera causa* – namely, that it is real, universally sufficient and adequate.[3] Herschel accepted Lyell's geological theory, however, mainly because "neither of these (rival theories to Lyell) can be regarded as real causes in the sense here intended; for we do not *know* that the globe has so cooled from fusion, nor are we sure that such supposed greater activity of former than of present volcanoes really did exist" (Herschel 1987, p. 146). Even though Herschel thought that Lyell's climate theory was not complete, in the sense that it did not explain drastic climate change of the geological past, Herschel accepted it because it was *methodologically and ontologically superior with respect to its rivals*. In Herschel's words "the elevation of the bottom of the sea to become dry land has really been witnessed so often, and on such a scale, as to qualify it for a *vera causa* available in sound philosophy" (Herschel 1987, p. 144). In contrast to alternative explanations, he maintained that in Lyell's "we have an evident real cause, of sufficient universality, and acting *in the right direction*, to account for the phenomenon" (Herschel 1987, p. 147).

Talking about the usual speculations included in several theories before Lyell, Herschel said that "these, indeed, were favorite speculations with a race of geologists now extinct; but the science itself has undergone a total change of character, even within the last half century, and is brought, at length, *effectually within the list of the inductive sciences*" (Herschel 1987, p. 282, my emphasis). But, was there another reason in Herschel's

[3] The *vera causa* issue has been sufficiently studied; I should concentrate only on his reasons to reject the hypothesis.

view to put geology on the list of inductive sciences, apart from the fact that it did not recur to hypotheses? I think so, and it is useful to establish a fundamental difference with Whewell's ideas about historical causation, as we shall see. That reason was supported by Herschel's conception of how true causes operate. According to him, the legitimate explanations must be causal, and there is a clear sense of causal explanation in Herschel.[4] He asserted that "the first thing that a philosophical mind considers, when any new phenomenon presents itself, is its *explanation*, or reference to an immediate producing cause" (Herschel 1987 p. 144). Herschel traces here a distinction between proximate causes and ultimate ones. In the former, the immediate cause of any phenomena is any other which goes before. Herschel maintained that scientific research must not stop when proximate causes had been found, but that "in our investigation of nature, we shall ever be able to arrive at a knowledge of ultimate causes" (Herschel 1987, p. 87). These kind of causes are what we find in phenomena which are not susceptible of further analysis; we can consider them as simple ones. Even though scientific investigation can be characterized with two kinds of causes – proximate and ultimate ones –, the causal relation is the same in both cases. Herschel expressed this relation in following roles "1st, invariable connection . . . 2d, invariable negation . . . 3d, increase or diminution of the effect . . . 4th, proportionality, and 5th reversal of the effect" (Herschel 1987, pp. 151-152). The causality expressed in these rules involves the ideas of invariability and necessary connection. The point that I wish to underline is the ahistorical character of Herschel's idea of natural law; i.e., that the law's effects do not vary with time. In other words, the action of the cause is independent of the specific moment of its application; in an important sense, this kind of causality is timeless. In cases when a certain cause appears, and if there is nothing to impede its natural action, a specific effect could be expected. As we will see, Whewell called this kind of causality "permanent causality."

The atemporal and ahistorical character of Herschel's notion of causality is the ground of the relation between causes and the laws of nature. Even though this relation is not very clear, it is of great importance to clarify the issue. Herschel maintained that "we may therefore regard a law of nature either, 1st, as a general proposition, announcing, in abstract terms, a whole group of particular facts relating to the behavior of natural agents in proposed circumstances; or, 2dly, as a proposition announcing that a whole class of individuals agreeing in one character agree also in another" (Herschel 1987, p. 100). The latter guides us, if not to an

[4] Black had determined three senses in which Herschel understood the notion of *cause*. He emphasized the third as the main use which Herschel assumed for his theory of science. I follow Black's (1960) interpretation here.

ultimate cause, at last to a proximate one[5], in the sense that whenever we observe two phenomena invariably connected to one another, we conclude that in fact they are interrelated either in a cause-effect relation or as the common effect of the same cause. Thus, for Herschel, laws of nature are statements which express constant relations between cause and effect, relations exist in the world. It is typical that the example which Herschel used to show the double aspect of a law of nature was the double refraction phenomenon, in which there are constant relations involved. This kind of causality presupposes an ontology of *events*, not of *processes*. That is, every time an event occurs, it is expected that another specific event occurs. Processes are made by a series of events and there could be several ways in which they relate to each other in order to articulate processes; as we shall see, Whewell will construct his idea of historical causation on the basis of this difference. The example Herschel gave to show causal relation was double refraction where, in many different substances that are impacted by a beam of polarized light, the periodic colors are expected. It is a clear example where in such a way that every particular event is governed by natural laws always in the same manner. Herschel expressed it in this way: "every law is a provision for cases which *may* occur, and has relation to an infinite number of cases that never have occurred, and never will. Now, it is this provision, *a priori*, for contingencies, this contemplation of possible occurrences . . . that impresses us with the notion of a *law* and a *cause*" (Herschel 1987, p. 36).

As we have seen, Lyell's theory was considered by Herschel as an inductive science because it incorporated true causes *and* these true causes operated as natural laws, in the same way that laws operate in optics. These reflections lead us to a final consideration which represents other epistemic differences with Whewell – namely, Herschel's idea of unity of science. Herschel maintained a certain thesis of the unity of the sciences, and that made it possible to apply the idea of causality in a general form to all sciences, *including geology*. To a first approximation, that unity consists in the existence of mutual relations between several branches of science. For example, mineralogy is an illustrative case; he said that "there is no branch of science which presents so many prints of contact with other departments of physical research, and serves as a connecting link between so many distant points of philosophical speculation, as this." (Herschel 1987, p. 290). Mineralogy offers several such links to geology, chemistry, optics, crystallography, physics, and to others. According to Herschel, the unity of science is articulated by the interrelations of different types of knowledge, which were obtained in various areas. It is

[5] Proximate causes are the most immediate to the phenomenon we investigate; ultimate causes are founded in phenomena which cannot be analyzed in further.

an epistemic unity partly because the manner in which different phenomena of diverse disciplines are explained is the same: "our business is to disentangle, to arrange, and to present them in a separate and distinct state: and to this end we are called upon to resolve the important but complicated problem, – Given the effect, or assemblage of effects, to find the causes" (Herschel 1987, pp. 221-222). The subdivisions of Natural Philosophy, into several branches and its mutual relations, are solidified by search for true causes.

## 4. *The Palætiological Sciences*

William Whewell evaluated Lyell's geological theory in his *History of Inductive Sciences* of 1837 and in his *Philosophy of Inductive Sciences* of 1840. In the *History*, he maintained geology was one of the sciences he called palætiological:

> Geology, forming one of the palætiological class of sciences, which trace back the history of the earth and its inhabitants on philosophical ground, is thus associated with a number of other kinds of research, which are concerned about language, law, art, and consequently about the internal faculties of man, his thoughts, his social habits, his conception of right, his love of beauty. Geology being thus brought into the atmosphere of moral and mental speculations, it may be expected that their investigations of the probable past will share an influence common to them; and that she will not be allowed to point to an origin of her own, a merely physical beginning of things; but that, as she approaches towards such a goal, she will be led to see that it is the origin of many trains of events, the point of convergence of many lines. (1857, vol. 3, p. 384)

But, what exactly were the palætiological sciences for Whewell? Did he refer to a subgroup of the sciences or to an autonomous group of sciences? Whewell affirmed that these sciences are "those in which the object is to ascend from the present state of things to a more ancient condition, from which the present is derived by intelligible causes" (Whewell 1857, p. 637).[6] One of the fundamental epistemic features of palætiological sciences is to outline the causal history of certain phenomena. The crucial point, insofar as it is a difference with respect to Herschel, is the way Whewell understood the "intelligible causes." In order to throw some light on this expression, it is necessary to analyze certain basic ideas of Whewell.

Whewell distinguished two kinds of causality; *mechanistic* and *historical*. The former is typically the mechanical sciences and the latter of palætiological ones. We will see the differences between them below. Whewell constructed the word 'palætiology' from two Greek terms: *ætio*

[6] What Whewell meant by 'to ascend' had already been partially analyzed by Quincy (1988, p. 293).

that means *a cause* and *paleo* that means *ancient*. About ætiology he maintained that

> The sciences which treat of causes have sometimes been termed *ætiological*, from *αιτια*, *a cause*: but this term would not sufficiently describe the speculations of which we now speak; since it might include sciences which treat of Permanent Causality, like Mechanics, as well as inquiries concerning Progressive Causation. (Whewell 1837, p. 397)

and in his *Philosophy* insisted that

> while *Palæontology* describes the beings which have lived in former ages without investigating their causes, and *Ætiology* treats of causes without historical from mechanical causation; *Palætiology* is a combination of the two sciences; exploring, by means of the second, the phenomena presented by the first. (Whewell 1857, p. 638)

It is important to mention that Whewell did not distinguish between "progressive causality" and "historical causality," unlike the crucial difference between historical and mechanical causation. In order to understand the latter difference, it is useful to note that the goal of palætiological sciences is making intelligible the change of certain kinds of phenomena. Progressive causality is *progressive* just because it makes intelligible the responsible causes which have carried a particular historical process to its current state.[7] Compatibly with this goal, what should be done is to present a plausible series of processes by means of which the present state was able to be come what it is. To Whewell "actual past events" determine progressive causality. These kinds of events are the facts through which the historical reconstruction is accomplished; they tell us how things reached their current constitution. In geology, for example, the fossil record, the Serapis temple, the stratographical divisions, etc., are the actual past events in the construction of explanations. Later, I will return to this point, because it is one of the Whewell's central criticisms of geology.

We find the explicit and clear difference between mechanical and progressive causality in a paragraph of his *Philosophy*:

> We have already considered sciences which depended mainly upon the Idea of Cause, namely, the Mechanical Sciences. But it is obvious that the Idea of Cause in the researches now in consideration (palætiological sciences) must be employed in a very different way from that in which we applied it formerly. Force is the *cause* of motion, because force at all times and under all circumstances, if not counteracted, produces motion; but the cause of the present condition and elevation of Alps, whatever it was, was manifested in a series of events of which each happened but once, and occupied its proper place in the series of time. The former is *mechanical*, the latter *historical*, *cause*. (Whewell 1857, p. 654)

[7] A "historical process" entails an enormous amount of time by human standards. For example, even some chemical reactions need a certain time to carry out their processes; they could not been called "historical processes" in the sense defended here.

Mechanical causality is deterministic and ahistorical, in the sense that any time and under any circumstances – that do not prevent it – it produces the expected effect. This is just Herschel's meaning of causality. By contrast, Whewell understood progressive causality as a causal chain that embraces events since the beginning of things until the present moment; the past has been a series of connected events by means of this causality and the last term of the series is the present. Every phenomenon that had occurred in any historical process – for example, in the solar system, the history of Earth or its creatures – had been cause and effect, i.e., the effect of precedent event and cause of further one, so the idea of a causal chain is not only a fruitful metaphor but the fundamental manner in which historical causation acts.

Here we find what Whewell thought was the fundamental problem for the palætiological sciences – namely, how one determines the way in which every link of the chain acts upon another one and if it is possible to calculate backwards the succession to the very beginning. Determining the effect of an historical or progressive cause is one way of expounding and solving this problem, but it is a difficulty that presupposes the problem of individuating an historical cause. Whewell explained this general problem in geology as follows:

> For instance, among the currents of lava which we traces as part of the shores of Italy and Sicily, *Which* shall we select as belonging to the existing order of things? In going backwards in time, where shall we draw the line? and why at such particular point? These questions are important, for our estimate of the efficacy of known causes will vary with the extent of the effects which we ascribe to them. (Whewell 1857, p. 658)

The epistemological problem of individuation emerges from the impossibility of establishing the efficacy of the known causes. In others words, the knowledge of some past causes requires them to be individuated as well as to know what are the effects we legitimately attribute to them. Whewell's point was that these sciences had no solid grounds for assessing effects when we reason from phenomena to causes. Determining the effects of the geological causes constituted the core of different geological controversies: "The controversies to which the various theories of geologists gave rise, proceeding in various ways upon the effects of the existing causes of change, led men to observe . . . the actual operation of such causes" (Whewell 1837, p. 452). It looks like Whewell was inclining to actualism, but at the same time he implicitly suggested that actualism was insufficient to consider any geological theory reliable, insofar as the problem of determining exactly the effect of a geological cause was a more fundamental problem and recurring to actually existing causes occasionally aid not resolve it. It is a difficulty that produced controversies in rival geological theories.

A problem directly related to the previous one was the fact that every new consecutive step in the development of a historical phenomenon becomes more complex each time. Whewell expressed this second great problem of the palætiological sciences as follows:

> In all of them (palætiological sciences) we have the same kind of manifestations of a number of successive changes, each springing out of a preceding state; and in all, the phenomena at each step become more and more complicated, by involving the results of all that has preceded, modified by supervening agencies. (Whewell 1837, p. 399)

The history of languages is a clear example of this growing or accumulating complexity. These languages are conglomerations of other ones, dead or alive, which vary through the time. They are not fixed entities; here, unlike in the case of mechanical causation, we cannot say that whenever specific causes exist, there will be particular effects. Languages, social customs, political institution, all of them palætiological sciences like geology, undergo the superposition of formations through time, one above the other. This turns every step into a more complex one. A process in the palætiological sciences experiences consecutive changes and it is articulated by events – links of the causal chain – which undergo something we could call "profits" and "losses." These historical processes are not constituted by *the same phenomenon* changing through the time, but what Whewell was implicitly affirming was that these historical processes produce phenomena which have a new identity with every step. Whewell offered English language as an example of a phenomenon which had gradually changed its characteristics. In geology it occurs in the same way.

The problem of complexity was closely linked to the problem of individuation, since growing complexity does not permit one to establish clear criteria of individuation to describe the phenomena governed by historical causation. A description of current English is different from one made three-hundred years ago and increasingly impossible as we go backwards in time.

Thus, according to Whewell, the palætiological sciences – faced with the problems that historical or progressive causality presents – have a *hypothetical* character. Even though these sciences deal with actual past events, the problem is that we do not have most of them because they are lost in the past, therefore we could not construct reliable theories. In the case of geology, it forms a kind of causal progression which has "acted *in succession* in the hypothetical history of the earth and its inhabitants" (Whewell 1857, p. 656). But the very fact that it is an *hypothetical* history is not an insuperable methodological difficulty for Whewell, because he accepted that science works with hypotheses, provided they have fulfilled rigorous methodological criteria. In one place in his *Philosophy*, he maintained, as one of the more important criteria of hypothetical choice, that: "it is a test of true theories not only to account for, but to predict phenomena" (Whewell 1958, p. 70). But palætiological sciences did not satisfy this fundamental test.

According to Whewell, there are three interrelated phases in the historical development of palætiological sciences. For geology, these phases are as follows: *descriptive* or *phenomenological* geology, causal or *geological dynamics*, and finally the phase of theoretical maturity, *systematic* geology. Referring to geologists, he maintained that "before they could obtain any sound Geological Theory, they must carefully cultivate Geological Ætiology" (Whewell 1857, p. 653). The descriptive phase is classified as the methodological root "for no description of a large and varied mass of phenomena can be useful or intelligible without classification" (Whewell 1857, p. 645). In the classification the discoverer has "to detect the true principles of natural classes, and to select marks by which these may be recognized" (Whewell 1857, p. 645).

The second phase, the geological dynamics, is a bridge between the descriptive phase and the theoretical one. In geological dynamics, the general empirical laws which govern phenomena are determined. This second phase is a transition and it basically is

> a science which shall investigate and determine the laws and consequences of the known causes of changes such as those which Geology considers; – and which shall do this, not in an occasional, imperfect, and unconnected manner, but by systematic, complete, and conclusive methods; – shall, in short, be a Science, and not a promiscuous assemblage of desultory essays. (Whewell, 1837, p. 451)

It is crucial to observe that even though geological systematics should be a systematic, complete, and conclusive science, *it is only a phase of transition to conform a theoretical geology*. This second phase is still an ætiological science insofar as it refers to particular facts, for example, aqueous causes, igneous causes, elevation of continents, etc. For Whewell, Lyell's theory is in just this phase: "Mr. Lyell's work (*Principles of Geology*) in particular, has eminently contributed to place Geological Dynamics in its proper prominent position" (Whewell 1837, p. 451). ("Prominent position" in the sense that Lyell appealed to known causes to construct a causal geological science; that is a fundamental characteristic in the constitution of a theoretical or physical geology). We will see later why Lyell's work could not claim a theoretical status according to Whewell.

The third phase is one in which palætiological sciences reach their theoretical maturity, i.e., sciences constituted by true theories. In his *Philosophy* Whewell maintained that

> We can account for known facts by intelligible causes, we can infer latest fact from manifest effects, so as to obtain a distinct insight into the whole history of events up to the present time, and to see the last result of the whole in the present condition of things. The term *Theory*, when rigorously employed in such sciences as those which we

> here consider, bears nearly the sense which I have adopted: it implies a consistent and systematic view of the actual facts, combined with a true apprehension of their connection and causes. (Whewell 1857, p. 663)

Only in that developing phase can we rely on a broad point of view of really occurring events. When this phase has achieved a complete development, it will be a demonstrative science and will involve general cases, i.e., real and permanent geological knowledge like all other forms of *physical* knowledge. Whewell thought, however, that geology still had not reached this phase of development (1840). Even more, in the *History* he maintained that the beginning of physical geology had not even commenced: "the history of Physical Astronomy almost commences with Newton, and few persons will venture to assert that the Newton of Geology has yet appeared" (Whewell 1837, p. 497). Whewell did not maintain that there are no geological theories up to 1840 which pretended to be true and sound; on the contrary, he claimed that geological theories have been abundant and diverse. (He was talking about catastrophism, uniformitarianism, plutonism and neptunism.) The central point here is that, although these general theories pretended to be true, "we do not trace this character" (Whewell 1837, p. 497) in most of them.

Whewell's principal criticisms against the existence of a true *physical* geology are basically epistemological, which he derived from the development of the sciences. First of all, geological theories aiming to be *physical* theories, like those of the neptunists and plutonists, have been premature ones, insofar as only limited cases and not general circumstances support them. To establish the division of rocks or their origin, although it constitutes a geological truth, is not enough to establish a general geological theory, despite the pretension of those groups to have achieved the latter. According to Whewell, these two general theories attempted to give a complete and simple explanation for every phenomenon of Earth history using only some limited facts. Werner, the father of the neptunism, elaborated his geological theory examining only a small region of Germany while Hutton, leading advocate of the plutonists, analyzed only a few strata of England and Scotland (Whewell 1837, p. 503). The underlying argument in these criticisms of Whewell is that even before these theories had developed a *descriptive* geology, they pretended to have established themselves as true and permanent geological knowledge.

Whewell accepted some points of catastrophism, insofar as it was possible to explain the total change of the animal and vegetable formations in the fossil record.[8] Whewell's inclination in favor of catastrophism is

---

[8] The favorable opinion of Whewell with respect to the relation between catastrophists and Natural Theology, and the creation of new species, had been analyzed in Cannon (1950).

due to the fact that he considered the species to be fixed and he was an advocate of the design argument in biological matters (Ruse 1977). Even though Whewell found some virtues in catastrophism, for example a clear connection between Natural Theology and species extinctions, he did not consider catastrophism as a *physical* geology, in spite of the fact that it constituted "the first steps in geological dynamics" (Whewell 1837, p. 507).

## 5. *Some Methodological and Epistemological Divergences between Whewell and Herschel on Lyell's Theory*

It is necessary to analyze carefully Whewell's criticisms of uniformitarianism, particularly of Lyell's theory, since here there are some clear methodological and epistemological divergences relative to Herschel. First of all, Whewell maintained that Lyell advocated an uniformitarianism "*by reference to Causes now in Operation*" (Whewell 1837, p. 511), thereby emphasizing the subtitle that Lyell utilized in his *Principles of Geology*. What is the possible reason for Whewell to remark: "causes now in operation"? The answer could be twofold. Either Whewell positively emphasized Lyell's reference to known causes, unlike rival theories, or he negatively underlined Lyell's attempt to elaborate geological explanations directing attention *only* to causes *now* in operation. There are two strong reasons to reject the former and to accept the latter. Firstly, Whewell believed in catastrophes and they are not known causes. Secondly, Whewell rejected Herschel's association between catastrophism and the hypotheses method, on the one hand, and uniformitarianism and the *vera causa* method, on the other. Both catastrophism and uniformitarianism are global and epistemologically unsatisfactory theories to Whewell. Let us see this issue in detail.

Whewell asserted, referring to Lyell's uniformitarianism, that "it must be granted at once, to the advocates of this geological uniformity, that *we are not arbitrarily to assume* the existence of catastrophes" (Whewell 1837, p. 513, my emphasis). According to Whewell, there are no reasons *a priori* to put aside catastrophism as an empirically unsupported hypothesis; which was a strategy followed by Lyell to reject catastrophism (Whewell 1837, pp. 511-512). For Whewell, huge catastrophes are not *directly observed* causes but inferences strongly supported by a very different sort of geological evidence. Whewell was referring here to the epistemic principle of the palætiological sciences. As we saw, it is through the knowledge of effects that we determine causes, i.e., effects show us which true causes we should invoke. The methodological principle Lyell followed was to avoid the mention of geological forces different in kind and degree from those acting in the present. But

> when Mr. Lyell goes further, and considers it a merit in a course of geological speculation that it *rejects* any difference between the intensity of existing and past causes, we conceive that he errs no less than those whom he insures. "An *earnest and patient endeavour to reconciliate* the former indications of change" (Lyell 1830-33, vol. 3, p. 328) with any restricted class of causes, – a habit which he enjoins, – is not, we suggest, the temper in which science ought to be pursued. (Whewell 1837, p. 513, Whewell's emphasis)

Whewell considered that Lyell's procedure was a wrong way to develop science, particularly if a palætiological science, insofar as effects indicate to us the nature and intensity of causes. According to Whewell, Lyell violated basic methodological principles of geological dynamics with such an idea. Whewell added one further consideration – namely, the accumulation of causal efficacy during *time* constitutes doubtless a great favor to the theorist of geology; but *force* is also a geological power which we should not underestimate, even though we could not measure its limits nor understand its nature (Whewell 1837, p. 513).

Whewell emphasized Lyell's methodological inaccuracy in the following passage:

> In truth, we know causal only by their efforts; and in order to learn the nature of the causes which modify the earth, we must study them through all ages of their action, and not select arbitrarily the period in which we live as the standard for all other epochs. The forces which have produced the Alps and the Andes are known to us by experience, no less than the forces which have raised Etna to its present height; for we learn their amount in both cases by their results. Why , then, do we make a merit of using the latter case as a measure for the former? Or how can we know the true scale of such force, except by comprehending in our view all the facts which we can bring together? (Whewell 1837, p. 514)

It is interesting to observe that Whewell accused Lyell on the same grounds that the latter criticized and rejected catastrophism – namely, Lyell was *arbitrarily* choosing a period of time – the present – to explain the geological past. Choosing "arbitrarily" means that if he would have lived in a catastrophic geological period, then he would have referred to other uses; in other words, it was an accident that Lyell lived in a quiescent geological period. Lyell criticized the arbitrary choice of huge catastrophes by the catastrophists; but his own choice of the causes *now* in operation equally was likely to be an accident. Another of Whewell's criticisms of Lyell was that he constructs his geological theory without a previous strong geological dynamics, although Whewell considered Lyell's work *only* as a significant advance in geological dynamics. It is crucial to notice that Whewell did not deny the possibility that Lyell's theory might be true, but he thought it was impossible to know if the Lyellian scale of force was right: There was not enough evidence to know the dimension of geological forces, nor enough knowledge about the limit of the catastrophic forces which must be admitted for a theory like Lyell's.

But Lyell's theory presented other problems according to Whewell. Recurring to known causes did not guarantee that one would reach reliable knowledge in palætiological sciences. Whewell explained that, according to Lyell's reasoning, it would be possible to trace similarities between the principles of the early phase of the formation of languages, governments, or societies, and the processes *now* in operation. But for Whewell it was clear that this kind of reasoning was wrong because "where do we now find a language in the process of formation, unfolding itself in inflections, terminations, changes of vowels by grammatical relations, such as characterize the oldest known languages? Where do we see a nation, by its natural faculties, inventing writing, or the arts of life, as we find them in the most ancient civilized nations?" (Whewell 1837, p. 516). The processes of formation at early phases of historical phenomena were different from those now in operation. From a certain paint of view, Lyell's attempt to extend the methodological standards of *vera causa* from Newtonian science to geology was not completely adequate. "Astronomy, as the science of cyclical motions, has nothing in common with geology" (Whewell 1837, p. 516), because the former is not a palætiological science.

Whewell was implicitly suggesting that the causal notion of mechanical sciences could not be utilized in palætiological theories. Whewell's second argument against the analogy between astronomy and geology pointed out certain misunderstandings of the way in which known causes were applied. This time the mistake consisted in the historical development of astronomy. Whewell asked if the main merit of the founders of astronomy was really to assume that the celestial motions were consequences of the choice of *known* causes. Certainly not. As an example he referred to the occasion when Newton tried for the first time to explain the movements of the Moon by the effect of gravitational force and he failed because the measurements he employed were wrong.[9] Whewell asked if in the light of this failure it would have been right for Newton to have overlooked the differences he found out, since otherwise it would be necessary to recur to causes different from these we usually observe in action; "or was there any praise due to these who assumed the celestial forces to be the same with gravity, rather than to those who assimilated them with any other known force, as magnetism, till the calculation of the

[9] Newton tells us that in 1666, a year in which he lived in Woolsthorpe because of the plague, he had already found that gravity extends from the Earth to the Moon's orbit, from which he deduced that the force that maintains the Moon in its own orbit varies in inverse relation to the square of its distance. Newton, however, took 21 years to publish these results. In a conversation with Pemberton in his old age, Newton declared that without necessary books in Woolsthorpe, he estimated an arch degree to be 60 miles instead of the more exact measure of 69.5 miles.

time and amount of these forces, from the celestial phenomena, had clearly sanctioned such an identification?" (Whewell 1837, p. 515). Whewell maintained that it was not possible to recur to known causes at every moment, even though in past time some of the causes now considered as known causes were supposed to be unknown causes. Finally, Whewell (1837, p. 515) said that "we are not to select a conclusion now well proved, to persuade ourselves that it would have been wise to assume it anterior to proof, and to attempt to philosophize in the method thus recommended."

## 6. *Conclusion*

We have seen that there were important methodological differences between Herschel and Whewell with respect to the methodological status of Lyell's theory. Such differences go well beyond the dichotomy between rationalists and empiricists. Whewell's main problem was to answer the question of how to elaborate epistemically reliable sciences which integrated historical causality, taking very seriously the historical processes by which every science is constructed. By contrast, Herschel maintained an ahistorical point of view about both the causes and the natural laws through which explanations are made.

*Universidad de Guanajuato*
Facultad de Filosofía y Letras
Ex-Convento de Valenciana S/n
Guanajuato, Gto. 36040
México
godfrey@servidor.unam.mx

## REFERENCES

Cannon, W. (1960). The Uniformitarian-Catastrophist Debate. *Isis* **51**, 38-55.
Cannon, W. (1976). Charles Lyell, Radical Actualism and Theory. *British Journal for the History of Science* **9**, 104-20.
Cannon, S. (1978). *Science in Culture*. New York: Science History.
Black, R.M, C. Ducasse and F. Madden (1960). *Theories of Scientific Method: The Renaissance through the Nineteenth Century*. Seattle: University Washington Pres.
Buchwald, J.Z. (1989). *The Rise of the Wave Theory of Light. Optical Theory and Experiment in the Early Nineteenth Century*. Chicago: University of Chicago Press.
Gillispie, C. (1951). *Genesis and Geology*. Cambridge: Cambridge University Press.
Greene, M. (1983). *Geology in the Nineteenth Century*. Ithaca, NY: Cornell University Press.
Good, G. (1987). John Herschel's Optical Researches and the Development of His Ideas an Method and Causality. *Studies in the History and Philosophy of Science* **18**, 1-41.

Gould, S.J. (1965). Is Uniformitarianism Necessary? *American Journal of Science* **263**, 223-228.
Gould, S.J. (1987). *Time's Arrow. Time's Cycle*. Cambridge, MA: Harvard University Press.
Herschel, J. (1987). *A Preliminary Discourse on the Study of Natural Philosophy*. Chicago: University of Chicago Press.
Hodge, J. (1991). The History of the Earth Life, and Man: Whewell and Palætiological Science. In: M. Fisch and S. Schaffer (eds.), *W. Whewell: A Composite Portrait,* pp. 255-288. Oxford: Clarendon Press.
Hutton, J. (1788). Theory of the Earth; or an Investigation of the Laws Observable in the Composition, Dissolution, and Restoration of Land upon the Globe. *Transaction of the Royal Society of Edinburgh* **1**, 209-304.
Laudan, L. (1981). *Science and Hypothesis: Historical Essays of Scientific Methodology*. Dordrecht: Reidel.
Laudan, R. (1992). What's So Special about the Past? In: M. Nitecki and D. Nitecki (eds.), *History and Evolution,* pp. 55-67. New York: State University of New York Press.
Laudan, R. (1987). *From Mineralogy to Geology. The Foundations of Sciences,* 1650-1830. Chicago: University of Chicago Press.
Lyell, Ch. (1830-33). *Principles of Geology*. 3 vols. London: J. Murray.
Quincy, C. (1988). Robert Chambers and William Whewell: A Nineteenth-Century Debate over the Origin of Language. *Journal of the History of Ideas* **49**, 287-300.
Rudwick, M. (1972). *The Meaning of Fossils*. London: Macdonald.
Rudwick, M. (1992). Darwin and catastrophism. In: J. Bourriau (ed.), *Understanding Catastrophe,* pp. 57-82. Cambridge: Cambridge University Press.
Ruse, M. (1976). Charles Lyell and the Philosophers of Science. *British Journal for the History of Science* **9**, 121-131.
Ruse, M. (1977). William Whewell and the Argument from Design. *The Monist* **69**, 244-268.
Ruse, M. (1979). *The Darwinian Revolution. Sciences Red in Tooth and Claw*. Chicago: University of Chicago Press.
Whewell, W. (1837). *History of the Inductive Sciences*. 3 vols. London: Parker.
Whewell, W. (1857). *The Philosophy of Inductive Sciences. Founded upon Their History*. London: Parker.
Whewell, W. (1958). *Novum Organon Renovatum*. London: Parker.

**Joseph Margolis**

# A SMALL APPRECIATION OF WILLIAM WHEWELL

ABSTRACT. The paper explores the possibility of a third option, beyond rationalism and empiricism, in the theory of induction and the analysis of necessary paths in science, by way of Whewell's running dispute with Mill.

The disagreement between William Whewell and John Stuart Mill, variously described in terms of the logic and methodology of science, more pointedly instructive – beyond the adequacy of Millian inductivism – with regard to the deeper differences between induction and deduction, is one of those delicious controversies that suddenly expose, quite serendipitously, the perennially skin-thin nature of the confident assertions on either side of important doctrinal divides. There is an exaggeratedly disjunctive assumption among the champions of famous "fundamental" disputes, which the history of their initial disagreement (down to our own day) almost always betrays through the dawning discovery of the inherent limitations of the intransigent vocabularies and presuppositions of both sides. Certainly, the instinct to polarize the conceptual disagreement about the adequacy of an inductivist treatment of science has seriously and adversely affected the reception of Whewell's canny grasp of important difficulties that Mill could not really meet by way of his canonized account of induction.

Whewell was himself inclined to favor certain extreme possibilities; but the validity of his perception of the weaknesses of the Millian canon (by which I do not mean merely the famous inductive "methods" but rather the empiricist explanation of inductive generalization and its relationship to necessary truths) need not be finally cast in terms of whatever systematizations Whewell favored. That would merely extend the *folie* of the argument by hopelessly disjunctive extremes. It is also true that (as in adapting Kant's transcendentalism) Whewell himself was often critical of the improbable extremes to which a recuperative theory of

In: John Wettersten, *Whewell's Critics: Have They Prevented Him from Doing Good?*, edited by James A. Bell (*Poznań Studies in the Philosophy of the Sciences and the Humanities*, vol. 85), pp. 377-384. Amsterdam/New York, NY: Rodopi, 2004.

rational methodology might be drawn.

I am impressed by John Wettersten's scrupulous and indefatigable efforts to set the record straight in Whewell's favor.[1] I am not, strictly speaking, a party to the dispute as it has played itself out and is still being played out: chiefly, if I may say so, because the alternatives seem to me to be still too primitive and too disjunctive – and too much committed to the nimble defense of tendentious options that are not likely to be tenable as they stand. But I don't think there can be much doubt that Whewell deserves a better inning than he's received, and I'm happy to acknowledge that better theories of the methodology of science (whatever they may prove to be) are likely to owe a considerable debt to Whewell's proposals and conjectures. Certainly his sense of the conceptual inseparability of the history of science and the analysis of what the method of science is – impossible on the empiricist grounds favored in Mill's and Herschel's inductivism – is a very early anticipation of the central puzzle of the late twentieth-century philosophies of science. In my mind, that would be enough to redeem Whewell, though there is much more to Whewell's theory than that.

I cannot contribute in a fresh way to a close reading of Whewell but I am very much struck by Whewell's courage and ingenuity in refusing (if I read him aright), precisely when he appeals to Kant's transcendental opposition to empiricism, to endorse Kant's sort of transcendentalism (at least in strict "application," if not in ultimate "hope") in fashioning a middle ground between empiricism and Kantianism, a ground that, informed by the controversies made familiar in our own century by the French sociologists of science, by Fleck, by Kuhn, by the recent *rapporteurs* of "laboratory life," might now be formulated in terms of societally entrenched but quite provisional practices. This is certainly not distant from Whewell's own inclination, which draws him with some misgiving in the direction of Hegel and Schelling (as Herschel contends) – possibly in a way that is also not very distant from the labors of that other isolated figure in nineteenth century philosophy of science, Charles Sanders Peirce. I press the point, of course, because both Whewell and Peirce play an instructive role in the development of Karl Popper's anti-inductivism, which Wettersten has in mind in recovering Whewell.

In any case, I intend no more than a small contribution: primarily to strengthen certain distinctions *sine qua non* by which the middle ground I'm attributing to Whewell might more robustly attract participants in the late continuation of the original dispute. That is, participants who (I should expect) are increasingly persuaded that *no* extreme position like the

[1] I have benefited from a reading of Wettersten's *Whewell's Critics: Have They Prevented Him from Doing Good?* as well as chapters 1-2 of his (1992).

empiricism leading from Locke to Mill (or, from Bacon to Mill) could possibly legitimate inductivism; or that, for similar reasons, no "rationalist" opposition that inclined toward a full-dress Kantianism could possibly fit the actual progressive history of science or (as a consequence) any reasonable replacement of a primitive inductivism. But, in saying all that, I mean to suggest that the opposition between the extreme positions is itself responsible for keeping the debate on the wrong track. Not in the sense that entirely different questions would have to be considered, but only that the important questions (posed in the historical debate) have been inappropriately construed in terms of polarized disjunctions that must be favored or rejected in the exclusionary way in which they were first presented. It may well be the best way to accept Wettersten's good work to suggest what finally to favor when we are clearer about Whewell's contribution, even if it is not precisely what Whewell would have made of his own work and even if it does not lead to the final victory of a "critical rationalism."

The lesson in all of this is, to put it simply, that the cognitive resources servicing the sciences cannot be disjunctively sorted in terms of sensation and reason. To put the matter thus is to put the matter crudely. But it is no cruder than Mill's own account of inductivism or, frankly, Kant's facultative division between sensation and thought that runs, despite important changes in the direction of transcendentalism, from the *Inaugural Dissertation* to the famous letter to Marcus Herz just preceding the publication of the *Critique of Pure Reason* to the *Critique* itself. In challenging Mill, Whewell was echoing the rationalist critique of empiricism that begins in Leibniz and extends to Kant: and in distancing himself from Kant's immensely attractive transcendental version of rationalism, Whewell moves to a relatively untried course of speculation that admits the *priority* of reason but not its *apriority*!

I find Whewell's novelty there. If we begin with Mill's conception that the whole of knowledge depends on inductive inferences from original sensations – that this encompasses mathematical knowledge as well as natural science: hence, necessary truth – then Whewell's objection seems unanswerable. We have the gist of it in Whewell's reply to Herschel's review of his own *Philosophy of the Inductive Sciences* (1969). For, Mill's own *Logic* completes in effect Herschel's empiricist conception of induction in science – and, effectively, in mathematics, which Herschel had not fully formulated. In any case, in his reply to Herschel's review,[2] Whewell argues that the necessity of a geometric theorem requires axioms governing figures (parallel lines, for instance), that cannot be derived from

[2] I draw here on Wettersten (1992, pp. 34-35), who summarizes Whewell's reply "Remarks on a Review of the Philosophy of the Inductive Sciences."

experience to argue that those same axioms are themselves derived from experience would, Whewell claims, lead to a regress that could only be resolved by appeal to further "ideas" not themselves derived from experience. All that is familiar but devastating (as far as it goes) against classic empiricism. But the picture is much more complicated, and Whewell was aware that it was. That is what I see in Whewell's resistance to Kantian apriority.

It is in such contexts that Whewell appears to insist that "ideas" must be imposed on the data of experience, if necessity is to be found in natural science; although, for Whewell, mathematics is not itself a form of natural science. Mill demurs, because, for him, deductive necessity is essentially a shorthand summary of inductive regularities over increasingly inclusive sets of observations, which therefore do not require necessity to be introduced *ab extra*. Whewell sees the advantage, here, of Kant's distinction between sensibility and understanding, which easily accounts for necessary truths by way of ampler resources than Mill's empiricism can provide. But Whewell is also prepared to replace Kant's apriority with the unexceptional priority of leading ideas drawn from the progressive history of science itself. That is, "priority" means nothing more than the temporal antecedence of orienting conjectures drawn from prior science and "imposed" (critically) on further empirical inquiry. Where these same "ideas" are thought to impose an "ideal element" on experience, as regarding "propositions which are known to be necessarily true," Whewell inclines in Kant's direction – as with space, time, causality, number, and similar "fundamental ideas." They do account for necessity in empirical knowledge, but they are not (for that reason), Whewell believes, transcendental *in Kant's sense*.

Whewell does indeed treat "fundamental ideas" as the "source of axioms" – hence, the source of necessary truths. But he believes that they are *not* the necessary, enabling, *a priori* concepts on which the very "possibility" of science depends. They are instead clarified, revised, strengthened, hence also capable of being discarded, by the ongoing work of the history of science. To the "informed" mind, their necessity, whenever valid, is apparent enough; hence, to that same mind, the necessary truths of the various sciences become apparent – where science requires additional "fundamental ideas" that Kant never considered. Hence, what appear to be the axioms of the various sciences may well be false and may need to be revised or replaced without, of course, disabling the "possibility" of science itself. (In the way, for instance, in which Kant's treatment of Euclid's geometry proved to go contrary to his own claims in the preface to the second edition of the First *Critique*.) Whewell lays this out very clearly in a response to a "Kantian" criticism of his own doctrine, advanced by a contemporary critic, Henry Longueville Mansel. Wettersten provides the relevant texts (see Wettersten, this volume, §1.5).

But the decisive fact remains: Whewell, who follows Kant in a way, by accepting the disjoint sources of knowledge that Kant treats facultatively, opposes the apriorism of the latter's transcendental idealism. I regard that as signifying both a serious limitation in Whewell and the beginning of a correction in Whewell himself, a weakness that vestigially continues the Cartesian impulse that runs through Descartes and Locke in their seemingly opposed ways and, therefore, may be found again in Kant and Mill. It only appears that the Cartesian assurance has been replaced by Kant – and, accordingly, opposed by Mill. The truth is, rather, this: affirming a disjunction between sensation and reason as between the disjoint sources of our cognitive powers has become, ever since Locke, the essential continuation of the Cartesian strain itself. So that Kant *and* Mill – *and* Whewell – remain Cartesians in their respective treatment of necessary truths, even where they differ from one another (in the radical way they do) regarding the local legitimation of necessity. The irony is that it is only when we reach Hegel that the disjunction between the sensory and the rational is ever exorcised (though of course Hegel himself seems at times to view his own undertaking in Kantian terms).

Speaking of Whewell, the important lesson remains as before: one cannot find a ground for the substantive necessities of natural science and mathematics, in Whewell, unless one subscribes to a facultative division between the sensory and the rational. But if that is so, then, in a curious way, Whewell has not really escaped either Kant or Mill! He *has* avoided Kantian transcendentalism and Millian empiricism, but he has not recovered (as both Herschel and Mill complain – rightly but for the wrong reasons) an account of scientific proof. And so, there must be an element of Kantian transcendentalism in Whewell's account, though his best intuitions deny it.

It is at this point, precisely, that I think it useful to suggest a larger view of the significance of the disagreement between Mill and Whewell. In fact, Wettersten (this volume, § 1.2.3.1) is persuaded that Whewell finds himself obliged to treat truth, even necessary truth, as changeable, so that a theory might be true at time $t$ and, with a change in science or "fundamental ideas," false at $t'$ ! I cannot assess the textual evidence, but the supposed blunder is easily reinterpreted in an illuminating and plausible way: give up the disjunction between the "sensory" and the "intelligible" (in Kant's general way), and concede that assignable truth-values, not truth itself, may change under the pressure of changing science and changing fundamental ideas – no paradox then remains. But that requires paying a certain price: namely, *the would-be necessity of the necessary truths of science will have to be relativized to the contingent state of science itself (to the salient "ideas" we regard as "fundamental"), with respect to which we judge what to regard as the operative axioms of the particular inquiry in question.*

It was perhaps Whewell's facultative disjunction which made it difficult for him to admit that seemingly clear and distinct axioms might, finally, *not be* axioms at all – hence, his paradox. But that is still Cartesianism pure and simple. For his part, Mill continues to insist that inductive processes do not presuppose or require "theory" – what, in Whewell's idiom, comes to the imposition of "fundamental ideas" upon sensory observation. (That was the supposed charm of Mill's Methods.) Perhaps this clarifies the sense (suggested earlier) in which, although Whewell has a very strong argument against the empiricist's treatment of necessity, that argument is hardly enough to confirm that Whewell himself satisfactorily recovers necessity – or the ground of scientific proof, which (he thinks) requires the recovery of necessity.

My own suggestion is: first, that, contrary to the rationalist tradition, there is no disjunction between natural science and mathematics, though Mill and Herschel were wrong to treat mathematics as "empirical" in the empiricist's way; second, that "necessary" is an epithet, when applied to non-logical and non-analytic truths, that cannot fail to be relativized to the historical phase of inquiry at which certain "fundamental ideas" are thought to lead to the axioms of the inquiry in question, (subject to unforeseen revisions in the ongoing work of any such inquiry); and, third, that the rationale for any proposed such necessities is not attributable to the distinct exercise of any faculty disjoined in principle ("metaphysically" distinct, in the empiricists' pejorative idiom) from sensory experience itself.

The quarrel between Whewell and Mill, I suggest, has been hopelessly deformed by the acceptance, on both sides, of a disjunction between sensory experience and intelligible theory (or "ideas") imposed (in some prior way) on the determinate data of sensory observation. Mill disjoins sensory "description" from "theory" – which, of course, is impossible to defend; Whewell insists that description is already oriented in a cognizing way by executive theories – which affects the logic of induction, but which also reserves (quite untenably) a separable source for such theorizing "ideas" – for "imposing" them on sensations.

In my opinion, the disjunction between "thought" and "sense" is classically linked to one or another form of objectivism, rationalist or empiricist, which runs from Descartes to Mill (and Whewell) and includes Kant (who, I suggest, is a Cartesian in the sense intended). The resolution of that completely indefensible disjunction is deflected, in the dispute between Mill and Whewell, by the puzzle of necessary truth and scientific proof. The fact is, *if* predicative similarity cannot be confirmed by any empiricist or rationalist means unless some version of platonism can be sustained (which I personally take to be impossible), then whatever objectivity may be ascribed to predicative uniformities is (i) affirmable only within the indissoluble space in which what "belongs" to our cognized, "independent" world can be assigned only by way of some constructivist account; and that (ii) to go the way of (i) entails a constructivist reading of otherwise disjunctive distinctions that would adversely affect what it means to be objective in the way of science. In particular, there can be no principled disjunction between induction and deduction (involving necessary truths about nature), between the *a priori* and the *a posteriori*, between the necessary and the contingent, between the subjective and the objective, between "ideas" (theories) and "facts" (observations), or (even) between the logical and the metaphysical.[3]

This is not the place to pursue the preferability of the "third" option I am interposing between rationalism and empiricism. My thought is, rather, that Whewell seems to have taken a very strong initial step in its direction, by linking the history and philosophy of science as he does and by distancing himself, given his rationalist proclivities, from Kant's form of transcendentalism. He seems to have found himself obliged to draw conclusions about the changeability of necessary truth itself – *a fortiori*, about scientific proof – that (if I'm right) could have been easily avoided by merely rejecting the facultative divisions that belong to empiricists and rationalists alike. In the spirit of a small appreciation, then, I suggest that we can now make the required move for Whewell and reinterpret what he might have said regarding his dispute with Mill and Herschel. He would certainly not have appeared as vulnerable (to our eyes) as Mill and Herschel supposed. But to make such a move would be to bring Whewell forward to our time, where, although the matter is still hardly settled, the indissoluble union of the subjective and the objective (sometimes called antirealism) is at least no longer drummed out of play as incoherent.

Wettersten cites, at the close of his own discussion, a remarkable anticipation in Whewell's own words of what (I suggest) might have resolved his peculiar vulnerability in the exchange involving Herschel and Mill:

> That necessary truth is progressive – that science is the idealization of facts and that the process goes on from age to age and advances with the advance of scientific discovery – these are the doctrines which I have endeavored to establish and to elucidate (Whewell, 1971, p. 354).[4]

---

[3] An excellent illustration of the violation of the last may be found in Dummett (1991).

[4] I am not familiar with the recent book by Fisch (1991). But I find it interesting that the interpretation of Whewell remains to this day more or less settled in terms of the question whether he was a (failed) empiricist or a (failed) Kantian. The passage just given shows

There is nothing here that requires the facultative disjunction or even the assurance of necessary truths about nature: nothing that requires transcendentalism. On the contrary, the formula given is reconcilable with the sense that, only at the end of history (which is to say, never in our time), can we expect to exceed the notion that would-be necessities in nature are *more* than what appears to be *a priori* within one historical age or another.[5] But that, I think, is to place Whewell beyond rationalism and empiricism.

*Temple University*
Department of Philosophy
Philadelphia, PA 19122
U.S.A.

## REFERENCES

Dummett, M. (1991). *The Logical Basis of Metaphysics*. Cambridge, MA: Harvard University Press.
Fisch, M. (1991). *William Whewell*. Oxford: Clarendon Press.
Foucault, M. (1966). *Les mots et les choses*. Paris: Gallimard.
Wettersten, J.R. (1992). *The Roots of Critical Rationalism*. Amsterdam: Rodopi.
Whewell, W. (1967). *The Philosophy of the Inductive Sciences, Founded upon Their History*. Second edition. London: Frank Cass. Reprinted from 1847 edition.
Whewell, W. (1971). *On Philosophy of Discovery*. New York: Burt Franklin. Reprinted from 1856 edition.

---

how reasonable it is to judge him to be neither. I may also say that Whewell's elaboration of the issue (in the passage cited) confirms an extraordinary convergence between his own views and those of Peirce. But that is an entirely different matter.

[5] It would be wrong to press the point, but this is not altogether unlike what Michel Foucault (1966) had in mind in speaking of the "historical *a priori*" (against both Kant and Husserl). Certainly, it converges with Peirce's pragmatism.

# REPLIES TO COMMENTARIES

**John Wettersten**

# REPLIES TO COMMENTARIES

In their commentaries on my monograph Michael Segre and Joseph Agassi have raised interesting methodological questions about how to study the history of the philosophy of science. Segre finds my approach wanting, because I do not let Whewell speak enough for himself, and thereby fail to satisfy Segre's desire to know just what Whewell's chief contribution was. Agassi discusses the place of critical rationalist studies of the history of philosophy among current alternatives. He suggests that we may find there a needed substitute for Popper's attempt to demarcate science from non-science. We may, he points out, demarcate good from bad philosophical work as by demarcating pretense; as I have tried to do. This shift in problems is called for by the change in studies of rationality from the exclusive study of scientific methods to the piecemeal study of methods of rationality. He discusses two problems raised by such studies: How can we avoid judging our predecessors by our standards rather than theirs? and, How can we study pretense?

Two of the six comments on my monograph take issue with my interpretation of Whewell. Ron Curtis points out that Whewell's theology is "hard-nosed" and played an important role in the development of his philosophy of science. He questions whether Whewell thought and/or reasonably defended the view, that, because his historical justification was scientific and sufficient, he did not need a theological justification of his philosophy of science. Maurice Finocchiaro illustrates my observation that Whewell's *History* moved significantly away from inductivist standards, but argues that his philosophy of science was inductivist in some sense.

Godfrey J. Guillaumin contends that the differences in the appraisals of Lyell's geology, offered by Herschel and Whewell, are largely independent of their other philosophical disputes.

Joseph Margolis suggests that we can appreciate Whewell today by (1) seeing how incredibly far he came in overcoming the dichotomy between

In: John Wettersten, *Whewell's Critics: Have They Prevented Him from Doing Good?*, edited by James A. Bell (*Poznań Studies in the Philosophy of the Sciences and the Humanities*, vol. 85), pp. 387-3??. Amsterdam/New York, NY: Rodopi, 2004.

rationalism and empiricism, by (2) seeing that his attempt to provide theory of scientific proof failed, because he did not do this thoroughly enough, and by (3) improving his view today by removing the remnants of this dichotomy from his view with the proposal that assignments of truth values are historically relative.

In reply to Segre I explain why I wanted to let my interpretation of Whewell grow out of responses to his critics, why my appreciation of him lies more in the problems which his view raised than the virtues of his system, and why the study of the former is more important than the latter. In response to Agassi I suggest, on the one hand, that the study of our forerunners in terms of their standards needs to include social analyses of what was demanded of them at the time and that the most coherent portrayal of their views is not necessarily the best, and, that the study of pretense can often be carried out with an objective methodological approach not requiring insights into motivation.

Curtis offers an appreciative and interesting perspective on the role of Whewell's theology in the development of his philosophy and corrects an impression I may have left that I disparaged it. But, I reject his suggestion that, presuming Whewell's own view of scientific proof is correct, he needed a theological justification for his philosophy of science. Finocchiaro illustrates nicely my suggestion that Whewell's history was not inductivist in the way Agassi describes inductivist histories. But, I find that his suggestion that his philosophy of science was in some sense inductivist vague and quite unconvincing, because he ignores both the question of the role of empirical research in Whewell, which is to make ideas clear and distinct, as well as his theory of justification by intuition.

Guillaumin, I suggest, fails to make his point that the differences between Whewell and Herschel go beyond their differences in methodology, because he has virtually no overview of the differences between the methodologies of Herschel and of Whewell. Margolis' appreciation of Whewell is concise and to the point. He finds important problems growing out of the attempt to improve him by rejecting his untenable view of proof, which allegedly reveals a remnant of the dichotomy between rationalism and empiricism. But, I cannot accept Margolis' proposal, that we may best solve them, if we avoid the demand for proof of necessity and accept the view that the assignment of truth-values can only be made relevant to an historical context. Since both Margolis and I see rationality as historically based and capable of being improved, our views lead to interesting problems about how to criticize and improve (socially accepted) standards of rationality, which need to be solved to avoid a mere success philosophy. These are touched on in Agassi's commentary and my reply to it.

## Michael Segre: The Method of My Study; System vs. Historical Dialectic

Michael Segre finds my discussion of the reaction to Whewell interesting and valuable, at least as a description of the dialectic, which followed Whewell's path breaking research. But he is nevertheless not very happy. He maintains that (1) although my style follows such thinkers as Galileo in using a dialogue form, I do not measure up to their standards, because I do not let Whewell speak for himself. As a consequence of this first defect I (2) do not explain just what Whewell's contribution was (Segre suggests it was due to his break with the radicalism of the Enlightenment). Segre further complains that I (3) ignore Whewell's history, that I (4) do not explain his view of induction, and that I (5) fail to say more about Duhem. I will first answer Segre's five objections in turn; I will then offer an additional brief explanation of what I was about, as this does not seem to have been sufficiently explained in my text.

### 1. *Traditional Complaints about the Effectiveness of My Approach*

Segre's five complaints are all concerned that my portrayal is not comprehensive enough and that it does not show distinctly enough the virtues of a Whewellian view.

Segre compliments me by comparing my style to that of Plato, Giordano Bruno, and Galileo. It is, he says, nearly a dialogue in which part of the meandering of philosophical argument is portrayed. But, he quickly complains, I do not do what my far more illustrious predecessors succeeded in doing: I do not make either Whewell's view or the reasons for his importance sufficiently clear. I do not do so, he says, because, contrary to these illustrious models for presentation, I do not let Whewell speak for himself enough. I give, he says, too much space to his critics – as if Galileo had given hardly any space to Salviati as a defender of the Copernican system and merely let his critics have their say.

When compared with the artistry and revolutionary character of Galileo's dialogues, my essay must, of course, be judged inferior. But leaving that aside, my intent was by no means the same as Galileo's. My method is different. Galileo was fighting for the Copernican theory. In sharp contrast with this aim, I am not fighting for Whewell's philosophy. Rather, I am pleading for a reappraisal of the history of the philosophy of science and especially of Whewell's role in it. I am also seeking to change the way we do philosophy of science: It should be done historically. Segre recognizes that my aims are different, when he notes that I have written a study of philosophical dialectics and the development of an aspect of the philosophy of science. But then he quickly ignores this observation and returns to his plea that I present Whewell's view in a more conventional and systematic way, one which separates it from this dialectic.

This criticism and proposal is not new, as I have heard it from editors before. A defense of Whewell is not, however, what is called for today. We all agree that it is false. I did pose the question of what Whewell's philosophy said. But this was necessary in order to show how the debate in the philosophy of science over the last one hundred and fifty years had gone awry, because it ignored him. Social conformity and adherence to established views have led to the neglect of important views and problems. The task is to show how this has taken place in order to straighten things out a bit today. A change in our view of current problems and how they have arisen is needed. Even though Whewell's view is false and in spite of strenuous efforts to suppress it, it has had enormous impact. The main story today of Whewell's view is how this took place.

I had no intention of presenting Whewell's system as a system. Rather I hoped to explain his view by showing how he met criticism to it. I ask, were there good reasons for rejecting Whewell's view? And I review the reasons given and explain that Whewell had answers to them. These answers build his theory. I summarized the main points of his view at the end of the essay. Now this is not the normal way philosophical views are presented. But that was not my intention. My intention was to show how far Whewell could go in answering his critics. At times he had no good answers. But he was even then at least as good as, if not better than, his critics. I think it is useful to look at the history of philosophy in this way, as it makes more sense out of developments, even if it does not meet some traditional standards for the portrayal of philosophies as systems. This traditional method has its problems, since the history of philosophy is not first and foremost the history of comprehensive and coherent systems. It is the history of conjectures, which are laden with problems. If we are to understand this history, we must see the problems and how our predecessors responded to them. I think my style is appropriate for this purpose – however much it may be improved upon – and the traditional demand Segre places on my text counter-productive.

Segre secondly complains that I do not make clear enough what Whewell's contribution was. He seeks to do the job for me by saying that Whewell broke with the radicalism of the Enlightenment. I do not find this suggestion enlightening or useful and will, therefore, ignore it. It smacks all too much of an historiography, which tries too hard to give medals – as Agassi described it – by saying just what someone (essentially) did. We do not need to use this outdated approach: We can look instead at how problems changed after interesting ideas were introduced. And I have explained how this happened, how problems in psychology, in the psychology of research, in methodology, in epistemology, in method in the philosophy of science and in the history of science were changed. This is what I find important.

I can mention some central points. Whewell pointed out that scientists learn by trial and error. He explained that no inductive method, either one of discovery or of proof is possible, but that an empirical method designed to render ideas clear and distinct by tests, refutations, changes and refinements remains. He pointed out that the philosophy of science should explain the history of science and that it should be self-reflexive; it should be a science of science. The history of science, he pointed out, consists of stages of conjecture, proof and refinement. He rejected sensationalism without rejecting empiricism, as Duhem then also did.

Segre's third complaint is that I ignore Whewell's history. Now I do point out that he gave history a new purpose: It should be a test of the philosophy of science. Any philosophy should be capable of explaining it. This was a revolutionary development, because it opened up historical speculation about scientific progress, which quickly encouraged others – especially Duhem. Secondly, Whewell pointed out that speculation and criticism were important historical aspects of science. Thirdly, he pointed out the important role of refinement of theories. Whewell said that this was the third of three stages in the development of any fundamental idea. Duhem, I conjecture, took over this idea of the importance of the refinement of theories in scientific progress from Whewell but made it all of science. He could do this by abandoning the search for truth, which is the purpose of the first two stages. Thirdly, I discuss and criticize Agassi's view of Whewell's history and point out that Agassi treats historians as black or white just as he criticized historians for treating scientists as black and white and therefore ignores the progress that Whewell made. (See Finocchiaro's discussion and my reply to it.) I should further note that in Whewell's researches there can be no sharp separation between history and philosophy of science.

Segre's fourth complaint is that I do not make Whewell's theory of induction clear enough. I think I have done that so I have trouble understanding his complaint. But hardly anyone today wants to accept the view that Whewell was no inductivist, but an intuitionist, in spite of the fact that Whewell's contemporaries were more in agreement with my interpretation, as they condemned him for being a Platonist. The difference between Whewell and inductivists lies in their view of the proper aim of empirical research. According to inductivists we gather evidence and use this evidence as a basis for (inductive) inference to true ideas. Some say evidence must be gathered first; others that theory formation may precede the gathering of evidence, but that the evidence serves as premises in an argument for the proof of theory. Whewell said that we cannot gather evidence before we have an idea. And, by the time we have gathered sufficient evidence to infer that a theory is true, the inference is superfluous. The aim of empirical research for Whewell is not gathering evidence for use as premises in arguments for the truth of theories. The aim of empirical research is to refine and correct theories until they are so clear and distinct that we see their truth by intuition. Only empirical research, that is, deducing statements from theories and making observations to see if they are true, Whewell claimed, can render ideas sufficiently clear that their truth can be seen intuitively. And we know empirically that ideas are proven in this way. We have learned it from Whewell's effort to refine his theory of science by testing and improving it through studies of the history of science.

Segre's fifth complaint is that I should say more about Duhem. I have said briefly above what I think the relationship between Whewell and Duhem is. I have said more in an article on *ad hoc* hypotheses (Wettersten 1998). Duhem follows Whewell's historical method, his anti-inductivism, and his third stage of the development of science, (the retirement of Proven theories) thereby removing the chief complaint against Whewell – he had no theory of proof – but also sacrificed his realism. Agassi has a few interesting comments on this subject in his commentary.

## 2. *New Aims for the Study of the History of Rationality: How Can We Choose Our Heritage?*

In undertaking my study of the development of Whewell's ideas the central problem which concerned me was: How can one find a good alternative to the uncritical acceptance of the up-to-date intellectual fashions as the determining factors in the formulation of today's problems? Descartes' solution to this problem, that is, to start totally afresh, is obviously unfeasible for any modern version one might care to pursue. Piecemeal criticism brings us forward, but we are in danger of losing the forest for the trees. We need new methods for shaping the intellectual context in which we work. In a quite general sense we may think better if we autonomously recast our historical and social context in some coherent, useful and interesting way. I came to the conclusion that the problem of choosing one's heritage can be best solved with historical-cum-sociological inquiries. Useful inquiries are designed not merely to trace the development of problems up to the present, but also to critically appraise the various turns discussions have taken as a matter of historical fact. These turns may have been the result of good arguments we may accept, or good arguments we may improve, or poor arguments that gained influence, or mere politics. It is thus useful to see how our heritage has been shaped, before one simply accepts it as a framework for today's problems. When we have gained this knowledge, we may improve our ability to make corrections where we find they are called for and to shape our heritage autonomously and rationally.

Critical rationalist methods can be used to shape one's views of his social-intellectual heritage. They allow one to choose one's heritage by rationally re-appraising the way it has been presented. Critical rationalist methods can do this because critical rationalism is not radical, as are traditional approaches that seek the truth, but yet offers methods for the pursuit of truth. Traditional approaches do not recognize the importance of background knowledge, because they presume that each piece of knowledge will be justified by its own special evidence. If background knowledge is conceded to play some important role in the pursuit of truth, it is presumed that it is the most stable part of this pursuit as Mario Bunge suggests. It is that which one may take for granted, because it has been proven; there is no need to treat it skeptically. Because traditional approaches incorporate this attitude, they offer no methods for dealing either systematically or effectively with problems posed by defective background knowledge. Because traditional approaches do not acknowledge the importance of background knowledge, they are condemned to either accept it uncritically or to handle the problems it raises ineffectively with ad hoc methods. Conventionalist views recognize the importance of background views, but only as a basis for moving further uncritically. The recognition of the role of background knowledge in forming problems, its fallibility, combined with new theories of rationality offer tools for the construction of methods for dealing with it appreciatively and critically.

The way we choose our philosophical heritage determines to a significant degree how we see other fields. It is thus fundamental; it is not a mere accouterment to the study of contemporary problems. The background knowledge we presume shapes the problems we formulate in all fields. It determines in large part the aims we follow. We determine with background knowledge where weak spots in current views are and where hope may lie for further progress. If we accept as background knowledge the view that scientific progress demands empirical proof and/or the that any alternative leads inevitably to irrationalism, we are led to restrict our views of live alternatives in quite unhappy ways. If we find alternative background assumptions, we may think better. But this cannot be done arbitrarily. We seem to be like a dog trying to catch its tail. We want to select background knowledge well and not uncritically. But the selection of background knowledge seems to depend on the very traditional approaches we are seeking to overcome: How can we choose background knowledge, if not on the basis of that which has been previously settled, i.e. proven or merely institutionally established?

Critical rationalist methods can break out of this circle, because we may use them to work back and forth, from background knowledge to problem and back, treating now one aspect and then the other as problematical. But if this process is to be progressive, it must be done in a controlled manner. A historical method is the key to finding the needed control. If we are to choose and reshape this background knowledge we must see how it was developed and where and why it became successful. Which problems have been solved? Which have been ignored? Which have been pushed aside for political reasons? Where can we find new ones which have not yet been noticed?

A general answer to the question of how we may intelligently choose our intellectual heritage is as follows. Historically influential arguments as well as arguments which might usefully have had influence, but which as a matter of fact did not, should be re-examined so that we may: (1) learn from good – sometimes mistaken – arguments, (2) learn how confusions have been passed on, so that we may overcome their influence on our thinking, (3) learn how pretense, politics etc. have shaped current views of the value of past contributions with the same aim as in (2), (4) find problems, past and present, and appraise them under the assumption that they are the real motor of real intellectual history, regardless of which views have been established. They do this either by leading to new views or by demanding some kind of suppression of the logics of situations. These tasks should be carried out while (5) keeping in mind the distinction between the views of those studied and the views of the inquirer. There are unlimited perspectives from which such inquires may be carried out so that (6) competing appraisals from competing points of view should be deemed an enrichment to our heritage and an aid to choosing our own.

I presume that the social-cum-intellectual environment at any time is shaped with more or less precision by loosely defined sets of established views. By established I simply mean those views which are deemed by editors of journals as representatives of various intellectual audiences, university departments, sponsors of conferences, etc. to be proper and interesting subjects of debate. Those views most widely deemed proper subjects of debate are for the moment the most established – which does not mean that they are the most stable, since revolutions do occur and just before a revolution there may be remarkable unanimity about the superiority of the view which is about to be overthrown. Established views have their own logical and social credentials. They are merely logical, when taken to be arguments for the acceptance of such views. They are social and historical insofar as they have become parts of the background knowledge widely shared by any particular intellectual group. In order to understand one's intellectual context, it is necessary to understand this social-historical setting of the arguments, which set the tone of debates at any given time.

This task is not merely descriptive. If we are to use the identification our heritage has an effective background for formulating interesting problems and for setting a fruitful agenda today, we will need to shape our heritage more critically than a mere description of various trends allows for. In order to identify ones heritage, one first needs some theory (or theories) about which arguments and which background assumptions are important today. When these seem unproblematic, there seems little reason to investigate their history to see how and why they have become so influential. One may do this, of course, merely to document the correctness of current fashions. In my opinion the history of analytical philosophy, *The Semantic Tradition from Kant to Carnap* by J. Alberto Coffa (1991) is an example of a failed attempt to do just that. When we find the dominant stream not to our liking, when we are dissatisfied for whatever reason with the intellectual context we happen to find ourselves in, there is more reason to investigate critically how that situation came about. What arguments, what turns of intellectual affairs led to the present situation? The quality of the arguments used in this development will have much to say about whether the standard background knowledge should be accepted, rejected, or revised. A critical analysis not merely of current views, but also of the situation these views are responding to, is needed when one is sufficiently at odds with the intellectual environment one happens to find oneself in. Being at odds with one's environment may be due to the fact that something is sorely amiss or it may be merely due to a personal quirk or an intellectual mistake. An historical study may go far toward clearing this up.

How then may we proceed to conduct such an historical-cum-sociological-cum-normative inquiry? We may look for the antecedents of those views we find interesting as well as of those views that we find not so interesting but that seem to stand in the way of progressive research programs from our point of view. These are, of course, mere conjectures. They may then be tested by spelling out how current theories have been developed over some longer time period.

My special interest was, of course, awakened by Popper's view of science and rationality on the one hand and the overwhelmingly negative reaction to his philosophy on the other. I asked: Why is this so? What intellectual and social forces are here at work, which, in my opinion, leads to this unhappy situation? My inquiries began with attempts to understand the development of Popper's view. While studying methods in psychological research as a graduate student, I had noticed that Popper's views were quite close to some psychologists. I first noticed similarities with the Gestalt school, but then more importantly learned, what others had discovered before me, that it was above all the Würzburg school, which had laid the groundwork for the development of a modern deductivist theory of science. While investigating this history I found that Whewell – who I kept an eye on due to suggestions by Agassi – often appeared. It seemed to me clear that he had played a far more important role in the shaping of our intellectual heritage than he was given credit for. At first I described this role insofar as it had laid the groundwork for Popper's view. This particular influence came above all through psychology. In the monograph under discussion here I studied the reaction to him, how it forced his influence underground, and how this fact shaped the debates about science ever since. How did this happen? Why had he come to play such a small role in the standard versions of our intellectual heritage? I sought to find out and to tell this story.

The place to begin, I presumed, was the initial reaction to him, and to proceed to show how this reaction set the stage for what followed. In order to correct what followed, I presumed, it would be quite useful to look at the sources of those views which followed in the original reaction and to trace their development, the arguments or strategies that were used to come to terms with Whewell and/or to suppress his views. I was encouraged by the fact that Whewell seemed a good case to use to illustrate a method. The commentaries on him were so few that it seemed that a virtually comprehensive overview of the debates about his theories could be reconstructed. This reconstruction could show not only how, but also where, the philosophy of science and rationality had gone quite awry on some quite important points due to intellectual mistakes, confusions and pretense as well. This could explain why discussion of important problems and alternatives had been blocked.

## Joseph Agassi: Two Methodological Problems for Critical Rationalist Studies in the History of Philosophy

Agassi contrasts my approach to writing the history of philosophy not only with Hegelian and Wittgensteinian alternatives, but also with Popper's, to which it is so indebted. The departures I have made from Popper's philosophy raise new problems and he discusses some of them. My central thesis is that critical rationalist studies in the history of philosophy may be used to choose rationally one's philosophical heritage, to formulate problems well, to set an agenda intelligently and to retain one's autonomy. Jose Luis Borges, Agassi notes, pointed out that we need to solve this problem, as, I might add, did Alexander Koyré as well. My central problem is: How may we do this well? One central aim of my study of Whewell, then, is to develop methods for doing this.

My study, Agassi says, is an interesting extension of critical rationalism to the study of the history of philosophy. His sympathetic interpretation of the place of my study in current research explains some advantages of historical critical rationalist studies over the now traditional, dominant Hegelian and Wittgensteinian alternatives. He asks what we gain by the switch from traditional justificationist historical studies of philosophy to critical rationalist ones, as well as what problems this switch might raise. He moves smoothly from the discussion of traditional approaches to the study of the history of philosophy – the irrationalist view of Hegel, the traditional dogmatic view and its modern Wittgensteinian version – to Popper's view and on to more recent modifications of it, including mine. His portrayal of these alternatives is fine indeed although it plays down his own exceedingly important contributions to it. It not only sets the context well for current developments, enabling one to see their import against the background of traditional alternatives, but begins a discussion of the methodology of critical rationalist alternatives. Agassi's portrayal goes far toward formulating well some of those problems which are now important, thereby enhancing an agenda for current research.

I will focus on Agassi's discussion of the development of critical rationalism and the problems it raises, not because it is more important than that which accompanies it and sets the stage for it, but because it is there that it will be most useful to look at differences of opinion. In this way we explore various possibilities not only for setting today's agenda, but also at the methods for doing so, which is Agassi's central concern.

## 1. *Two Problems for the Critical Rationalist Study of the History of Philosophy*

The study of the history of philosophy from a critical rationalist perspective, Agassi says, is one more interesting development brought forth by this school. He finds both advantages and problems in this endeavor. Its chief advantages are that it enables us (1) to critically appraise thinkers without being dogmatic and (2) to understand the rationality of the development of views without appealing to an Hegelian, irrationalist view of the development of thought. We may look at past thinkers critically, as embodying normal human frailties, but we may also see them as rational to this or that degree and the work of some of them as worthy of study. We are not limited in our criticism by acceptance of current views, nor must we condemn, when we find disagreements and mistakes.

But Agassi also sees problems that arise when one uses critical rationalism as a framework for studying the history of philosophy. These problems, he thinks, are even more difficult to overcome than those we encounter in writing the histories of other fields, such as the history of science, where the problems are severe enough. Agassi focuses on two of these problems. The first is the problem of how we treat intellectually non-serious work: How may we identify work that suffers from pretense or is intellectually dishonest? This problem arises out of the sociological approach that I follow. I seek to understand how ideas have been developed against a social background, which Popper seeks to block out, in order to concentrate on the pursuit of truth alone.

The second problem Agassi poses is: How may one fairly and without distortion apply a critical rationalist framework to appraise the views and practice of individuals, who were seeking to follow quite different standards? How may we impose a critical rationalist framework on the history of philosophy and still evaluate our predecessors in accord with their own, non-critical rationalist standards? He points out that we not only need to appreciate historical figures insofar as they meet their own standards, but also that we need to appreciate the progress they made in accord with the standards of critical rationalism. This may indeed be fairer to them, he explains, since it allows us to see their errors *and* appreciate them, whereas on traditional views it is very difficult to appreciate anyone one finds to have been mistaken.

These two problems are new, because they presume that we have to take into account the differences in standards between the historian and the person studied: We cannot simply condemn on the basis of our standards, nor can we simply accept prior standards as the working of reason. In specific cases we may also find a deviance between the rational practice of a philosopher and either his otherwise accepted standards or the results of his inquiries, but we do not wish to condemn as wrong and/or dishonest all such deviance. It is too easy, Agassi thinks, to explain behavior which appears to conflict with some standard as mere misbehavior. So, we need new standards and approaches, which help us see them in their context following their standards as best they can. But what should these be?

## 2. *Pretense in the History of Thought*

Agassi points out that Popper implored us to employ a rule when studying thinkers, which I reject. Popper said that we should always and exclusively look at those real intellectual problems thinkers have tried to solve. In this way, he thought, we may not only better understand them, but also treat them fairly, by avoiding other considerations such as sociological ones, which might lead us to view their thought, say, from the perspective of the interests it served rather than its truth alone. But I do not restrict myself in this way. I do not look only at the theoretical problems thinkers tried to solve, but also at the strategies, techniques and arguments they have used for overcoming political difficulties, even when these have little or nothing to do with the solutions to theoretical problems. I look at their practice of rationality in a social context. The use of this practice to explain the history of thought seems to harbor just those dangers, which Popper with good reason warned us about. Agassi suggests that a further central departure of mine from Popper lies in a proposal for the study of bad intellectual work. Popper demarcates pseudo-science from science with the criterion of non-testability, but has no theory of demarcation outside of science. (This is not quite true, since he did propose that all good philosophy must seek to solve problems that arise outside of philosophy; I reject this proposal as well.) I propose instead that we distinguish bad from good (philosophical) work by the identification of pretense. He suggests that I thereby fill a lacuna left by Bartley and Hattiangadi: Bartley proposed that we study rationality instead of science and Hattiangadi suggests that we view rationality as a set of techniques. But neither thinker offers a substitute for Popper's theory of pseudo-science. A proposal such as the one he attributes to me, Agassi suggests, is rendered necessary by these developments in critical rationalism. This sets a central problem, then, for Agassi's discussion of critical rationalist philosophy: If we apply this proposal, how can we study, identify and/or demarcate pretense effectively, without judging our predecessors unfairly or improperly?

What rules should we use for identifying and/or demarcating pretense? Agassi says he does not know; but he explains why it is important to identify it. We can only avoid this task, he explains, by employing traditional dogmatic views, which serve as blinders when we turn to historical studies. Furthermore there are serious portrayals of historical events, which indicate that pretense is present. We need, in short, a solution to the problem of how we are to judge pretense, if we are to realistically describe the history of thought. Critical rationalism, he explains, opens up new problems such as these. He points out that one inadequacy of Hegelian irrationalism – history thinks for us – or of Wittgensteinian demands that we think only within the context of (established) common sense is that they rule out such studies.

Agassi sees the problem of identifying pretense as quite serious; he does not want to judge historical figures too quickly or easily as violating this or that standard. We need to look seriously, he emphasizes, at the logic of their situation. It might be all too easy, he says, to find thinkers violating some standard of "honesty" which we endorse but they do not. How can we appreciate differences and distinguish honest work from that which is less so? How can we avoid being too harsh on just those people who have benefited humanity with their work? This difficulty is still further compounded when we find excellent views defended by thinkers we highly respect for their accomplishments using defenses we find less than admirable.

We do not want to be sympathetic only to those who have made honest mistakes and less so to those who have behaved mischievously, but we should also appreciate those thinkers whose general quality is high, but who may have used defenses we find unacceptable from time to time. Agassi suggests that we view the history of philosophy and science much more like political historians view the subjects of their inquiries. Political historians find no difficulty in seeing the good that even tyrants may have done. They appraise this good without offering any excuse – explicit or implicit – for the points of view or behavior of those they study. This proposal is certainly a good part of the solution. Before discussing it further, however, I will turn to the second problem first; they are closely connected.

### 3. *The Problem of Using a Critical Rationalist Framework to Study Thinkers Holding Different Views*

The second problem Agassi poses is: How may we apply critical rationalist standards to the study of those thinkers who have applied quite different standards to shape and judge their own work? Agassi shows how the imposition of the traditional framework on the history of science does not work: It does not value criticism and insists that no mistakes be deemed valuable. This leads to all sorts of contortions. But finding standards for the imposition of a dogmatic framework is still easier than that for a critical one, since one may simply evaluate past work as in conformity with one's own perspective or not.

Before proceeding with the discussion of this problem, let me mention one caveat. Insofar as the history of science has been influenced by philosophy, and I think this influence is considerable, finding standards for the imposition of a framework on the history of science cannot be easier than that for the history of philosophy, since the former must include to some degree the latter, as Agassi has so elegantly shown with his studies of the impact of metaphysics on science. These studies need, I think, to be supplemented by the studies of the impact of changing methodologies on science, so that the history of science is not portrayed as the result of the use of a single methodology – as Popper and perhaps Agassi view it – but as the result of steady interaction between methodology and scientific practice – as Agassi's bootstrap theory would have it on my reading of it.

Agassi suggests that one aspect of the task of writing history from a critical rationalist point of view might include attempts to look at the "proofs" of those we study from their points of view, that we seek to understand why they accepted arguments as conclusive, which we find so faulty. Why did they defend the points of view they did and why they found arguments convincing, which we do not find so today?

Agassi praises a program that Lakatos sought to carry out for a while. This program is to reconstruct the "proofs" offered by historical figures, so as to show why they were deemed by these figures to be "proofs" at all. In order to do that, we need to study their standards. This program, Agassi suggests, is much better than the Hegelianism, which Lakatos later defended: It explains the logic of the reasoning in the thought of historical figures; it does not appeal to the mysterious workings of Reason in order to say why some view was accepted, and it thereby avoids the obscurantism of such an appeal.

I find the transition, however, from Lakatos' first historical program to his later Hegelianism so natural that the first program bothers me as well. For, on both programs we presume that there is some proof theory, which allows some arguments to be deemed valid, which we later have found to be invalid on the basis of new proof theories. But there need be no such theory and in most cases I would be surprised to find one. Even Lakatos' first program has a Hegelian presumption of the validity of historical arguments, according to conceptions of rationality that change over time.

Agassi recommends that we study proofs in order to render them valid according to their standards. He suggests that we might be able to find those standards of proof, which Newton and/or Marx implicitly used. Now it may be that Newton or Marx were implicitly applying standards of proof, which it may be possible to reconstruct. I doubt however that this is the case. They presented arguments, Agassi points out, which their contemporaries could not effectively criticize. But criticism which came later may very well illuminate confusions, which led to this inability, rather than to any coherent set of standards for proof under which their arguments would have to be deemed valid.

Does this mean that the program which Agassi describes and which, according to him, Lakatos once tried and failed to implement is misguided, that it presumes that a level of (implicit) rationality may be found in the work of those studied, which is much too high? I think not. One can try to carry out the program, but only under this proviso: There may be no solution to it at all and any coherent view one constructs may or may not be that held by the thinkers studied. It may, indeed, often be more fruitful to study the connection between even confused and contradictory social standards, on the one hand, and specific arguments on the other, than to try to reconstruct a rationality of standards, which, if it was there, was hardly even conscious to the actors themselves. The former program comes much too close to the Hegelian view of reason which Lakatos later, according to Agassi, endorsed in conversations and for which Agassi rightly expresses contempt.

Let me suggest an alternative to Lakatos' first program which does not share his nearly Hegelian presumptions. When actions seem not to follow from the views of those who act, we have difficulty in explaining them. In the case of critical rationalist histories of thought this problem arises quite definitely: We attribute, for example, to the actors intentions to solve problems raised by criticism, even when they themselves deemed themselves to be going about matters in some other way, say, by gathering information before proceeding to find the true explanation of some phenomena. Critical rationalism offers here a solution. It presumes, namely, that no system is complete and consistent. Inconsistencies allow some person to act now one way and now another, since thinkers will understandably tend to appeal in any particular situation to that view which happens to be accepted by his colleagues as appropriate. We can still however explain views partially by explaining how thinkers acted this way or that, on this or that view because it was seen by their contemporaries as the appropriate thing to do in that specific situation. This may be dissatisfying to some. But the alternative is to construct rationales when none are – or were – present. Of course we may always ask if some explanation might be improved, whether the actions of some thinker may be seen to have exemplified a higher degree of rationality than has hitherto been recognized. But in each case we must be just as critical of these attempts to improve our appreciation by testing them to see if they attribute too much rationality to the actors. We cannot view the fact that some view makes some action appear more rational than some other view as proof of the superiority of that view.

The partial explanations we give of the rationality of those we study will also have to rely on social analyses. What role did they play in society and how did this effect the views they defended and the methods they used to do so? In this way the personal and the social explanation cannot be separated, as historians have often sought to do, by explaining systems as if they were responses merely to problems of explaining events in accord with timeless standards of justification. In fact the rules change over time and such changes need to be taken into account in any history of thought. In the case of *avant garde* thinkers the social standards of justification, which they use to defend their theories in discussions with opponents, may differ from those standards, which flow from their own views. This is the case, indeed, with Whewell. And my study of Whewell may be looked at as carrying out a task in regard to Whewell, which is much like that which Lakatos tried and failed to carry out in the case of Newton. I try to explain, namely, how Whewell both proposed new standards of proof and applied these standards to his own view. The new standard was that when an idea had been shown to explain empirically that material relevant for some field – whereby relevance is determined by the explanation itself, a fundamental idea defines a discipline – in such a way that the educated intuition sees it to be necessarily true, it is so. Now this view explains why Whewell thought his view was justified. For, he thought he had done that for the history of science. No alternative could. This shows also why his opponents claimed he had no theory of proof, because he had no Kantian or empiricist theory of proof.

### 4. *The Subsidiary Role of Personal Investigations in the Identification and/or Demarcation of Pretense in Critical Rationalist Historical Studies*

The approach to the understanding of the history of thought which I have sketched above as an alternative to, or if you prefer, addition to that proposed by Agassi; that is, that we look at the social standards which thinkers have attempted to meet when we seek to understand why they adopted and defended the views they did, allows one to see the problem of identifying and/or demarcating pretense in a new light. This latter problem turns out primarily to be one of the identification of arguments revealing pretense. Examples are the portrayal or programs as adequate theories or the criticism of opponents for falling to solve problems which one cannot solve oneself without noting that. The identification of such dubious acts as deliberately concealing debt to others turns out to be quite subsidiary. It is often not the case that one needs to look for personal or external evidence of pretense in order to make analyses of pretense in the history of thought. If can explain the moves thinkers made in terms of the standards of the day, one may learn which standards were inappropriate from this or that point of view and when standards have been violated from our or their point of view. Whether they were violated is a problem for methodological critique. It often has no special, personal character. If one finds a potential violation, one may even seek standards under which it would not be a violation. The identification of pretense involves piecemeal analyses, which may determine where and when standards were violated, or, perhaps, where two standards conflicted, making the violation of at least one unavoidable. We may also learn in this way to better identify moves involving pretense and thereby raise our level of rationality.

## Ron Curtis: Can We Separate Whewell's Task of Justifying His Philosophy of Science from His Theology?

Ron Curtis defends an interesting and challenging thesis: Whewell not only needed his theology to complete and to justify his philosophy of science – as commentators have long held – but also that this theology does not constitute a special, religious appeal. Whewell's theology, he says, is a "hard-nosed," respectable, intelligent and integral part of his philosophy. He further points out that his theology and his idea of proof precede his philosophy of science and guided its development. This interesting perspective on Whewell was missing in my presentation. The central disagreement between Curtis and me is whether Whewell needed his theology to complete and/or to justify his philosophy of science. I claim he did not, but this disagreement can be separated from Curtis' other interesting and valuable observations.

### 1. *Are There Two Mistakes in My Reading of Whewell's Theology?*

Ron Curtis begins his interesting comment by reminding us, as he has previously and insightfully pointed out, that Whewell was no anti-scientific creationist. Whewell thought, Curtis explains, that the world, including man, should be explained in law-like or scientific terms. So, Curtis continues, insofar as Whewell was an opponent of evolutionism, he was no simple-minded anti-scientific creationist, which the literature all too often makes him out to be. Rather he sought alternative explanations, which better satisfied his high scientific standards.

Curtis identifies mistakes in the interpretation of Whewell's view of evolution as due to inductivist attempts to write – or rewrite – the history of science in accord with up-to-date views. He then uses this analysis to explain alleged mistakes in the interpretation of his theology. He finds that his observations about Whewell's theory of evolution have consequences for the traditional interpretations of the role of Whewell's theology in his philosophy: The causes of the misinterpretations of Whewell's views on evolution and of his views on theology are the same. In both cases efforts are made to paint his philosophy in all black terms, his view of creation and his view of theology as beyond the bounds of science or respectable philosophy of science. Curtis here argues that all previous commentators, myself included, have got it wrong. Many, myself included, have got it wrong because they deem Whewell's theology something external to his philosophy, as a creationist view, which Whewell could not defend as part of philosophy. I have got it doubly wrong, he says. I not only assert that Whewell, from his own point of view, did not need his theology to complete or justify his philosophy of science. I also make too many concessions to the traditional view of this theology as external to his philosophy. He thinks I seek to keep Whewell's theology and his philosophy of science separate, because I share the popular view of his theology to a much too high degree. Let me turn first to the question of my motivation, of whether I share the disparaging popular view of his theology too much.

## 2. *Curtis Corrects a Mistaken Impression*

Curtis cites examples of misunderstandings of Whewell's theology in the work of Stoll and Butts as well as in my discussion of these thinkers. The problem he finds lies in the false interpretation of theology as a mere anti-scientific creationism, rather than an integral part of his admirable philosophy. Curtis contends that we should not be embarrassed either by Whewell's need for his theology or his theology. He thinks I am. As a consequence he thinks I unhappily seek to downplay the importance of Whewell's theology throughout my book. I portray it, he claims, as a position that emerges late in the game and one that Whewell's readers may accept or reject as they choose. I mistakenly assert, he suggests, that it is independent of his philosophy of science. Still further I allegedly defer to the notion that the *Bridgewater Treatise* is a bad book and that the theological deduction is the sort of "ploy" one might expect from the man who wrote it. Curtis then quotes me, allegedly documenting this last claim: "should we really be surprised that the author of the *Bridgewater Treatise,* who explained such facts as the distance of the Earth from the Sun as signs of God's plan, should now explain the possibility of knowledge with a theology?" – says Wettersten (this volume, p. 191). I do not say that this was "ploy" or that the book was bad. I only say that one can see from one context what would, in another context, be convincing to him.

Curtis suggests further that he senses my unease, because I keep returning to the issue, repeating my arguments. Now I do return to the issue and repeat my arguments as the situation demands, but this does not reflect unease. It reflects the simple fact that various authors presented Whewell's view erroneously. Since I did not want to make each discussion contingent on the others, I repeated some arguments which happened to be relevant for several of those commentators I discuss. I did that with other issues as well. This fact merely reflects the style of the book.

I did in fact treat Whewell's theology gingerly, and Whewell certainly took it to be a more important aspect of his philosophy than I do. So, the tenor of my remarks might portray a somewhat misleading picture as Curtis claims. The picture of Whewell which emerges from my analysis may not make clear enough how important theology was to Whewell personally and/or how well thought out it was. If so it is good to have this picture corrected by Curtis. But it has little to do with my attitude or my analysis of his philosophy.

I do *not* say that the *Bridgewater Treatise* is a bad book, nor do I say that Whewell's move in introducing theology is a mere ploy. I did say that given his religious perspective – which so far as I know I nowhere belittled – we might expect him to develop a religious interpretation of science. This interpretation would not necessarily be so important to those who did not share his religious inclinations, as I took him to recognize. This is not, I think, belittling and the remark about the *Bridgewater Treatise* was not in any case *meant* to be disparaging. I am, of course, grateful for Curtis' criticism and correction. His emphasis on the rational nature of Whewell's theology and the background it may have offered for the development of his view is useful for understanding Whewell. The fact that he has pointed this out allows me to correct any false impression about this I may have left.

### 3. *Did Whewell Think He Needed His Theology to Complete and/or Justify His Philosophy of Science?*

We come next to two questions, which Curtis alleges I conflate. These are (1) the question of whether, as a matter of fact, he needed a theology to complete and/or justify his philosophy of science and (2) the question of whether he thought he needed it for one or both of these purposes. Let us look at these two questions more closely. I do not think that Whewell's philosophy of science is in fact justified by his history. I think, indeed, that it is false. I do think that Whewell thought he could justify his philosophy of science with his history alone. I can only say that if one presumes that Whewell's method is correct, and that Whewell's history of science, that is, his science of science, meets his standards. One may deduce from these (false) assumptions that he did not need it.

I say that Whewell thought he justified his view of science without theology because he in fact did not need it to meet his own standards of proof, that is, to construct a science of science. Given his standards of proof, he did not need his theology to justify his philosophy of science. I do not think that to look at things this way is to confuse the questions Curtis poses. As Curtis formulates the first question, it is in any case hardly of interest, since it is clear that his theory needs something more than history to be justified. Since Curtis and I are both fallibilists, it would be rather pointless to seek what, from our point of view, he needs to justify his theory.

Curtis continues his criticism by saying that I have trouble finding textual evidence to support my thesis that Whewell did not need his theology to complete and/or to justify his philosophy of science. If he means by textual evidence direct textual evidence and not such evidence that comes from interpreting various aspects of Whewell's philosophy, he is right. But this comment is nevertheless terribly misleading. It does not show, as Curtis apparently intended to show, that my view in this regard suffers from some peculiar difficulty: It is difficult to find direct textual evidence for *any* view of this matter. In defense of his view Curtis does no better than to reinterpret the passages I discuss. I did not find any direct commentary on the issue outside of his discussion of the issue in the *Philosophy of Discovery*.

Curtis disagrees with my interpretation of two passages. In the first passage Whewell says, supposing a reference to God be admitted, let us see what consequences flow from it. Curtis and I have two disagreements about this passage. I say, since Whewell says "supposing," he means you need not suppose a reference to God, that the following considerations apply only to those who do, in particular to those who are dissatisfied with the incomplete view of man offered by his philosophy of science. I base my interpretation of this text on what went before, which Curtis does not mention. Whewell mentions that his view may be not be deemed satisfactory for some, since it does not offer insight in regard to some matters. The second disagreement we have about this passage is whether Whewell hopes to prove here the truth of his philosophy of science. Curtis says yes and I say no. Whewell does not say that his philosophy of science is not proven. Rather, he says that he will trace the consequences of his philosophy of science. If we take this comment at face value, Whewell is tracing the consequences of his view of science, which is independently proven. This is how Whewell, on my view, presents the logic of the situation.

Let us turn to the second passage on which Curtis disagrees with me. Whewell there states that those who reject his theological deduction would contend that man's thoughts and ideas agree with the world because they are borrowed from them, that is, they would side with Herschel. Curtis' reading is quite plausible, that is, that if you reject his theology, you must also reject his philosophy of science. But this reading is not the only possible one. It can also be read as the mere assertion, that those who reject his theology do in fact assert that ideas correspond to the world because they are borrowed from it. There is a further possibility: Whewell thought that his theological argument was based on his philosophy of science, which was independently established. But he thought that no one who accepted his argumentation in regard to the philosophy of science could refrain from going one step farther, because it was the same type of argument. Now, if all things were equal, I would tend to Curtis' reading which only allows for the two alternatives of (1) accepting his philosophy of science plus his theology or (2) of rejecting both together and endorsing inductivism. But all things are not equal. I find his reading of this passage incompatible with the above passage, which on my reading definitely presumes both that the philosophy of science is established without the theology and that interest in theology is something above and beyond interest in the philosophy of science. If I am to reconcile these passages, I find that my proposal is the better option. Whewell, I suggest, either intended to portray the contrast between himself and his opponents, which existed in fact, or he intended to suggest that it would be scarcely reasonable to accept his philosophy of science without accepting his theology as well. Neither of these readings of this passage requires any change in my analysis of the logic of Whewell's view.

We may review the logic of Whewell's view by turning to the second question: Did Whewell's philosophy of science require his theology to complete and/or to justify it as a matter of fact? Now, as I have noted above, the obvious preliminary answer to this question as it stands is that of course it needed something because it is not proven, as we all agree. But this is not or should not be at issue here. What might be at issue is whether, according to Whewell's theory of the proof of a philosophy of science, it needed any further proof. We must look more closely to see if Whewell's justification of his philosophy of science is based on his historical study alone, as I claim, or rather if some other interpretation more favorable to Curtis' is better.

Curtis suggests we consider the possibility that, even if Whewell thought he needed his theological interpretation, he may in fact not have needed it. He presumes that I would suggest this alternative. But this is not quite right. I do not think his proposal can be so defended, revised or whatever so that it would yield a justification. I merely claim that Whewell thought his theory of science was justified according to his own standards, because he thought it fully explained the history of science and no alternative could. On my view both Whewell and his competitors took as fact that science found and justified the truth. There was no need to prove it and no party to the debate called it into question. I read Whewell as a Kantian in the sense that he presumed that scientific method existed and that science found necessary truth. The task of the philosophy of science on this view is to explain how science does that. When the philosophy of science has done that, its task is completed.

Now Curtis seems to agree that Whewell's philosophy of science is, accordingly to Whewell, a science of science. But he calls into question the view that this science of science needed no other justification than its ability to explain science when no other theory could. I find his view difficult to understand. According to Whewell any specific scientific theory needs no other justification than that it explains the facts in its field and no other theory does. If the theory of science is a science of science, then how can the theory of science need some further justification than that it explains the facts of the history of science when no other theory can? Why must it be better than science?

Curtis nevertheless insists that Whewell needs some further justification for his philosophy of science. He insists that as a matter of fact he must answer Cartesian questions. I cannot see why he must do that. Curtis says, namely, that Whewell needs to exhibit some guarantee that our intuition does not deceive us.

Curtis agrees with my reading of Whewell according to which, on Whewell's view, science proves ideas necessary by refining then until the educated intuition of scientists sees them to be true. He agrees with me that this view is a science of science, that is, it explains the history of science. He also says it is incomplete. It leaves open, namely, the question of why that should be the case. I agree that this is true. The question that arises then is only whether Whewell had to complete his theory by explaining why this is the case and by showing in some other way than by an historical demonstration that it is true. I have argued that, although Whewell's theory is in fact not proven, the fact that it is not complete does not mean that it is not proven. Whewell's theory of how science attains the truth does not need to form a complete or systematic account to be proven, if we accept Whewell's standards.

Now, if one is to argue with Whewell, one has to either accept his standards and say he does not meet them, or one has to say what his standards are and show them to be faulty. My complaint with the tradition is that it insists on using non-Whewellian standards to evaluate his view without ever taking into account Whewell's own standards. I do not see in Curtis' reading any special consideration of Whewell's standards or any realization that they may have been different from his competitors. And thus I find his insistence that Whewell needs something more than the appeal to the capability of his theory to explain the history of science to be arbitrary.

Curtis contends that Whewell needs his theology to complete and justify his philosophy of science. In defense of this view Curtis claims that the "same problem" of justification that arises for theories of science arises as well for the science of science, i.e. Whewell's philosophy of science. Like other sciences, he suggests, the philosophy of science uses ideas not arising out of the facts in its domain but rather ideas that have an independent, mental origin. Why, then, he asks, have these ideas led to the truth of the matter? An explanation for science as well as for the philosophy of science is, he says, needed. Has our intuition or the view that only Whewell's theory can account for the history of science misled us? How can we know that God is no deceiver?

Whewell's philosophy of science raises problems of explaining why our ideas correspond to the world and why ideas seen to be true by the educated intuition are true. But it does not raise questions about whether ideas correspond to the world or whether ideas seen to be true by the educated intuition are so or whether God might deceive us. Following Kant Whewell presumed that science found the truth. He did not seek to prove that science did this but to explain how it did it. The problems of proving this did not arise for him, any more than they did for Kant. If he accepts as a fact that science finds necessary ideas and proves them to be so, he may, as he did, ask and answer the question why this is so. But in answering this question he does not use God to argue for the truth of science. Rather, he uses the truth of science and the nature of man's special capacity to do science as the foundation for arguments for God. He uses these facts to argue for God just as he had earlier used properties of the universe such as its very fortuitous distance from the sun to argue for God and his benevolent plan for the world. We might even say, in sympathy with Curtis' emphasis on the importance of theology for Whewell, that his primary aim was religious, that he cared more about proving the existence of God and his goodness than he did about proving anything about science. But this is a personal matter and not a matter of the logic of his view. Indeed, if that were the case he might be especially interested in finding independent grounds for the belief in God.

Curtis continues his argument with the proposal that Whewell developed his theology as he developed his philosophy (of science). It was not, he says, a late epicyclic addition to his view. Now it is not my view, as Curtis suggests, that Whewell's theology was epicyclic, if he understands by epicyclic something ad hoc or something that does not fit in his philosophical system. And it is obvious that in point of time Whewell offered theological proofs earlier than his philosophy of science. I merely assert that the logic of his specific argument for God which he offered in the *Philosophy of Discovery* – and this is what we have been talking about – presumed the truth of his philosophy of science.

In order to develop his view of the role of theology in Whewell's thought, Curtis attempts to explain the development of his theology in terms of his philosophy. This is a fine contribution to our understanding of Whewell. But I do not think that it will help him show that Whewell needed his theology to justify his philosophy – or science – of science. Curtis shows how the idea of Design has, according to Whewell, become so clear and distinct that he cannot deny its truth. Curtis' suggestion that Whewell developed his view of design through criticism is interesting. But he does not explain this development. He only notes that he developed some new arguments based in his view of science. But I do not see that this has any consequence for the question of whether Whewell on Whewell's standards needed his theology to complete and/or justify his philosophy of science.

In summing up his view Curtis says that Whewell is much more the hard-nosed theologian than I would like to admit. This is not so. But my only concern was to separate his theology from his philosophy of science, that is, to show that the philosophy of science needed no theology for its justification. His theology may be as hard-nosed as you please. Curtis then suggests that Whewell's theology need not be any cause for embarrassment. This is fine. There is no reason to identify with him and no reason to be embarrassed by his theological concerns.

Curtis then suggests that Whewell gave no explanation of the origin of Fundamental Ideas. This is not true. Whewell said that they were innate – implanted there by God – but not in a clear form. We have the God-given capacity to render these ideas clear and to see their necessity if we do science. Curtis then notes that Whewell's criticism challenged Darwin to find a better theory of the origin of Fundamental Ideas. For that reason alone his view cannot be read as black. This is interesting, but beside the point of the disagreement between Curtis and myself.

Curtis has read me all too much as accepting presumptions of a tradition, which first finds that Whewell needs a theology to complete his philosophy of science and then rejects him for this need or for the theology itself. He finds also that the need is there, but finds it fulfilled in a "hard-nosed" way. I find no need, but do not deny that what is there is hard-nosed. I find Curtis' view of the standards which Whewell sought to fulfill too traditional. Whewell proposed and fulfilled new, historical standards for the philosophy of science. These standards were not Cartesian. They did presume that the central problem of the philosophy of science was to show that science finds the truth. They were Kantian and presumed that this task was to explain how it found the truth.

Curtis maintains that Whewell needed his theology to justify his philosophy of science. But Whewell only says that without it his view is incomplete. And furthermore Whewell did not demand completeness for truth. This is clear from his theory of the antimony of ideas. He says that this cannot be explained, but must be presumed. Curtis has fully ignored this argument for my view, which I discuss in that passage he notes in my discussion of Stoll. The tradition has demanded from Whewell a complete system and Whewell rejected this demand. Curtis never discusses this issue, because he never discusses the difference between Whewell's standards and those of his competitors. I find, then, no reason in Curtis' comment that would lead either to the conclusion that Whewell thought his theology was necessary for his philosophy of science or to the conclusion that it was necessary given his standards of proof.

### Maurice Finocchiaro: Whewell Was No Inductivist!

In spite of my efforts to show that Whewell was not an inductivist, Maurice Finocchiaro proposes that in some sense he was. Finocchiaro hopes to clarify matters by examining various senses of the word "inductivist" and then to discuss first, whether Whewell's history is "inductivist" and secondly what his view was. The main task of his essay is to examine one part of Whewell's history to determine whether he is not, as I say he is not, an inductivist and to say what, on the basis of this study, we might say that he is. Finocchiaro agrees with most of my analysis, but we do differ on Whewell's view of justification – which I deem to be intuitionist and he something else – and on whether his view should be characterized as some form of inductivism. Before turning to a discussion of his central task of showing that Whewell's methodology was inductivist in some sense or other and his *History* was not, I will respond to some of his preliminary remarks, which he uses to introduce and justify his more central task.

1. *Why Was Whewell Rejected?*

Finocchiaro complains that he is still unclear as to why Whewell should have been so frequently misinterpreted and so unfairly criticized. In my essay I observed that Whewell had been decisively rejected as confused and wrong-headed. I then looked for adequate intellectual grounds for these charges. Although my search has been comprehensive, I have not found them; I have suggested that there were none. Finocchiaro apparently accepts this thesis. Now, if there are no adequate intellectual grounds for such rejection, then, I presume, there may very well be other ones. Among these are simple confusion and the difficulty of coming to terms with radically new ideas. And these played, in my opinion, a crucial role in the shaping the poor reception that Whewell received. But they were not, I think, the only ones. There was also a fear that the new ideas might undermine science, as I have pointed out. Now my proposals that Whewell was rejected because of dogmatic commitments to inductivism and/or Kantian views, difficulty of coming to terms with new ideas, and fear for the welfare of science, might not be adequate. But it is, at any rate, more than a beginning. I do not claim to have said the last word on this issue, but only to have shown how acute it is. The reasons I have given, if accepted, open up new and interesting problems. Some of these are sociological, concerning the response to new ideas within the scientific community.

2. *Do Negative Theses Require More Documentation than Positive Ones?*

Finocchiaro's next observation is that I deal too harshly with some of those criticized; he seems to mean Agassi. It may be the case that I have been too harsh on my former teacher and close friend of many years. I did not want to seem harsh on those far away and too lenient on those close and tried to keep the standards the same. Perhaps I overcompensated a presumed tendency to go too lightly on my friend. But I am not sure that I can accept Finocchiaro's case for my allegedly excessive harshness. He suggests, namely, that a favorable interpretation of any thinker does not need as strong a documentation as an unfavorable one requires. Agassi's view of Whewell is more than sympathetic, so it does not need as strong a documentation as a hostile one; but my criticism of it, by Finocchiaro's standard, being a negative view of sympathetic interpretation, requires more documentation.

I do not think that Finocchiaro's rule can be accepted. Whether an interpretation is interesting or not is the major though not the only standard for criticism. And how much documentation is needed is a question of context and of style, etc. Some favorable and some unfavorable views can be dismissed without further ado; and some of both varieties may deserve serious study. But we should also not confuse seriousness with documentation: Much documentation is humbug and much discussion without documentation is quite serious. If my documentation of my discussion of Agassi was too sparse for his taste, Finocchiaro has been so kind to make up for my laxness by improving it without finding any need to change it.

But there is another important reason for rejecting Finocchiaro's suggestion. When authors try to interpret "positively," they often mean more in accord with what they – and not the person being interpreted – think is correct. And this often does injustice to the author as happened to Popper, for example, when he was interpreted as offering a theory of meaning. Finocchiaro's own reading of Whewell may seem more "positive" because it makes him more modern, but it does him injustice by robbing his view of the strength it had in the 19$^{th}$ century.

### 3. *Should One Raise Problems One does Not or Cannot Solve?*

Finocchiaro also complains that, although some of my criticisms are plausible and sound as far as they go, they do not go far enough. The reason for this is that they raise important and interesting issues that I do not explore. Now in my view any group of criticisms that raise so many important and interesting issues that they cannot be handled in a monograph of nearly three hundred pages is very good indeed. If Finocchiaro is right, I have done this, and I thank him warmly for this high compliment. But the standard Finocchiaro here implicitly uses, i.e. that one should not raise problems which one does not solve, is horrible. This is often done, but only at the expense of intellectual honesty. The existence of unsolved problems is hidden. I say the opposite: Whenever we notice an unsolved problem we should note it and especially then when we find it too difficult to solve or even when we simply do not have the space or time.

Finocchiaro mentions one of the problems, which is raised in my criticism of Laudan, though he could have mentioned others as well. I suggest that Laudan and others failed to understand the justification, which Whewell claimed for his own theory. He claimed, namely, that his theory was justified by the facts of science. We should ask further, Finocchiaro says, whether the fact or facts cited by Whewell really are facts and whether his theory – in the particular discussion the theory of the consilience of inductions – accounts for the putative fact or facts.

I find Finocchiaro's standard too demanding. The issue of whether the claim for a justification has been rightly interpreted and whether it is in fact a justification are two quite different questions. The discussion of the first does not necessarily demand a discussion of the second. As a matter of fact the further questions did not interest me, since I presume that some of the facts he cites as facts are not facts – we do not know necessary truths – and, if they were his account is inadequate – we could not know them by intuition. Whewell's methodology cannot be justified by his history: We obviously do not, as Whewell claimed he could, prove that scientists intuit the necessity of fundamental ideas on the basis of his historical studies. Even a study which follows Whewell's own method, that is, that seeks to explain the history of science so as to make it intuitively clear to historians that his thesis is necessarily true, could not do this. Whewell's interesting and innovative theory of empirical research as the only possible education of scientific intuition cannot be true.

## 4. *How Should We Use the Term 'Inductivist'?*

In the second section of his essay Finocchiaro turns to a discussion of the meaning of the term 'inductivist' and various aspects Whewell's thought which may or may not be "inductivist." He thinks clarifications are needed because, he alleges, I have not been sufficiently precise. I need to discuss four points to cover his critical comments. To begin with Finocchiaro correctly notes that the claim that Whewell is not inductivist could apply to historical or philosophical views as well as to his own procedures of discovery and justification. I do not know why Finocchiaro complains, since I covered all bases in my monograph. But, if for some reason I failed to make myself clear enough, let there be no further doubt: On my view he was not an inductivist in any aspect of his theory or practice.

Secondly, Finocchiaro quite correctly notes that I am not using the word 'induction' in the sense in which Whewell used it. Whether his claim that I do not make this sufficiently clear is correct is hard for me to judge. I did discuss this issue. I explained that I was following the best Whewellian scholar that I came across – Robert Blanché – who first proposed that Whewell was not an inductivist. Blanché and I, as I explained, are quite aware that Whewell's use of the word 'induction' is different from ours. I said that clearly; perhaps not prominently enough. Finocchiaro correctly notes that when I refer to induction I am referring to (1) inference from facts to theory by generalization or (2) discovery by observation of facts or reduction of theory to facts alone. He further notes that my use is traditional. But he still is puzzled as to why a defender of Whewell would want to "concede" to Whewell's critics that this is a proper use of the term induction. I should note in passing that I am not a defender of Whewell's theory as a live option. I am only a defender of the view that it was a very important and innovative development, whose value has been much too much ignored. Following Whewell himself Finocchiaro wishes to call Whewell's theory inductive. Whewell tried to do that and no doubt with a motivation similar to Finocchiaro's, that is, he wanted to have his view deemed a new theory of induction rather than a new non-inductivist theory, since all respectable theories of science were called inductive. But this usage merely led to confusion and did him no good politically. It may have been one cause for the common and erroneous charge leveled against him that he had no theory of proof: If he offers a theory of induction, where is the (inductive) theory of proof?

In contrast to Finocchiaro, I think it is unwise to use a word with a clear meaning in a new way simply to suggest some sort of connection with a tradition. But if we are determined to do that, the minimum we must do is to make quite clear what this new meaning is. Finocchiaro does not do that, even though his major complaint was, that I did not make my point clear enough. In fact I offer a simple and clear view of induction. Finocchiaro introduces a number of alternative aspects and calls for new definitions. This has the effect of confusing a simple analysis. It hardly brings us forward.

Finocchiaro has another reason for not following my usage of the term: The conception of induction, which I use, is, he says, not up-to-date, since the up-to-date version is probabilistic. Finocchiaro misreads me: This conception is also inductive in my use of the term, since it also involves inference from facts to theory. The fact that the conclusion is only probable is of no consequence for my usage. I may have not said this explicitly but he knows my Popperian sympathies and that all Popperians see the matter in this way. My usage is quite up-to-date and quite normal. There is no need for misunderstanding me in this way.

Thirdly, Finocchiaro incorrectly describes the view, which I attribute to Whewell. He says correctly that I say Whewell is anti-inductivist, because he denies that scientific theories are inferred by generalization from facts: I also say, as Finocchiaro notes, that on Whewell's view (1) scientific theories are explanations from which facts are deducible and (2) all observation is theory laden. But then Finocchiaro continues by falsely maintaining that I say on Whewell's view scientific theories have an inherently hypothetical element. This is not true, since I claim that Whewell thought that scientific theories were necessary and known to be so. If hypothetical means not proven even to any slight degree, Whewell denied that any fundamental idea is hypothetical. All theories go through a hypothetical stage to be sure. But established scientific theories have no hypothetical element in them. Finocchiaro seems to be misled by the inductivist tradition into thinking that any theory not reducible to facts must be hypothetical, i.e. not certain.

Fourthly, Finocchiaro discusses what he deems another sense of inductivism. He takes this idea from Agassi, which is that the inductivist presentation of the history of science paints events in the history of science in overly black and white terms. I suggest, and Finocchiaro is in agreement with me, that Whewell is not inductivist in this sense. I do not deem this view a different sense of inductivism than that which has gone before, but merely a corollary of it, which has been shown to be so by Agassi. Finocchiaro thinks that whether Whewell is inductivist in this sense of the word is of great import. He endorses my criticism of Agassi according to which Agassi treats historians in the same way that he objects to historians treating scientists, that is, as black or white. But he thinks my documentation insufficient and wishes to examine the claim. But he wants to do more. He wishes to find out what Whewell's own methodology is by looking at his practice. We come back to Whewell's justification of his view, that is, his claim that he is doing a science of science. His methodology should be justified by its ability to account scientifically for scientific practice. What then, Finocchiaro asks, is Whewell's scientific, i.e. historical, practice?

### 5. *What Is Whewell's Historical Practice?*

There are three aspects to Finocchiaro's discussion of Whewell's scientific, i.e. historical, practice as exhibited in his treatment of Ptolemy, Copernicus, Galileo and Kepler. The first consists of a documentation of my claim that Whewell did not fall prey to those sins analyzed by Agassi in his study of the historiography of science. I can only say, if you want more illustration, fine! The second aspect of his analysis consists of a portrayal of Whewell's portrayal of scientific practice. In accord with my view Finocchiaro points out the importance, which Whewell gives to ideas, imagination, and conjecture. He does not, as I have explained and as Finocchiaro further illustrates, hold what Finocchiaro calls an empiricist view and what I call an inductivist view; rather he emphasizes the role of ideas in scientific research and theory, we agree. Finocchiaro offers further illustration.

The third aspect of Finocchiaro's discussion of Whewell's practice is his own portrayal of Whewell's philosophy of science, which he claims to have culled from his study of his history and which he deems to be an improvement over mine. His prime reason for deeming his view an improvement seems to be that on his view Whewell may be deemed an inductivist of sorts. On my view this is no advantage and, indeed, a source of confusion, as I have already explained. If one chooses to deem Whewell an inductivist, the least one must do is to clearly explain just what this amounts to. But the best Finocchiaro can offer in this regard is a tautology: Whewell is an inductivist in the sense of the term used by Whewell. But Finocchiaro does not ask what this sense is. It is, as one sees from the second quotation in § 811 above, that some conception is superinduced upon the facts. And this shows that Finocchiaro's claim is false.

But what then is Finocchiaro claiming, when he claims that Whewell is an inductivist? Is he claiming that Whewell is Whewellian? I presume that this is not all he would like to say after his careful portrayal of Whewell's historical practice. What he does say is that Whewell exhibits empiricist leanings in addition to intellectualist ones and that this does not exhibit any incoherency. I have pointed this out in my monograph. Up to this point there is no difference between us other than a name. What puzzles me further, however, is that Finocchiaro, after complaining that my analysis is not careful enough, offers a vague and sketchy view of the relation between empirical research and ideas without ever asking what my view of the matter is. He gives the erroneous impression that I merely claim that Whewell is not an inductivist, without offering an interpretation of Whewell's view of empirical research. But I have not only offered, but repeatedly emphasized, an interpretation which goes far beyond that which Finocchiaro offers. I explained that Whewell viewed empirical research as a process by which the intuition was educated. Theories are tested in science in order to discover difficulties. They are subsequently refined to improve their accuracy and to increase their explanatory power. This process enables scientists to "see" the necessity of theories. On his view empirical research is necessary in order to find and refine the correct idea, but we make no inference from fact to theory: We intuit the necessity of idea after we have conducted a sufficient amount of empirical research; and only then! This is Whewell's theory of justification by intuition educated by empirical research. We cannot make sense of Whewell, if we ignore this aspect of his view, as Finocchiaro does. Because Whewell's view of justification is a form of intuitionism and because he rejects inference from facts to theory, I think it is quite misleading to insist on calling him an inductivist. And to do so to give him some honor is simply foolish: There is no particular honor or dishonor in being or not being an inductivist.

At the beginning of his essay Finocchiaro said that the still did not understand why Whewell had been so systematically misunderstood. One reason for this is the determination by virtually all commentators to either insist that he is confused, because he is not an inductivist, or to interpret him as an inductivist for external reasons – such as that being an inductivist is an honor or to bring him in line with up-to-date – which Finocchiaro has here once again tried to do. But Whewell is neither confused nor an inductivist in any clear sense of that word other than his own, which is a sense no one shares with him.

## Godfrey Guillaumin: To What Extent do Herschel's Narrow Empiricism and Whewell's Deductive Methodology Influence Their Respective Appraisals of Geology and Its Problems?

Godfrey Guillaumin hopes to show that "the core of his (Whewell's) idea of scientific enterprise (is) a difference between mechanical and historical causation" (p. 349). This view of science differs from Herschel's, Guillaumin maintains, since Herschel denies this distinction. Although Guillaumin makes no attempt to show how this "core" is relevant to Whewell's methodological theory, he does propose that this difference between Herschel and Whewell is crucial for their differing analyses of Lyell's geology. He says that some scholars have maintained that discrepancies between Herschel and Whewell in regard to geology depend on differences in their interpretations of *vera causa*. But Guillaumin argues their differences go beyond their disagreements about *vera causa* or between "rationalism and empiricism." Guillaumin hopes to correct an allegedly misleading interpretation of the important debate between Herschel and Whewell, advocated by Ruse, which attributes their disagreements in geology to their disagreement about *vera causa* (or "rationalism vs. empiricism"). But he nowhere explains Ruse's view. He also says that he will discuss an important question I have raised – Were there significant methodological and philosophical differences between Herschel and Whewell? In conclusion he says, and I quote because I cannot summarize what he wants to say, "Whewell's main problem was to answer the question of how to elaborate epistemically reliable sciences which integrated historical causality, taking very seriously the historical processes by which every science is constructed. By contrast Herschel maintained an ahistorical point of view about both the causes and the natural laws through which explanations are made" (p. 366).

Whewell distinguishes, Guillaumin says, those sciences, which are historical from those, which are mechanical, whereas Herschel allows for only mechanical sciences. This distinction is allegedly based on their analyses of the specific science of geology and extends beyond (is not a result of?) differences in their methodologies and/or epistemologies of science.

Guillaumin claims that Herschel and Whewell disagreed about whether all science should be conducted the way physics is. Herschel, he claims, says yes, and therefore finds Lyell's theory good, whereas Whewell says no, and therefore finds Lyell's theory unacceptable. In his attempt to make his case Guillaumin presumes that the method of physics equals Newton's method, which equals the so-called *vera causa* method, which equals Herschel's method (which is summarized as the view that all knowledge is based in direct observation and the methods of science are those which later became known as Mill's four (or five) methods of induction), which equals Lyell's method. On Guillaumin's reading Whewell accepts this view of *vera causa* but adds to the methods of physics an historical approach suited for geology.

This analysis will not do because the assumptions it makes about physics and the so-called *vera causa* method are false. Whewell and Herschel agreed that Newton's observations about *vera causa* apply to all sciences. Guillaumin ignores entirely Whewell's discussion of Newton's comments on method. Whewell and Herschel disagreed about what the method of physics was, and Whewell rejected the theory that all knowledge was derived from direct observation – it also required ideas – and he found Mill's methods useless. Whewell did not seek to add additional sciences with different methods. He rejected the methods of both Herschel and of Lyell on the basis of his own methodology. He rejected "actualism" (the theory that only geological causes now acting have produced geological changes in the past) and "uniformitarianism" (the theory that these causes have had the same degree of impact on geological change in the past that the now exercise) not simply because geology was historical, which all sides to the disputes admitted, but also because these assumptions were contrary to the spirit of science as revealed in its history, especially the history of physics: They arbitrarily preclude real possibilities.

Guillaumin also thinks that Whewell found a special problem in geology of identifying the right events to be explained. But Whewell saw this as a problem for all sciences. And the task which Guillaumin's conclusion obscurely hints at, that of finding the proper clear ideas, was central for all sciences.

Guillaumin further says that because Whewell extended the sciences to historical ones he allowed for hypotheses, whereas Herschel, because he held that the method of physics, i.e. direct observation and Mill's methods, should be used in all sciences, did not allow them. But Herschel did allow for hypotheses, albeit apologetically, and in an ad hoc way as Agassi and I have pointed out. And the method of physics has little to do with direct observation and Mill's methods, as Whewell demonstrated. Whewell not only allowed for hypotheses, but encouraged their use in all sciences including physics, but only as a research tool. The final, fundamental idea, should be proven. His criticism of Lyell was in part that he had not proven his theory.

Whewell criticized "actualism" – those forces acting today have always been those forces, which have produced geological change – and "uniformitarianism" – these forces have always acted with the same degree of impact – on the grounds that they are hypotheses. Herschel and Lyell accepted them on the basis of their narrow empiricism. These views only went beyond their general views in the sense that they were applications of them – I do not say correct or incorrect ones.

Guillaumin hoped to find differences between Whewell's view of geology and Herschel's, which went beyond their more general views. But he has no grasp of their general differences and thus cannot judge either what goes beyond them or how. He thus remains unprepared to carry out the task he has set himself of explaining to what degree the differences between Whewell and Herschel in geology go beyond their more general philosophical differences. Until he has undertaken a more general analysis, there is little point in saying more about this premature attempt to go beyond such an analysis.

## Joseph Margolis: The Dichotomy between Ideas and Sensations Cannot Be Overcome

Joseph Margolis poses the question of what we may especially appreciate in Whewell. His answer differs somewhat from mine, because we look at him from differing perspectives. But by and large it does not conflict with my attempt to revise our picture of Whewell's place in the history of philosophy. It shows how he may also be appreciated from a different perspective. Margolis observes that Whewell paved the way for the overcoming of some of those dichotomies that he finds harmful to current philosophical debates. One such dichotomy is that between rationalism or deduction or a priori truths on the one hand and empiricism on the other. Another is that between subjective and objective. Although I do not share some of the philosophical observations Margolis makes about current research, I do think that he is right to read Whewell in the way he does, that is, as moving toward overcoming those dichotomies he finds harmful. I think, indeed, that Whewell goes farther toward this goal than even Margolis gives him credit for. Let me first say, then, why I think that is so.

This common ground leads us to see the central philosophical problems of today in quite similar ways. Finding such common ground is always important, because it allows for fruitful debate. We disagree about solutions: Is it better to be skeptics in search of truth, hoping to increase our (speculative) knowledge of it, or should we assign truth values to theories on the basis of today's social standards, while making clear that these assignments change when standards change?

Margolis finds his sympathy for Whewell above all in his rejection of both a Kantian view of ideas as given a priori and his rejection of empiricist reductionism: Both ideas and observation are needed and inextricably intertwined, Whewell is never tired of saying. Margolis appreciates Whewell's view that even necessary truths grow. This is just that point on which so many thinkers have taken him to task – with notable exceptions such as de Morgan. Margolis notes that Whewell's theory of the development of ideas places empirical research in the forefront. Whewell gives, indeed, empirical research a new aim which is, as I have suggested in my monograph, the clarification of ideas until their necessary truth may be seen by the educated intuition. This aim is achieved, according to Whewell, by critically carrying out empirical tests and by using the results to correct ideas, refine them, and to make them more precise. Indeed, he claims that no idea can be rendered clear without critical empirical research, thereby merging inseparably the rationalist pursuit of clarity and distinctness with the empiricist's pursuit of knowledge based on observation while changing both significantly.

Margolis finds, however, that in the end Whewell falls back into that dichotomy, which he did so much to undermine. He falls back into it, says Margolis, due to his theory of the identification of necessary truth. Necessary truths must be rationalist, distinct from empirical research, transcendental. But, Margolis says, if the dichotomy between rationalism and empiricism is to be fully overcome – a task we need to accomplish, if we are to achieve an adequate epistemological theory today – we cannot allow for it there either: Necessary truth must always be allowed to develop, to change.

Whewell is indeed in agreement with Margolis' claim that we cannot completely separate ideas from facts; perhaps more than Margolis notes. The progress of science lies in the idealization of facts – as Margolis quotes Whewell saying – but, Whewell adds, this idealization can never be complete. If it were, we would no longer have knowledge of the world. So, even though we see the necessary truth of ideas only by intuition and only after they have been rendered so clear that the educated mind sees that they are necessarily true, this education can only be achieved by critical empirical research on some idea. Furthermore *some* empirical remnant must remain when facts are idealized; if it did not, we would no longer have knowledge of the world. There is an antinomy, Whewell says, between facts and theories, observation and ideas, which can never either be fully explained or fully sorted out: Ideas are about the world because they are intertwined with observation. If we could separate fully the ideal from the empirical elements, the ideas would no longer be about the world. A consequence of this impossibility of separating mental and empirical elements is that we can never explain how ideas and observation are united. This view of the so-called antinomy of philosophy raises a difficulty for Whewell's theory of proof, which is the task of reconciling the view that knowledge is necessary and seen by intuition to be so with the view that it is to some, even quite small, degree empirical. I will not discuss how Whewell might get out of this difficulty, for, I do not want to be here, as Margolis observes, tendentious in defense of Whewell.

Now Margolis maintains, as the empiricist critics of Whewell have also maintained, that Whewell fails to offer a theory of proof. This is, in may opinion, mistaken. Whewell did offer a theory of proof, albeit one which no one could accept today and even for reasons offered in the 19th century. But this fact should not lead us to underestimate the power his theory had then. For, he argued that all philosophers of science are agreed, as they then in fact were, that science achieves necessary truth. It cannot be by induction, he convincingly argued. But we do see that all scientists see the truth without induction. They must, then, see necessary truth intuitively. It was then possible to reject this argument for reasons such as those offered by de Morgan. But one could do this only at the cost of opening up deep problems in the philosophy of science, which no one wanted to do. So, they erroneously said he had no theory, when they really meant – or should have meant – that his theory was inadequate. They excused themselves thereby from responding to his arguments. Even de Morgan, who had much more sympathy for Whewell than most and who criticized Whewell's theory of intuitive proof, did not respond to Whewell's transcendental arguments in favor of them. He did not explain where Whewell's argument had gone awry.

Now Margolis continues by observing that in place of Whewell's much too transcendental view of necessary truths, we must place a view which allows for the change of (properly?) assignable truth values. And this seems to fit nicely with Whewell's view of the growth of necessary ideas. But a caveat is called for here. When Whewell said necessary truth grows, he meant that it was learned, but also that it was cumulative. He did not allow it to change – until he noticed that Kepler's theory contradicted Newton's, so that some weakening of this view was, regrettably, necessary. Margolis offers a modern rendition of how Whewell might have responded today. Margolis notes Whewell's observation that the best of science changes and that this historical observation forced Whewell to say that truth changes and thereby dangerously close to Hegel, just that position Whewell wanted to avoid. Margolis says that this move can now be seen to have been unnecessary: We may hold that truth does not change but that assignable truth-values do. Now here my philosophical views part company with those of Margolis, since I prefer the critical rationalist alternative of viewing knowledge as conjectural and steadily improving to the view that properly assignable truth-values change.

Margolis seems to think that the overcoming of the dichotomy between rationalism and empiricism or subjectivism and objectivism and the view that assignable truth values change are necessary bedfellows: If we reject the dichotomy, then we must accept the theory of changing truth values as well. But this is not the case. We can accept the Whewellian view that all ideas have both (1) a mental, rational side on the one hand and an empiricist side on the other and (2) that these are inseparable without accepting (3) the view that properly assignable truth values change. If we wish to overcome the dichotomy between subjective and objective, we need merely to say that all ideas, all theories are invented by us. They have, as Whewell said, a source in ideas which is in us as well as a source in the world outside of us but which comes to us through observation. But no observation occurs without ideas and no idea can refer to the world without observation. We cannot today plausibly add, as Whewell did, that ideas with their source in us will also be refined until they are necessary, if we only use the scientific method to criticize, correct and refine them. We need, then, to be skeptics: We cannot claim to have justified by any standard whatsoever the claim that any idea is true. But we can claim to improve our conjectures and thereby learn more about the world. We may also add that on both views standards for good theories may change and be improved. But on one view these standards justify the assignment of truth-values and on the other view they do not.

On some points those who have wished to overcome the dichotomy between rationalism and empiricism and critical rationalists move rather close together. Joseph Agassi has, for example, in his development of a philosophical anthropology on the basis of critical rationalist theory of rationality, made the overcoming of some dichotomies – those, e.g. between truth and falsity, nature and convention, reality and appearance rationality and irrationality – the major theme of his study in philosophical anthropology. And the competing views of which problems are important today come in some respects rather close together. Both sides see subjective elements flowing into "knowledge" that are inextricably mixed with external ones. We need, then, to explain how knowledge is possible, even while acknowledging that it does not simply grow cumulatively: The appraisals of fundamental propositions change and should do so. The issue that divides camps is whether a skeptical realist position, on the one hand, or a justificationist position with accommodation of historical change, on the other hand, is a better response to this situation.

Those who say that "assignable truth values change" have to come to terms with the well-known problems of reconciling the claim that some truth value may be (justifiably?) assigned to some statement with the recognition that tomorrow a different truth value may be (justifiably?) assigned to the same statement. Since these may change, skepticism seems part and parcel of this position. On the other side of the fence no assignment of truth-values is deemed justified, but conjectures may be freely made and held provisionally. The crux of the matter, then, seems to be whether we may reasonably have, or should have, social standards for the (provisional) assignment of truth values or whether we may better work on standards for criticism and/or standards for institutionalization of theories without socially established standards for the assignment of truth values. Both sides, incidentally, emphasize the role of changing social standards. They both face the problem of how these may be profitably analyzed, criticized and improved. Agassi's commentary and my response to it touch on this problem. Central problems for the opposition is to explain why fallible social standards should be deemed a proper basis for the assignment of truth-values and which ones these are. A mere appeal to social success, which seems to be the alternative of some modern sociology of science for which Margolis expresses some sympathy, seems too vulgar to be taken seriously.

## REFERENCES

Coffa, J.A. (1991). *The Semantic Tradition from Kant to Carnap.* Edited by L. Wessels. Cambridge: Cambridge University Press.

# INDEX

## MONOGRAPHS-IN-DEBATE

## CONTENTS OF BACK ISSUES

### VOLUME 81 (2004)

**Evandro Agazzi**

RIGHT, WRONG AND SCIENCE
THE ETHICAL DIMENSIONS OF THE TECHNO-SCIENTIFIC ENTERPRISE
(Edited by Craig Dilworth)

VOLUME 83 (2005)

CONFIRMATION, EMPIRICAL PROGRESS AND TRUTH APPROXIMATION

ESSAYS IN DEBATE WITH THEO KUIPERS, VOL. 1

(Edited by Roberto Festa, Atocha Aliseda and Jeanne Peijnenburg)

R. Festa, A. Aliseda, J. Peijnenburg, *Introduction*; T.A.F. Kuipers, *The Threefold Evaluation of Theories: A Synopsis of From Instrumentalism to Constructive Realism. On Some Relations between Confirmation, Empirical Progress, and Truth Approximation (2000)*. **Confirmation and the HD Method** — P. Maher, *Qualitative Confirmation and the Ravens Paradox*; T.A.F. Kuipers, *Reply*; J.R. Welch, *Gruesome Predicates*; T.A.F. Kuipers, *Reply*; A. Aliseda, *Lacunae, Empirical Progress and Semantic Tableaux*; T.A.F. Kuipers, *Reply*. **Empirical Progress by Abduction and Induction** — J. Meheus, *Empirical Progress and Ampliative Adaptive Logics*; T.A.F. Kuipers, *Reply*; D. Batens, *On a Logic of Induction*; T.A.F. Kuipers, *Reply*; G. Schurz, *Bayesian H-D Confirmation and Structuralistic Truthlikeness: Discussion and Comparison with the Relevant-Element and the Content-Part Approach*; T.A.F. Kuipers, *Reply*. **Truth Approximation by Abduction** — I. Niiniluoto, *Abduction and Truthlikeness*; T.A.F. Kuipers, *Reply*; I. Douven, *Empirical Equivalence, Explanatory Force, and the Inference to the Best Theory*; T.A.F. Kuipers, *Reply*. **Truth Approximation by Empirical and Nonempirical Means** — B. Hamminga, *Constructive Realism and Scientific Progress*; T.A.F. Kuipers, *Reply*; D. Miller, *Beauty, a Road to the Truth?*; T.A.F. Kuipers, *Reply*; J.P. Zamora Bonilla, *Truthlikeness with a Human Face: On Some Connections between the Theory of Verisimilitude and the Sociology of Scientific Knowledge*; T.A.F. Kuipers, *Reply*. **Truthlikeness and Updating** — S.D. Zwart, *Updating Theories*; T.A.F. Kuipers, *Reply*; J. Van Benthem, *A Note on Modeling Theories*; T.A.F. Kuipers, *Reply*. **Refined Truth Approximation** — T. Mormann, *Geometry of Logic and Truth Approximation*; T.A.F. Kuipers, *Reply*; I.C. Burger, J. Heidema, *For Better, for Worse: Comparative Orderings on States and Theories*; T.A.F. Kuipers, *Reply*. **Realism and Metaphors** — J.J.A. Mooij, *Metaphor and Metaphysical Realism*; T.A.F. Kuipers, *Reply*; R. Festa, *On the Relations between (Neo-Classical) Philosophy of Science and Logic*; T.A.F. Kuipers, *Reply*; *Bibliography of Theo A.F. Kuipers*; *Index of Names*.

VOLUME 84 (2005)

COGNITIVE STRUCTURES IN SCIENTIFIC INQUIRY

ESSAYS IN DEBATE WITH THEO KUIPERS, VOL. 2

(Edited by Roberto Festa, Atocha Aliseda and Jeanne Peijnenburg)

R. Festa, A. Aliseda, J. Peijnenburg, *Introduction*; T.A.F. Kuipers, *Structures in Scientific Cognition: A Synopsis of Structures in Science. Heuristic Patterns Based on Cognitive Structures. An Advanced Textbook in Neo-Classical*

*Philosophy of Science (2001).* **Types of Research and Research Programs** — D. Atkinson, *A New Metaphysics: Finding a Niche for String Theory*; T.A.F. Kuipers, *Reply*; T. Nickles, *Problem Reduction: Some Thoughts*; T.A.F. Kuipers, *Reply*; M. Franssen, *Design Research Programs*; T.A.F. Kuipers, *Reply*; J.P. van Bendegem, *Proofs and Arguments: The Special Case of Mathematics*; T.A.F. Kuipers, *Reply*. **Types of Explanation** — E. Weber, H. De Preester, *Micro-Explanations of Laws*; T.A.F. Kuipers, *Reply*; E.R. Scerri, *On the Formalization of the Periodic Table*; T.A.F. Kuipers, *Reply*; J. Peijnenburg, *Classical, Nonclassical and Neoclassical Intentions*; T.A.F. Kuipers, *Reply*; A. Ruth Mackor, *Erklären, Verstehen and Simulation: Reconsidering the Role of Empathy in the Social Sciences*; T.A.F. Kuipers, *Reply*; A. Wouters, *Functional Explanation in Biology*; T.A.F. Kuipers, *Reply*; A. Grobler, A. Wisniewski, *Explanation and Theory Evaluation*; T.A.F. Kuipers, *Reply*. **Computational Approaches** — J. Kamps, *The Ubiquity of Background Knowledge*; T.A.F. Kuipers, *Reply*; A.P.M. Van Den Bosch, *Structures in Neuropharmacology*; T.A.F. Kuipers, *Reply*; P. Thagard, *Why Is Beauty a Road to the Truth?*; T.A.F. Kuipers, *Reply*; G.A.W. Vreeswijk, *Direct Connectionistic Methods for Scientific Theory Formation*; T.A.F. Kuipers, *Reply*. **Theories and Structures** — E. Ruttkamp, *Overdetermination of Theories by Empirical Models: A Realist Interpretation of Empirical Choices*; T.A.F. Kuipers, *Reply*; R.L. Causey, *What Is Structure?*; T.A.F. Kuipers, *Reply*. **Science and Ethics** — H. Zandvoort, *Knowledge, Risk, and Liability. Analysis of a Discussion Continuing within Science and Technology*; T.A.F. Kuipers, *Reply*; *Bibliography of Theo A.F. Kuipers*; *Index of Names*.

# THE JOURNAL OF THE BRITISH SOCIETY FOR PHENOMENOLOGY

An International Review of Philosophy and the Human Sciences

EDITOR: WOLFE MAYS

Volume 36, No.1, January 2005

Music, the Other, and Injustice


**Articles**




The *JBSP* publishes papers on phenomenology and existential philosophy as well as contributions from other fields of philosophy. Papers from workers in the Humanities and human sciences interested in the philosophy of their subject will be welcome. All papers and books for review to be sent to the Editor: Professor Wolfe Mays, Department of Politics and Philosophy, The Manchester Metropolitan University, Manchester M15 6LL, England. Subscription and advertisement enquiries to be sent to the publishers: Jackson Publishing and Distribution, 3 Gibsons Road, Heaton Moor, Stockport, Cheshire, SK4 4JX, England.

# PHILOSOPHIA

**PHILOSOPHICAL QUARTERLY OF ISRAEL**

**Edited by**

**Asa Kasher**

Laura Schwarz-Kipp Professor of Professional Ethics and Philosophy of Practice, Tel-Aviv University

## Volume 30, Nos. 1 – 4





Editorial addresses:
PHILOSOPHIA, Bar-Ilan Univrsity, Ramat-Gan 52900, Israel
Or:
Prof. Asa Kasher
L.Schwarz-Kipp Chair of Professional Ethics and Philosophy of Practice,
Tel-Aviv University, Tel-Aviv 69978, Israel
Fax: +972-3-6407953
E-Mail addresses: kasher@post.tau.ac.il or asakasher@hotmail.com